ALGEBRAIC CURVES IN CRYPTOGRAPHY

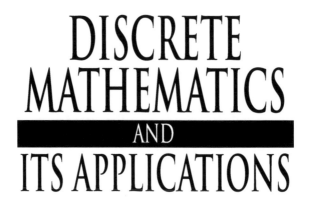

DISCRETE MATHEMATICS AND ITS APPLICATIONS

Series Editor
Kenneth H. Rosen, Ph.D.

DISCRETE MATHEMATICS AND ITS APPLICATIONS

Series Editor KENNETH H. ROSEN

ALGEBRAIC CURVES IN CRYPTOGRAPHY

San Ling
Huaxiong Wang
Chaoping Xing

CRC Press
Taylor & Francis Group
Boca Raton London New York

CRC Press is an imprint of the
Taylor & Francis Group, an **informa** business

A CHAPMAN & HALL BOOK

CRC Press
Taylor & Francis Group
6000 Broken Sound Parkway NW, Suite 300
Boca Raton, FL 33487-2742

First issued in paperback 2019

© 2013 by Taylor & Francis Group, LLC
CRC Press is an imprint of Taylor & Francis Group, an Informa business

No claim to original U.S. Government works

ISBN-13: 978-1-4200-7946-3 (hbk)
ISBN-13: 978-1-138-38141-4 (pbk)

Visit the Taylor & Francis Web site at
http://www.taylorandfrancis.com

and the CRC Press Web site at
http://www.crcpress.com

To my beloved wife Bee Keow

S. L.

To my wife Qun
and my daughter Anna

H. W.

To my wife Youqun Shi
and my children Zhengrong and Menghong

C. X.

Contents

Preface

Algebraic curves over finite fields have been studied by mathematicians at least as far back as Gauss more than two centuries ago. Tremendous progress was made on this topic in the 20th century, with the seminal work of Hasse and Weil in the 1930s and 1940s on the number of points on such curves, as well as many other important discoveries. A new milestone was reached in the 1980s when the use of algebraic curves led to major breakthroughs in some other domains: the discovery of algebraic geometry codes by Goppa that led to asymptotically good sequences of codes that outperformed the Gilbert-Varshamov bound, a classical benchmark for families of good linear codes; the invention of the elliptic curve public key cryptosystem independently by Koblitz and Miller; and the use of elliptic curves for the factorization of large integers by Lenstra.

The discovery of such unexpected relationships marked the beginning of a new chapter in cryptography, the study and practice of protecting data against adversaries. Intensified effort has been made to develop and refine cryptographic tools using algebraic curves, while potential applications have also motivated deeper theoretical studies of algebraic curves. Elliptic curve cryptography, in particular, has seen a surge in interest and research activities. A good number of books have also been written on this topic. However, the reach of algebraic curves in cryptography goes far beyond elliptic curve cryptography or, indeed, public key cryptography. Examples of other cryptographic applications of algebraic curves that have emerged in the past two decades include multiparty computations, secret sharing, authentication codes, frameproof codes, private information retrieval, key distribution, broadcast encryption, one-time signatures, and sequences for stream ciphers.

Having worked quite a bit in some of these areas in recent years, we felt that the richness of the use of algebraic curves in these other cryptographic applications has not been much covered systematically in the literature. The main intention of this book is to bridge this gap.

With graduate students and researchers intended as the primary readers of this book, we have tried to make the content as self-contained as possible. The book, therefore, begins with a chapter on the background knowledge of algebraic curves needed for the rest of the book. This is followed by a chapter on error-correcting codes, where a discussion on algebraic geometry codes is included. Indeed, in some of the topics covered in this book, algebraic curves enter into cryptography through algebraic geometry codes. Despite the wide availability of literature on elliptic curves and elliptic curve cryptography, it would still be odd for a book on algebraic curves in cryptography to omit this topic completely. Hence, a chapter is included to provide a brief introduction. Starting from Chapter 4, each chapter deals with a selected topic in cryptography (other than elliptic curve cryptography). The topics we have chosen to discuss in this book include secret sharing schemes, authentication codes, frameproof codes, key distribution schemes, broadcast encryption, and sequences. Each chapter begins with some introductory material for the topic before application of algebraic geometry is featured. It is quite possible for any of these chapters to be studied independently of the others.

We have received much support and help in the process of writing this book. For this, we are immensely grateful. We thank the Singapore National Research Foundation, Singapore Ministry of Education, and Nanyang Technological University, for the generous funding we have received for our research. We are also grateful to those who have given us precious feedback during the preparation, especially Carles Padró, and those who have provided invaluable technical help, in particular Jie Chen, Hoon Wei Lim, and Enver Ozdemir. Special thanks must go to Sze Ling Yeo, without whose help Chapter 3 would not have been possible. Part of this book was written when S. Ling was visiting École Normale Supérieure de Lyon and when H. Wang was visiting Macquarie University. It is a pleasure to thank these institutions for their hospitality.

San Ling
Huaxiong Wang
Chaoping Xing

List of Figures

List of Tables

Chapter 1

Introduction to Algebraic Curves

Algebraic curves have found many applications in coding and cryptography since the discovery of the Goppa geometry codes [71]. There are quite a few standard books on algebraic curves [61, 76, 75, 117], although most of these books discuss curves over algebraically closed fields. For our purpose, we are interested primarily in algebraic curves over finite fields, in fact, preferably algebraic curves with explicit defining equations. In this chapter, a self-contained introduction to algebraic curves is given.

1.1 Plane Curves

Throughout this chapter, and indeed throughout this book, we always assume that $\mathbb{F}_q$ is the finite field with q elements. We denote by $\mathbb{F}_q[x_1, x_2, \ldots, x_n]$ the multivariate polynomial ring with n variables.

A polynomial $f(x_1, x_2, \ldots, x_n) \in \mathbb{F}_q[x_1, x_2, \ldots, x_n]$ is called **absolutely irreducible** if it is irreducible over the algebraic closure $\overline{\mathbb{F}_q}$[1].

An **affine plane curve** over $\mathbb{F}_q$ is defined by $f(x, y) = 0$ with an absolutely irreducible polynomial $f(x, y) \in \mathbb{F}_q[x, y]$.

Example 1.1.1   (i) Let $\mathcal{E}$ be a curve over $\mathbb{F}_2$ defined by

$$y^2 + y + x^3 + x = 0.$$

It has four affine points $(0, 0), (0, 1), (1, 0), (1, 1)$ over $\mathbb{F}_2$. This curve is called an **elliptic curve**, a subject that will be extensively studied in Chapter 3.

(ii) Let q be an odd prime power and let $\mathcal{X}$ be a curve over $\mathbb{F}_q$ defined by

$$y^2 = f(x),$$

where $f(x)$ is a square-free polynomial over $\mathbb{F}_q$ of degree ≥ 3. This curve

---

[1]An algebraic closure K of a field F is an algebraic extension such that every univariate polynomial of degree ≥ 1 over K has at least one root in K. In the case where F is the finite field $\mathbb{F}_q$, an algebraic closure K can be taken to be the union $\cup_{i=1}^{\infty} \mathbb{F}_{q^i}$.

is called a **hyperelliptic curve**. When the degree of $f(x)$ is 3, $\mathcal{X}$ is actually an **elliptic curve** (see Chapter 3 as well).

(iii) An **affine plane line** is defined by $ax + by + c = 0$, where $(a, b) \neq (0, 0)$.

(iv) The curve over $\mathbb{F}_{r^2}$ defined by the equation

$$y^r + y - x^{r+1} = 0$$

is called a **Hermitian curve**.

If $f(x, y) \in \mathbb{F}_q[x, y]$ is not absolutely irreducible (though it is irreducible over $\mathbb{F}_q$), then it cannot define a plane curve. An illustration is given in the following example.

Example 1.1.2 Consider the polynomial $y^2 + x^4$. It is irreducible over $\mathbb{F}_3$. However, it is reducible over $\mathbb{F}_9$ (hence it is reducible over $\overline{\mathbb{F}}_3$). Over $\mathbb{F}_9$, it can be factored into a product $(y - \alpha x^2)(y + \alpha x^2)$, where α is a root of the polynomial $t^2 + 1 \in \mathbb{F}_3[t]$. Thus, $y^2 + x^4 = 0$ does not define a curve over $\mathbb{F}_3$. In fact, $y^2 + x^4 = 0$ implies that $y - \alpha x^2 = 0$ or $y + \alpha x^2 = 0$, i.e., $y^2 + x^4 = 0$ is the union of two other curves.

We are now ready to discuss projective plane curves. Before doing this, we need to introduce affine and projective planes.

Definition 1.1.3   (i) The n-dimensional **affine space** over $\mathbb{F}_q$, denoted by $\mathbf{A}^n$, is defined by

$$\mathbf{A}^n \overset{\text{def}}{=} \mathbf{A}^n(\overline{\mathbb{F}}_q) \overset{\text{def}}{=} \{(a_1, a_2, \ldots, a_n) : a_i \in \overline{\mathbb{F}}_q\}.$$

An element of $\mathbf{A}^n$ is called a **point**. A point in the subspace defined by $\mathbf{A}^n(\mathbb{F}_q) \overset{\text{def}}{=} \mathbf{A}^n \cap \mathbb{F}_q^n = \{(a_1, a_2, \ldots, a_n) : a_i \in \mathbb{F}_q\}$ is called $\mathbb{F}_q$-**rational**. In the case $n = 2$, $\mathbf{A}^2$ is called the **affine plane** over $\mathbb{F}_q$.

(ii) (Roughly speaking, the n-dimensional projective space is obtained by adding some "points at infinity" to the n-dimensional affine space. In order to do so, we need to go to the affine space of dimension $n + 1$.)

An n-dimensional **projective space** over $\mathbb{F}_q$, denoted by $\mathbf{P}^n(\overline{\mathbb{F}}_q)$ or $\mathbf{P}^n$, is the set of equivalence classes of nonzero $(n+1)$-tuples $(a_0, a_1, \ldots, a_n)$ of elements of $\overline{\mathbb{F}}_q$ under the equivalence relation given by:

$$(a_0, a_1, \ldots, a_n) \sim (b_0, b_1, \ldots, b_n)$$

if and only if there exists a nonzero element λ of $\overline{\mathbb{F}}_q$ such that $a_i = \lambda b_i$ for all $i = 0, 1, \ldots, n$. An element of $\mathbf{P}^n$ is called a **point**. We denote by $[a_0, a_1, \ldots, a_n]$ the equivalence class containing the $(n+1)$-tuple

$(a_0, a_1, \ldots, a_n)$. Thus, $[a_0, a_1, \ldots, a_n]$ and $[\lambda a_0, \lambda a_1, \ldots, \lambda a_n]$ stand for the same point if $\lambda \neq 0$. A point in the set

$$\mathbf{P}^n(\mathbb{F}_q) \stackrel{\text{def}}{=} \{[a_0, a_1, \ldots, a_n] \in \mathbf{P}^n : a_i \in \mathbb{F}_q \text{ for } i = 0, 1, \ldots, n\}$$

is called $\mathbb{F}_q$-**rational**.

When $n = 2$, $\mathbf{P}^2$ is called the **projective plane** over $\mathbb{F}_q$.

Proposition 1.1.4 (i) *The number of $\mathbb{F}_q$-rational points in $\mathbf{A}^2(\overline{\mathbb{F}}_q)$ is q^2.*

(ii) *The number of $\mathbb{F}_q$-rational points in $\mathbf{P}^2(\overline{\mathbb{F}}_q)$ is given by $q^2 + q + 1$.*

Proof. The first part is clear.

To prove (ii), we can list all the elements of $\mathbf{P}^2(\mathbb{F}_q)$. In fact, it is not difficult to verify that

$$\mathbf{P}^2(\mathbb{F}_q) = \{[a, b, 1] : a, b \in \mathbb{F}_q\} \cup \{[a, 1, 0] : a \in \mathbb{F}_q\} \cup \{[1, 0, 0]\}.$$

Counting the number of points in $\mathbf{P}^2(\mathbb{F}_q)$ gives the desired result. □

Remark 1.1.5 From the proof of Proposition 1.1.4, we can see that $\mathbf{P}^2(\mathbb{F}_q)$ can be regarded as consisting of two subsets. One consists of the points whose third coordinate is equal to 1. It has q^2 points and can be identified with $\mathbf{A}^2(\mathbb{F}_q)$. The other subset, consisting of all points whose third coordinate is equal to 0, has $q + 1$ points. The points in this subset are called "points at infinity."

Example 1.1.6 Consider the curve defined in Example 1.1.1(i). It has four points $(0, 0), (0, 1), (1, 0), (1, 1)$ in the affine space $\mathbf{A}^2(\mathbb{F}_2)$. To see that it has one "point at infinity," we have to homogenize this curve (i.e., creating a homogeneous equation from the one given, by introducing a new variable Z, such that specializing to $Z = 1$ returns us to the original given equation) to the following equation

$$Y^2 Z + Y Z^2 + X^3 + X Z^2 = 0.$$

Now, we can see that there are five points altogether in the projective plane satisfying this equation: $[0, 0, 1], [0, 1, 1], [1, 0, 1], [1, 1, 1]$, and $[0, 1, 0]$. The first four points are just the affine points. The last one is a new point. Hence, this point is sometimes denoted as ∞ if we only express the curve equation and points in the affine form. Indeed, in most cases, it is more convenient to work with affine curve equations and affine points.

From the above example, we know that, in order to find all the projective points of a given curve defined by an affine equation, we have to consider the homogeneous equation for the curve instead of the affine equation. A

homogeneous polynomial in three variables of degree m over $\mathbb{F}_q$ is of the form $\sum_{i+j+k=m} a_{ijk} X^i Y^j Z^k$, where $a_{ijk} \in \mathbb{F}_q$. If an affine plane curve is defined by a nonzero polynomial $f(x, y)$ of degree $m > 0$, then we can convert it into a projective plane curve by homogenizing it to $Z^m f(X/Z, Y/Z) = 0$ as we did in Example 1.1.6.

A **projective plane curve** over $\mathbb{F}_q$ is defined by $f(X, Y, Z) = 0$ with an absolutely irreducible homogeneous polynomial $f(X, Y, Z) \in \mathbb{F}_q[X, Y, Z]$.

Example 1.1.7   (i) The **Klein curve** is defined by the following homogeneous equation

$$X^3 Y + Y^3 Z + Z^3 X = 0.$$

  (ii) A **projective plane line** is defined by $aX + bY + cZ = 0$, where $(a, b, c) \neq (0, 0, 0)$. In the case $(a, b) \neq (0, 0)$, it is the projective line from the affine equation of Example 1.1.1(iii). However, compared with the affine line of Example 1.1.1(iii), it has one more point, i.e., $[b, -a, 0]$, which is viewed as a "point at infinity."

(iii) By homogenizing the affine equation of Example 1.1.1(iv), we obtain the **projective Hermitian curve** over $\mathbb{F}_{r^2}$ defined by $ZY^r + Z^r Y - X^{r+1} = 0$. Compared with the affine curve of Example 1.1.1(iv), it also has one more point, i.e., $[0, 1, 0]$, which is viewed as a "point at infinity" as well.

We now introduce the notion of smoothness of plane curves.

Definition 1.1.8 Let $\mathcal{X}$ be an affine curve defined by an equation $f(x, y) = 0$. A point P of $\mathcal{X}$ is called **nonsingular** or **simple** if

$$\left(\frac{\partial f}{\partial x}(P), \frac{\partial f}{\partial y}(P) \right) \neq (0, 0),$$

where $\frac{\partial f}{\partial x}(P), \frac{\partial f}{\partial y}(P)$ denote the formal partial derivatives of f with respect to x and y, respectively, at P. Otherwise, P is called a **singular** point. For a nonsingular point $P = (a, b)$, the **tangent line** of $\mathcal{X}$ at P is defined by

$$\frac{\partial f}{\partial x}(P)(x - a) + \frac{\partial f}{\partial y}(P)(y - b) = 0.$$

If all points of $\mathcal{X}$ are simple, then $\mathcal{X}$ is called **nonsingular** or **smooth**.

Note that all singular points of an affine plane $f(x, y) = 0$ can be determined by solving the system of equations

$$\begin{cases} f(x, y) = 0 \\ \frac{\partial f}{\partial x} = 0 \\ \frac{\partial f}{\partial y} = 0. \end{cases} \tag{1.1}$$

Previously, we converted affine curves to projective curves by homogenizing polynomials. Conversely, we can convert projective curves to affine curves by dehomogenizing polynomials (i.e., by setting one of the variables to be 1). In many cases, it is more convenient to deal with affine curves rather than projective curves.

Let $P = [a, b, c]$ be a point on the projective curve defined by $f(X, Y, Z) = 0$. As $(a, b, c) \neq (0, 0, 0)$, without loss of generality, we may assume that $c \neq 0$, thus $P = [a/c, b/c, 1]$ and the curve can be dehomogenized to the affine curve defined by $f(x, y, 1) = 0$. If $b \neq 0$ (respectively, $a \neq 0$), we can consider the affine curve $f(x, 1, y) = 0$ (respectively, $f(1, x, y) = 0$) instead.

Definition 1.1.9 Let $\mathcal{X}$ be a projective curve defined by an equation $f(X, Y, Z) = 0$. Let $P = [a, b, c]$ be a point of $\mathcal{X}$. We may assume that $c \neq 0$. Then, P is called **nonsingular** or **simple** if $(a/c, b/c)$ is a simple point of the affine curve defined by $f(x, y, 1) = 0$. Otherwise, P is called a **singular** point. In the case where P is simple, the **tangent line** of $\mathcal{X}$ at P is defined by

$$\frac{\partial f}{\partial x}\left(\frac{a}{c}, \frac{b}{c}\right)(X - aZ) + \frac{\partial f}{\partial y}\left(\frac{a}{c}, \frac{b}{c}\right)(Y - bZ) = 0.$$

(Here, in $\frac{\partial f}{\partial x}$ and $\frac{\partial f}{\partial y}$, we regard f as the function $f(x, y, 1)$ in two variables x and y.) If all points of $\mathcal{X}$ are simple, then $\mathcal{X}$ is called **nonsingular** or **smooth**.

Example 1.1.10   (i) Consider the projective Klein curve over $\mathbb{F}_q$ defined in Example 1.1.7(i). If $P = [a, b, 1]$ is a singular point, we dehomogenize the equation to $x^3 y + y^3 + x = 0$. By (1.1), we obtain a system of equations

$$a^3 b + b^3 + a = 0, \quad 3a^2 b + 1 = 0, \quad a^3 + 3b^2 = 0.$$

Solving this system gives no solution if $q \not\equiv 0 \pmod 7$, and one solution $(a, b) = (2, 4)$ if $q \equiv 0 \pmod 7$. For points of the types $[a, 1, c]$ and $[1, b, c]$, we can obtain similar results. Thus, the projective Klein curve is smooth if $q \not\equiv 0 \pmod 7$, and it has only one singular point $[2, 4, 1]$ if $q \equiv 0 \pmod 7$.

(ii) Using the same method as in (i), we can show that the projective line defined in Example 1.1.7(ii) is smooth, hence the affine line defined in Example 1.1.1(iii) is also smooth.

(iii) Again using the same method, we can show that the projective elliptic curve defined in Example 1.1.6 is smooth, hence the affine curve defined in Example 1.1.1(i) is also smooth.

(iv) The projective Hermitian curve defined in Example 1.1.7(iii) is also smooth, hence the affine curve defined in Example 1.1.1(iv) is smooth as well.

(v) Finally, we consider the affine hyperelliptic curve over $\mathbb{F}_q$ defined in Example 1.1.1(ii). Solving the system of equations (1.1) for $g(x,y) = y^2 - f(x)$, we get $f(x) = f'(x) = 0$, where $f'(x)$ denotes the derivative of $f(x)$. This is impossible since $f(x)$ has no multiple roots. This implies that the affine hyperelliptic curve in Example 1.1.1(ii) is smooth.

To look at the projective hyperelliptic curve, we first homogenize the affine curve to get a projective curve: $Y^2 Z^{d-2} - Z^d f(X/Z) = 0$, where $d \geq 3$ denotes the degree of $f(x)$. Apart from those "finite" points $[a, b, 1]$ corresponding to the points on the affine curve $y^2 - f(x) = 0$, there is one more point $[0, 1, 0]$. We dehomogenize this curve to an affine equation: $z^{d-2} - z^d f(x/z) = 0$ and consider $(0, 0)$. It is easy to check that $(0, 0)$ is a singular point if $d > 3$ and a nonsingular point if $d = 3$. In conclusion, the projective hyperelliptic curve has one singular point $[0, 1, 0]$ for $d > 3$, and it is smooth for $d = 3$.

1.2   Algebraic Curves and Their Function Fields

In the previous section, an affine plane curve has two variables and is defined by one equation. If we want to define a curve in an n-dimensional space, we need n variables. Roughly speaking, a curve in an n-dimensional space is defined by $n - 1$ equations (we will have a more precise definition later (see Definition 1.2.3)).

Example 1.2.1 Consider the affine curve $\mathcal{X}$ over $\mathbb{F}_2$ in the 3-dimensional space defined by

$$\begin{cases} y^2 + y = x^3 + x \\ z^2 + z = y^3 + y^2. \end{cases}$$

It has eight $\mathbb{F}_2$-rational points: $(0,0,0), (0,0,1), (0,1,0), (0,1,1), (1,0,0), (1,0,1), (1,1,0)$, and $(1,1,1)$. Note that, for an affine plane curve over $\mathbb{F}_2$, it has at most four $\mathbb{F}_2$-rational points. This is because $\mathbf{A}^2(\mathbb{F}_2)$ has only four points.

To give a more precise definition of a curve, we have to study function fields first.

Let k be a finite field $\mathbb{F}_q$ or its algebraic closure $\overline{\mathbb{F}}_q$. For x transcendental over k, the **rational function field** $k(x)$ of one variable over k is the set of all $\frac{f(x)}{g(x)}$, where $f(x), g(x)$ are polynomials in $k[x]$. An extension K over k, denoted by K/k, is called a **function field** of one variable over k if, for any $x \in K \setminus k$, the field K is a finite algebraic extension over the rational function field $k(x)$.

Example 1.2.2   (i) Let $\mathcal{L}$ be the line defined by $ax + by + c = 0$ with $(a, b) \neq (0, 0)$, as in Example 1.1.1(iii). We may assume that $a \neq 0$. It is clear that y is a variable over $\mathbb{F}_q$. Since $x = -(by + c)/a$, x is algebraic over $\mathbb{F}_q(y)$. Hence, $\mathbb{F}_q(x, y)$ is a function field of one variable over $\mathbb{F}_q$. This function field is called the function field of $\mathcal{L}$, denoted by $\mathbb{F}_q(\mathcal{L})$. In fact, it is easy to see that $\mathbb{F}_q(x, y) = \mathbb{F}_q(y)$, i.e., $\mathbb{F}_q(\mathcal{L})$ is a rational function field.

(ii) In Example 1.1.1(i), x is a variable over $\mathbb{F}_2$, while y is algebraic over $\mathbb{F}_2(x)$. Thus, $\mathbb{F}_2(x, y)$ is a function field of one variable over $\mathbb{F}_2$. This function field is called the function field of $\mathcal{E}$, denoted by $\mathbb{F}_2(\mathcal{E})$.

(iii) In Example 1.2.1, x is a variable over $\mathbb{F}_2$, while y is algebraic over $\mathbb{F}_2(x)$. Since z is algebraic over $\mathbb{F}_2(x, y)$, z is also algebraic over $\mathbb{F}_2(x)$. Thus, $\mathbb{F}_2(x, y, z)$ is a function field of one variable over $\mathbb{F}_2$. This function field is called the function field of $\mathcal{X}$, denoted by $\mathbb{F}_2(\mathcal{X})$.

The above examples show that, for all the three curves, their function fields are function fields of one variable. This fact will be used to define a curve.

In order to define curves in higher dimensions, we need to use the n-dimensional projective space, for $n > 2$, defined in Definition 1.1.3.

Definition 1.2.3 An affine curve $\mathcal{X}$ over $\mathbb{F}_q$ in $\mathbf{A}^n$ is defined by a system of polynomial equations:

$$
\begin{cases}
f_1(x_1, x_2, \ldots, x_n) = 0 \\
\quad \vdots \\
f_m(x_1, x_2, \ldots, x_n) = 0
\end{cases}
\tag{1.2}
$$

with $f_i \in \mathbb{F}_q[x_1, x_2, \ldots, x_n]$ such that the following conditions are satisfied:

(i) the ideal of $\overline{\mathbb{F}}_q[x_1, x_2, \ldots, x_n]$ generated by $f_1, \ldots, f_m$ is a prime ideal (hence, the ideal of $\mathbb{F}_q[x_1, x_2, \ldots, x_n]$ generated by $f_1, \ldots, f_m$ is also a prime ideal);

(ii) the quotient field of the residue ring $\mathbb{F}_q[x_1, x_2, \ldots, x_n]/(f_1, \ldots, f_m)$ is a function field of one variable over $\mathbb{F}_q$, where $(f_1, \ldots, f_m)$ denotes the ideal of $\mathbb{F}_q[x_1, x_2, \ldots, x_n]$ generated by $f_1, \ldots, f_m$.

The quotient field in (ii) is called the **function field** of $\mathcal{X}$, denoted by $\mathbb{F}_q(\mathcal{X})$. A point P of $\mathbf{A}^n$ satisfying $f_i(P) = 0$, for all $i = 1, \ldots, m$, is called a point of $\mathcal{X}$. The point P is said to be an $\mathbb{F}_q$-**rational** (or **rational**) point of $\mathcal{X}$ if it belongs to $\mathbf{A}^n(\mathbb{F}_q)$.

A projective curve $\mathcal{Y}$ over $\mathbb{F}_q$ is obtained by homogenizing the polynomials in (1.2) satisfying the above two conditions. The function field of this projective curve is defined to be the function field of the above affine curve. A point P of $\mathbf{P}^n$ which is a zero for all the homogenized polynomials is called a point of $\mathcal{Y}$. The point P is said to be an $\mathbb{F}_q$-**rational** (or **rational**) point of $\mathcal{Y}$ if it belongs to $\mathbf{P}^n(\mathbb{F}_q)$.

Remark 1.2.4 (i) The first condition in Definition 1.2.3 cannot be changed to the weaker condition "the ideal of $\mathbb{F}_q[x_1, x_2, \ldots, x_n]$ generated by $f_1, \ldots, f_m$ is a prime ideal." For instance, the ideal of $\mathbb{F}_3[x, y]$ generated by $x^2 + y^4$ is a prime ideal, while the ideal of $\overline{\mathbb{F}}_3[x, y]$ generated by $x^2 + y^4$ is not a prime ideal. This requirement coincides with our definition of plane curves since $x^2 + y^4$ is not absolutely irreducible.

(ii) Since the ideal of $\mathbb{F}_q[x_1, x_2, \ldots, x_n]$ generated by $f_1, \ldots, f_m$ is a prime ideal, the residue ring $\mathbb{F}_q[x_1, x_2, \ldots, x_n]/(f_1, \ldots, f_m)$ is a domain and its quotient field can be defined.

(iii) It is not difficult to verify that the function fields defined in Example 1.2.2 coincide with the above definition of the function field of a general curve.

1.3 Smooth Curves

We generalize Definition 1.1.8 to curves in spaces of higher dimension.

Definition 1.3.1 Let $\mathcal{X} \subseteq \mathbf{A}^n$ be an affine curve defined by a set of polynomials $f_1, \ldots, f_m \in \mathbb{F}_q[x_1, \ldots, x_n]$. Then $\mathcal{X}$ is said to be **nonsingular** (or **smooth**, or **simple**) at a point P of $\mathcal{X}$ if the $m \times n$ **Jacobian matrix**

$$\left(\frac{\partial f_i}{\partial x_j}(P) \right)$$

at P has rank $n - 1$. Otherwise, $\mathcal{X}$ is said to be **singular** at P. If $\mathcal{X}$ is smooth at every point of $\mathcal{X}$, then we say that $\mathcal{X}$ is a **nonsingular** (or **smooth**) **affine curve**.

The above definition coincides with Definition 1.1.8 when $\mathcal{X}$ is a plane curve defined by one polynomial $f(x_1, x_2) = 0$.

For a projective curve in $\mathbf{P}^n$ and a point $P = [a_0, a_1, \ldots, a_n]$ with $a_i \neq 0$, we may dehomogenize all polynomials with respect to the index i to obtain an affine curve. Then P is called a **nonsingular** (or **smooth**, or **simple**) point if

$$(a_0/a_i, a_1/a_i, \ldots, a_{i-1}/a_i, a_{i+1}/a_i, \ldots, a_n/a_i)$$

is a smooth point on the affine curve.

Nonsingular points on a curve have a nice associated property, which we will see later in this section (Theorem 1.3.5). We first require the following notion.

Definition 1.3.2 Let K/k be a function field of one variable, where k is a finite field $\mathbb{F}_q$ or its algebraic closure $\overline{\mathbb{F}}_q$. A **discrete valuation** of K/k is a surjective map $\nu : K \to \mathbb{Z} \cup \{\infty\}$ satisfying the following conditions:

(i) $\nu(z) = \infty$ if and only if $z = 0$;

(ii) $\nu(yz) = \nu(y) + \nu(z)$ for all $y, z \in K$;

(iii) $\nu(y + z) \geq \min(\nu(y), \nu(z))$ for all $y, z \in K$;

(iv) $\nu(y + z) = \min(\nu(y), \nu(z))$ if $\nu(y) \neq \nu(z)$ for $y, z \in K$;

(v) $\nu(K^*) \neq \{0\}$, where $K^* = K \setminus \{0\}$;

(vi) $\nu(\alpha) = 0$ for all $\alpha \in k^* = k \setminus \{0\}$.

We now look at some examples of function fields of plane curves.

Example 1.3.3 Consider the plane line $\mathcal{L}$ over $\mathbb{F}_q$ defined by $ax + by + c = 0$ with $a \neq 0$. Then the function field $\mathbb{F}_q(\mathcal{L})$ is $\mathbb{F}_q(y)$ by Example 1.2.2(i). For any $y = \alpha \in \mathbb{F}_q$, we get a unique $\mathbb{F}_q$-rational point $P_\alpha \overset{\text{def}}{=} (-b\alpha/a - c/a, \alpha)$ on $\mathcal{L}$. Define a map ν_α from $\mathbb{F}_q(\mathcal{L})$ to $\mathbb{Z} \cup \{\infty\}$ by: $\nu_\alpha(0) = \infty$, and $\nu_\alpha(f(y)) = r$ if $f(y) = (y - \alpha)^r(g(y)/h(y)) \neq 0$ for some $g(y), h(y) \in \mathbb{F}_q[y]$ with $g(\alpha) \neq 0$ and $h(\alpha) \neq 0$. Then it is not difficult to verify that ν_α is a discrete valuation. This means that, for each "finite" $\mathbb{F}_q$-rational point P_α (by a "finite" point, we mean a point on the affine curve), we obtain a discrete valuation ν_α. For the "point at infinity" $[-(b+c)/a, 1, 0]$, we define another discrete valuation ν_∞ from $\mathbb{F}_q(\mathcal{L})$ to $\mathbb{Z} \cup \{\infty\}$ by: $\nu_\infty(0) = \infty$, and $\nu_\infty(g(y)/h(y)) = \deg(h(y)) - \deg(g(y))$ for $g(y), h(y) \in \mathbb{F}_q[y]$ with $g(y), h(y) \neq 0$. Again, one can verify that ν_∞ is a discrete valuation.

Now assume that β is an element in some extension field of $\mathbb{F}_q$ and let $p(y) \in \mathbb{F}_q[y]$ be the minimal polynomial of β with respect to $\mathbb{F}_q$. Define a map $\nu_{p(y)}$ from $\mathbb{F}_q(\mathcal{L})$ to $\mathbb{Z} \cup \{\infty\}$ by: $\nu_\alpha(0) = \infty$, and $\nu_{p(y)}(f(y)) = r$ if $f(y) = p(y)^r(g(y)/h(y)) \neq 0$ for some $g(y), h(y) \in \mathbb{F}_q[y]$ with $p(y) \nmid g(y)$ and $p(y) \nmid h(y)$. Then it is not difficult to verify that $\nu_{p(y)}$ is a discrete valuation. Thus, for each point $(-b\beta/a - c/a, \beta) \in \mathbf{A}^2$ of $\mathcal{L}$, we have a corresponding valuation $\nu_{p(y)}$.

In conclusion, we have shown the following:

(i) For every "finite" point $(-b\beta/a - c/a, \beta) = [-b\beta/a - c/a, \beta, 1]$, we have a corresponding discrete valuation.

(ii) If β and γ are conjugate, i.e., they have the same minimal polynomial over $\mathbb{F}_q$ (this is equivalent to $\gamma = \beta^{q^s}$ for some $s \in \mathbb{Z}$ (see [95, Theorem 2.13])), then the discrete valuations corresponding to $(-b\beta/a - c/a, \beta)$ and $(-b\gamma/a - c/a, \gamma)$ are equal.

(iii) Let $p(y) \in \mathbb{F}_q[y]$ be the minimal polynomial of β with respect to $\mathbb{F}_q$. Then $\nu_{p(y)}(u(y)) > 0$ if and only if $u(\beta)$ is well defined and $u(\beta) = 0$.

(iv) For the "point at infinity," we also have a corresponding discrete valuation.

In fact, all the discrete valuations of $\mathbb{F}_q(\mathcal{L}) = \mathbb{F}_q(y)$ are those corresponding to the points of $\mathcal{L}$ (see [151, Corollary I.2.3]).

Example 1.3.4 Consider the elliptic curve defined in Example 1.1.1(i).

(i) Define a map $\nu_{(0,0)}$ from $\mathbb{F}_2(\mathcal{E}) = \mathbb{F}_2(x,y)$ to $\mathbb{Z} \cup \{\infty\}$ by: $\nu_{(0,0)}(x) = 1$, $\nu_{(0,0)}(y) = 1$, and extend this map to the function field $\mathbb{F}_2(\mathcal{E}) = \mathbb{F}_2(x,y)$ so that it satisfies the conditions of Definition 1.3.2. Then it is a discrete valuation. We can see that $\nu_{(0,0)}(g(x,y)) > 0$ if and only if $g(0,0)$ is well defined and $g(0,0) = 0$.

(ii) Now for the "point at infinity" $O \stackrel{\text{def}}{=} [0,1,0]$, we define a map ν_O from $\mathbb{F}_2(\mathcal{E})$ to $\mathbb{Z} \cup \{\infty\}$ by: $\nu_O(x) = -2$, $\nu_O(y) = -3$, and extend this map to the function field $\mathbb{F}_2(\mathcal{E}) = \mathbb{F}_2(x,y)$ so that it satisfies the conditions of Definition 1.3.2. Then ν_O is a discrete valuation. We homogenize the equation of Example 1.1.1(i) to $ZY^2 + Z^2Y + X^3 + Z^2X = 0$ and then dehomogenize it to $z + z^2 + x^3 + z^2x = 0$ and consider the point $(0,0)$. We can see that $\nu_O(h(x,z)) > 0$ if and only if $h(0,0)$ is well defined and $h(0,0) = 0$.

The two examples above illustrate that we can define a discrete valuation for every nonsingular point of a curve.

Theorem 1.3.5 *Let P be a nonsingular point of an affine (or projective) curve $\mathcal{X}$ over $\mathbb{F}_q$ (P may not be $\mathbb{F}_q$-rational). Then there exists a unique discrete valuation of $\mathbb{F}_q(\mathcal{X})$, denoted by ν_P, such that $\nu_P(f) > 0$ if and only if $f(P)$ is well defined and $f(P) = 0$.*

We do not give a proof for this theorem. The reader may refer to [61, page 82, Corollary 4] and [117, Theorem 3.1.5] for the detailed proof.

For an $\mathbb{F}_q$-rational point P of $\mathcal{X}$, an element t of $\mathbb{F}_q(\mathcal{X})$ is called a **local parameter** at P if $t(P)$ is well defined and $\nu_P(t) = 1$. Note that a local parameter is not unique and always exists (see [151, pages 2–3] and [117, page 15]). For instance, both x and y in Example 1.3.4(i) are local parameters at $(0,0)$.

Let $\mathcal{X}$ be an affine curve over $\mathbb{F}_q$. For a point $P = (a_1, \ldots, a_n)$ (P may not be $\mathbb{F}_q$-rational), we denote by $P^{(q^i)}$ the point $(a_1^{q^i}, \ldots, a_n^{q^i})$. Two points P and Q of $\mathcal{X}$ are said to be **conjugate** if $Q = P^{(q^i)}$ for some $i \in \mathbb{Z}$. Note that, if $Q = P^{(q^i)}$, then $P = Q^{(q^j)}$ for some $j \in \mathbb{Z}$.

Now let $\mathcal{X}$ be a projective curve over $\mathbb{F}_q$. For a point P, not necessarily $\mathbb{F}_q$-rational, given by $P = [a_0, a_1, \ldots, a_{k-1}, 1, a_{k+1}, \ldots, a_n]$, we denote by $P^{(q^i)}$ the point $[a_0^{q^i}, a_1^{q^i}, \ldots, a_{k-1}^{q^i}, 1, a_{k+1}^{q^i}, \ldots, a_n^{q^i}]$. Two points P and Q of $\mathcal{X}$ are said to be **conjugate** if $Q = P^{(q^i)}$ for some $i \in \mathbb{Z}$. In this case, we also have that, if $Q = P^{(q^i)}$, then $P = Q^{(q^j)}$ for some $j \in \mathbb{Z}$. The collection of all points conjugate to a given point P is called a **closed point**. The **degree** of a closed point P, denoted by $\deg(\mathsf{P})$, is defined to be the cardinality of P.

Consider now the plane line $\mathcal{L}$ over $\mathbb{F}_q$ defined by $ax + by + c = 0$ with $a \neq 0$, and let $P_1 \overset{\text{def}}{=} (-b\beta_1/a - c/a, \beta_1)$ and $P_2 \overset{\text{def}}{=} (-b\beta_2/a - c/a, \beta_2)$ be two points on $\mathcal{L}$. From Example 1.3.3, we know that, if both β_1 and β_2 are roots of an irreducible polynomial $p(y) \in \mathbb{F}_q[y]$, then P_1 and P_2 define the same discrete valuation. Since β_1 and β_2 are roots of the same irreducible polynomial, we have $\beta_2 = \beta_1^{q^i}$ for some $i \in \mathbb{Z}$ and, hence, $P_2 = (-b\beta_2/a - c/a, \beta_2) = ((-b\beta_1/a - c/a)^{q^i}, \beta_1^{q^i}) = P_1^{(q^i)}$. Conversely, if $P_2 = P_1^{(q^i)}$ for some $i \in \mathbb{Z}$, then $\beta_2 = \beta_1^{q^i}$, so β_1 and β_2 are roots of the same irreducible polynomial. This means that two conjugate points correspond to the same discrete valuation of $\mathbb{F}_q(\mathcal{L})$. In fact, this is true for any curve.

Theorem 1.3.6 *Let P, Q be two nonsingular points of an affine (or projective) curve $\mathcal{X}$ over $\mathbb{F}_q$ (P, Q may not be $\mathbb{F}_q$-rational). Then these two points correspond to the same discrete valuation of $\mathbb{F}_q(\mathcal{X})$ if and only if they are conjugate to each other.*

We refer to [117, Theorem 3.1.15] for the detailed proof of this theorem.

Let P be a closed point. For a point $P \in \mathsf{P}$ and an element t of $\mathbb{F}_q(\mathcal{X})$ with $\nu_P(t) = 1$, we have $\nu_Q(t) = 1$ for all points Q conjugate to P since ν_P and ν_Q are identical. In this case, t is called a **local parameter** at P. Again, a local parameter is not unique and always exists (see [151, pages 2–3] and [117, page 15]) for any closed point. For instance, in Example 1.3.3, let $p(y)$ be an irreducible polynomial with two distinct roots β_1 and β_2. Then $p(y)$ is a local parameter of the closed point containing both $(-b\beta_1/a - c/a, \beta_1)$ and $(-b\beta_2/a - c/a, \beta_2)$.

1.4   Riemann-Roch Theorem

The Riemann-Roch Theorem is arguably one of the most fundamental results in the study of algebraic curves. We begin our discussion by first considering an example.

Example 1.4.1 (i) Consider the set $\mathcal{P}_k$ of all the polynomials over $\mathbb{F}_q$ of degree less than or equal to k. This set is actually a linear space over $\mathbb{F}_q$ of dimension $k+1$. We now interpret this space using discrete valuations. Let ν_∞ be the discrete valuation of the line $\mathcal{L}$ corresponding to the "point at infinity" discussed in Example 1.3.3. Then $\mathcal{P}_k$ is the set $\{f \in \mathbb{F}_q(\mathcal{L}) = \mathbb{F}_q(y) : \nu_\infty(f) \geq -k, \nu_Q(f) \geq 0$ for all other "finite" points $Q\}$.

(ii) Let $\mathcal{E}$ be the elliptic curve defined in Example 1.1.1(i). Let $O = [0, 1, 0]$ be the "point at infinity." By Example 1.3.4(ii), there is a discrete valuation

ν_O corresponding to O. For $k \geq 0$, consider the set

$$V_k \overset{\text{def}}{=} \{f \in \mathbb{F}_2(\mathcal{E}) = \mathbb{F}_2(x,y) : \nu_O(f) \geq -k, \ \nu_Q(f) \geq 0 \ \text{ for all other}$$
$$\text{``finite'' points } Q\}.$$

Then it is easy to verify that V_k is an $\mathbb{F}_2$-linear space. In fact, it has a basis

$$\{x^i y^j : i \geq 0, \ 0 \leq j \leq 1, \ 2i + 3j \leq k\}.$$

By counting the number of elements in the above basis, we know that the dimension of V_k is equal to k if $k \geq 1$, and $\dim_{\mathbb{F}_2}(V_0) = 1$.

The above examples suggest that, to a given point P of a smooth curve $\mathcal{X}$ over $\mathbb{F}_q$ and a nonnegative integer k, we can associate a set over $\mathbb{F}_q$, denoted by $\mathcal{L}(kP)$, i.e.,

$$\mathcal{L}(kP) \overset{\text{def}}{=} \{f \in \mathbb{F}_q(\mathcal{X}) : \nu_P(f) \geq -k, \ \nu_Q(f) \geq 0 \ \text{ for all points } Q \neq P\}. \tag{1.3}$$

By Definition 1.3.2, we can verify that $\mathcal{L}(kP)$ is a linear space over $\mathbb{F}_q$. This space is called a **Riemann-Roch space**. The following theorem shows that this Riemann-Roch space is a finite dimensional space over $\mathbb{F}_q$ and its dimension can be determined.

Theorem 1.4.2 *Let $\mathcal{X}$ be a smooth projective curve over $\mathbb{F}_q$. Then there exists a nonnegative integer g, which is an invariant of $\mathcal{X}$, such that, for every $\mathbb{F}_q$-rational point P and integer $k \geq 0$, the dimension of the Riemann-Roch space $\mathcal{L}(kP)$ is at least $k+1-g$. Moreover, the dimension is exactly equal to $k+1-g$ if $k \geq 2g - 1$.*

The above theorem is a special case of Theorem 1.4.7 and is called the **Riemann-Roch Theorem**. The invariant g in the above theorem is called the **genus** of $\mathcal{X}$. We do not prove this theorem here as we are only interested in its applications. There are many books to which the reader may refer for the proof of this theorem and the definition of the genus of a curve (see [151, Theorem I.5.15], [61, page 108] and [75, page 295, Theorem 1.3]).

To calculate the dimension of a Riemann-Roch space, it is essential to know the genus of a curve. However, it is in general not easy to find this invariant even when the curve is explicitly given. Fortunately, for a smooth plane curve, the genus can be determined in terms of the degree of the polynomial defining this curve.

Theorem 1.4.3 *Let $\mathcal{X}$ be a projective (respectively, affine) smooth plane curve defined by $f(X,Y,Z) = 0$ (respectively, $h(x,y) = 0$). Then the genus of $\mathcal{X}$ is equal to $(d-1)(d-2)/2$, where d is the degree of $f(X,Y,Z)$ (respectively, $h(x,y)$).*

Again, we direct the reader to the existing literature available, e.g., [61, page 102, Proposition 5], for the proof of Theorem 1.4.3. To illustrate Theorems 1.4.2 and 1.4.3, we look at some examples now.

Example 1.4.4 (i) The genus of a projective (or affine) plane line is $(1 - 1)(1 - 2)/2 = 0$. Thus, for any $k \geq 0$ and $\mathbb{F}_q$-rational point P, we have $\dim_{\mathbb{F}_q}(\mathcal{L}(kP)) = k + 1 - 0 = k + 1$. This result coincides with Example 1.4.1(i).

(ii) Let $\mathcal{E}$ be the elliptic curve defined in Example 1.1.1(i). It is smooth and hence its genus is $(3 - 1)(3 - 2)/2 = 1$. Thus, for any $k \geq 1$ and $\mathbb{F}_q$-rational point P, we have $\dim_{\mathbb{F}_q}(\mathcal{L}(kP)) = k + 1 - 1 = k$. This result coincides with Example 1.4.1(ii).

(iii) By Example 1.1.10(iv), the projective Hermitian curve defined in Example 1.1.7(iii) is smooth. Hence, its genus is $(r + 1 - 1)(r + 1 - 2)/2 = r(r - 1)/2$.

(iv) Let $\mathcal{X}$ be the hyperelliptic curve defined in Example 1.1.1(ii). If the degree d of $f(x)$ is bigger than 3, then it is not smooth by Example 1.1.10(v). Thus, we cannot use the formula in Theorem 1.4.3 to find the genus of this curve, although we know that the genus of this curve is $\lfloor (d - 1)/2 \rfloor$ (see [151, Proposition VI.2.3]). On the other hand, if the degree of $f(x)$ is 3, i.e., $\mathcal{X}$ is an elliptic curve, then $\mathcal{X}$ is smooth and hence its genus is $(3 - 1)(3 - 2)/2 = 1$.

Next, we generalize the Riemann-Roch spaces defined above.

Let $\mathcal{X}$ be a smooth curve over $\mathbb{F}_q$. A **divisor** is a formal sum $\sum_{P \in \mathcal{X}} n_P P$, with $n_P \in \mathbb{Z}$ for all $P \in \mathcal{X}$, and $n_P = 0$ for all but finitely many points $P \in \mathcal{X}$. For a divisor $D = \sum_{P \in \mathcal{X}} n_P P$, we also denote the coefficient n_P by $\nu_P(D)$. Note that the points in a divisor may not be $\mathbb{F}_q$-rational. A divisor $\sum_{P \in \mathcal{X}} n_P P$ is said to be $\mathbb{F}_q$-**rational** if $n_P = n_Q$ for any two conjugate points P, Q. In this book, except for Chapter 3 where divisors may not necessarily be $\mathbb{F}_q$-rational, we always mean an $\mathbb{F}_q$-rational divisor whenever a divisor is mentioned.

The **divisor group** of $\mathcal{X}$, denoted by $\mathrm{Div}(\mathcal{X})$, is the free abelian group consisting of all $\mathbb{F}_q$-rational divisors, with the obvious formal addition as the group operation. A divisor $D = \sum_{P \in \mathcal{X}} n_P P$ is called **positive** (or **effective**), written as $D \geq 0$, if $n_P \geq 0$ for all points $P \in \mathcal{X}$. For two divisors D and G with $D - G$ being positive, we write $D \geq G$ or $G \leq D$.

The **degree** of a divisor $D = \sum_{P \in \mathcal{X}} n_P P$, denoted by $\deg(D)$, is defined to be $\sum_{P \in \mathcal{X}} n_P$. It is clear that $\deg$ defines a group homomorphism from $\mathrm{Div}(\mathcal{X})$ to $\mathbb{Z}$. The kernel of this homomorphism is a subgroup of $\mathrm{Div}(\mathcal{X})$, denoted by $\mathrm{Div}^0(\mathcal{X})$, i.e.,

$$\mathrm{Div}^0(\mathcal{X}) \stackrel{\text{def}}{=} \left\{ \sum_{P \in \mathcal{X}} n_P P \in \mathrm{Div}(\mathcal{X}) : \sum_{P \in \mathcal{X}} n_P = 0 \right\}.$$

Lemma 1.4.5 *Let $\mathcal{X}$ be a smooth projective curve over $\mathbb{F}_q$. Let z be an element in $\mathbb{F}_q(\mathcal{X}) \setminus \mathbb{F}_q$. Then we have*

$$\sum_{\nu_P(z)>0} \nu_P(z) = -\sum_{\nu_P(z)<0} \nu_P(z) = [\mathbb{F}_q(\mathcal{X}) : \mathbb{F}_q(z)],$$

where $[\mathbb{F}_q(\mathcal{X}) : \mathbb{F}_q(z)]$ denotes the degree of the field extension $\mathbb{F}_q(\mathcal{X})/\mathbb{F}_q(z)$.

The reader may refer to [151, Theorem I.5.15], [61, page 108] or [75, page 295, Theorem 1.3] for the proof of the above lemma.

If $z \in \mathbb{F}_q \setminus \{0\}$, we define $\mathrm{div}(z) = 0$, and for $z \in \mathbb{F}_q(\mathcal{X}) \setminus \mathbb{F}_q$, we have, by Lemma 1.4.5, that $[\mathbb{F}_q(\mathcal{X}) : \mathbb{F}_q(z)]$ is finite (see the definitions of function fields of one variable and curves in Section 1.2), hence we can define a divisor

$$\mathrm{div}(z) = \sum_{P \in \mathcal{X}} \nu_P(z)P.$$

Such a divisor is called a **principal divisor**. If $\nu_P(z) > 0$, the point P is called a **zero** of z, and if $\nu_P(z) < 0$, the point P is called a **pole** of z. The **zero divisor** and **pole divisor** of z are defined as

$$\mathrm{div}_0(z) = \sum_{P \in \mathcal{X}:\ \nu_P(z)>0} \nu_P(z)P \quad \text{and} \quad \mathrm{div}_\infty(z) = -\sum_{P \in \mathcal{X}:\ \nu_P(z)<0} \nu_P(z)P,$$

respectively. By Lemma 1.4.5, a principal divisor has degree $\deg(\mathrm{div}_0(z)) - \deg(\mathrm{div}_\infty(z)) = 0$. Furthermore, as

$$\mathrm{div}(yz) = \sum_{P \in \mathcal{X}} \nu_P(yz)P = \sum_{P \in \mathcal{X}} (\nu_P(y) + \nu_P(z))P = \mathrm{div}(y) + \mathrm{div}(z)$$

for any two nonzero elements $y, z \in \mathbb{F}_q(\mathcal{X})$, all principal divisors form a subgroup of $\mathrm{Div}^0(\mathcal{X})$, denoted by $\mathrm{Princ}(\mathcal{X})$. The quotient group $\mathrm{Div}^0(\mathcal{X})/\mathrm{Princ}(\mathcal{X})$ is called the **divisor class group** of degree zero. Two divisors D and G are said to be **equivalent** if there exists a nonzero element $z \in \mathbb{F}_q(\mathcal{X})$ such that $D = G + \mathrm{div}(z)$.

Now, for any divisor D of $\mathcal{X}$ over $\mathbb{F}_q$, we define the **Riemann-Roch space** of D by

$$\mathcal{L}(D) \overset{\text{def}}{=} \{z \in \mathbb{F}_q(\mathcal{X}) \setminus \{0\} :\ \mathrm{div}(z) + D \geq 0\} \cup \{0\}.$$

Then it is easy to verify that $\mathcal{L}(D)$ is a vector space over $\mathbb{F}_q$. If $D = kP$ for an integer k and an $\mathbb{F}_q$-rational point P, then $\mathcal{L}(D)$ coincides with the Riemann-Roch space defined in (1.3). If G is equivalent to D with $D = G + \mathrm{div}(z)$, then it is easy to verify that $\mathcal{L}(D) = z\mathcal{L}(G)$. Hence, $\dim \mathcal{L}(D) = \dim \mathcal{L}(G)$. In this book, the dimension $\dim \mathcal{L}(G)$ is also denoted by $\ell(G)$.

Lemma 1.4.6 *Let $\mathcal{X}$ be a smooth projective curve over $\mathbb{F}_q$ and let D be a divisor of $\mathcal{X}$ over $\mathbb{F}_q$. Then*

(i) $\mathcal{L}(D)$ *is a subspace of* $\mathcal{L}(G)$ *if* $D \leq G$;

(ii) $\mathcal{L}(0) = \mathbb{F}_q$;

(iii) $\mathcal{L}(D) = \{0\}$ *if* $\deg(D) < 0$.

Proof. (i) is clear.

(ii) It is clear that $\mathbb{F}_q$ is contained in $\mathcal{L}(0)$. Now let a be a nonzero element of $\mathcal{L}(0)$, then $\mathrm{div}(a) \geq 0$ by definition. This means that a has no poles. On the other hand, by Lemma 1.4.5, a has at least one pole if $a \notin \mathbb{F}_q$. Thus, a must be an element of $\mathbb{F}_q$.

(iii) For any nonzero element z of $\mathbb{F}_q(\mathcal{X})$, we have $\deg(\mathrm{div}(z)) = 0$ (for $z \in \mathbb{F}_q(\mathcal{X}) \setminus \mathbb{F}_q$, this follows from Lemma 1.4.5, whereas, for $z \in \mathbb{F}_q$, it is obvious that $\mathrm{div}(z) = 0$). Hence, $\deg(D + \mathrm{div}(z)) < 0$. This implies that $D + \mathrm{div}(z)$ cannot be a positive divisor. Therefore, $z \notin \mathcal{L}(D)$, i.e., $\mathcal{L}(D) = \{0\}$. □

The following theorem generalizes the Riemann-Roch Theorem in Theorem 1.4.2.

Theorem 1.4.7 *Let* $\mathcal{X}$ *be a smooth projective curve over* $\mathbb{F}_q$ *of genus* g *and let* D *be a divisor of* $\mathcal{X}$ *over* $\mathbb{F}_q$. *Then*

$$\dim \mathcal{L}(D) \geq \deg(D) + 1 - g.$$

Furthermore, if $\deg(D) \geq 2g - 1$, *we have* $\dim \mathcal{L}(D) = \deg(D) + 1 - g$.

The reader may refer to [151, Theorem I.5.15] or [61, page 108] for the proof of the above theorem.

Example 1.4.8 Consider the projective line $\mathcal{L}$ defined by $aZ + bY + cZ = 0$ with $a \neq 0$. The function field of $\mathcal{L}$ is the same as that of the affine line defined by $ax + by + c = 0$, which is $\mathbb{F}_q(y)$ by Example 1.2.2(i). Let $p(y)$ be an irreducible polynomial of degree d over $\mathbb{F}_q$. Then it has d conjugate roots $\beta, \beta^q, \ldots, \beta^{q^{d-1}}$. Thus, all points conjugate to $[-(b\beta + c)/a, \beta, 1]$ are $P_i \stackrel{\text{def}}{=} [-(b\beta^{q^i} + c)/a, \beta^{q^i}, 1]$ for $i = 0, \ldots, d-1$ (cf. paragraph before Theorem 1.3.6). Let D be the divisor $\sum_{i=0}^{d-1} P_i$. Then D is an $\mathbb{F}_q$-rational divisor and

$$\mathcal{L}(D) = \left\{ \frac{f(y)}{p(y)} : f(y) \in \mathbb{F}_q[y], \ \deg(f(y)) \leq d \right\}.$$

It is easy to see that it is a vector space of dimension $d + 1$ over $\mathbb{F}_q$ since $\{y^i/p(y)\}_{i=0}^{d}$ is a basis. This result coincides with Theorem 1.4.7.

1.5 Rational Points and Zeta Functions

Let $\mathcal{X}$ be a projective curve over $\mathbb{F}_q$ defined by a finite set S of polynomials. We are interested in the number of $\mathbb{F}_q$-rational points on $\mathcal{X}$. In fact, there are only finitely many $\mathbb{F}_q$-rational points on $\mathcal{X}$. To see this, we assume that $\mathcal{X}$ is contained in the projective space $\mathbf{P}^n$. Then the set of $\mathbb{F}_q$-rational points is a subset of $\mathbf{P}^n(\mathbb{F}_q)$. Now our conclusion follows from the following result.

Proposition 1.5.1 *The cardinality of $\mathbf{P}^n(\mathbb{F}_q)$ is $(q^{n+1} - 1)/(q - 1)$.*

Proof. It is clear that there are altogether $q^{n+1} - 1$ nonzero elements in $\mathbf{A}^{n+1}$. By Definition 1.1.3(ii), two elements of $\mathbf{A}^{n+1}$ are in the same equivalence class if and only if they belong to the same vector space of dimension 1 over $\mathbb{F}_q$. Thus, there are altogether $(q^{n+1} - 1)/(q - 1)$ equivalence classes in $\mathbf{A}^{n+1}$, i.e., there are $(q^{n+1} - 1)/(q - 1)$ points in $\mathbf{P}^n$. $\square$

Let $\mathcal{X}$ be a smooth projective plane curve defined by $f(X, Y, Z) = 0$, where $f(X, Y, Z)$ is a polynomial of degree d. For each $x = \alpha \in \mathbb{F}_q$, there are at most d solutions for the equation $f(\alpha, y, 1)$, thus there are at most dq "finite points" on this curve. Together with those "points at infinity," there are at most $dq + q + 1$ $\mathbb{F}_q$-rational points (note that all the "points at infinity" of a projective plane are $[\beta, 1, 0]$, with $\beta \in \mathbb{F}_q$, and $[1, 0, 0]$). This implies that the number of points on $\mathcal{X}$ is controlled by its degree d. In fact, we will see in this section that the number of points on a plane curve is controlled by its genus, which is $(d - 1)(d - 2)/2$.

However, no matter how large the degree of a smooth projective plane curve is, it has at most $q^2 + q + 1$ $\mathbb{F}_q$-rational points since all $\mathbb{F}_q$-rational points of this curve belong to $\mathbf{P}^2$. Therefore, in order to get more points, we need to use curves in spaces of higher dimensions.

Let $\mathcal{X}$ be a smooth projective curve in $\mathbf{P}^n$ defined by polynomials $f_1, \ldots, f_{n-1} \in \mathbb{F}_q[X_0, X_1, \ldots, X_n]$. Then $\mathcal{X}$ is a curve over $\mathbb{F}_q$. Since these polynomials belong to $\mathbb{F}_q[X_0, X_1, \ldots, X_n]$, it is natural to view these polynomials over $\mathbb{F}_{q^m}[X_0, X_1, \ldots, X_n]$, for all $m \geq 1$. Thus, $\mathcal{X}$ is also a curve defined over $\mathbb{F}_{q^m}$. Hence, we can speak of $\mathbb{F}_{q^m}$-rational points of $\mathcal{X}$. It is clear that all $\mathbb{F}_q$-rational points are $\mathbb{F}_{q^m}$-rational points, for any $m \geq 1$. Let N_m denote the number of $\mathbb{F}_{q^m}$-rational points on $\mathcal{X}$, then $N_m \geq N_1$ for all $m \geq 1$. Define a power series associated with $\mathcal{X}$ as follows:

$$Z_{\mathcal{X}}(t) = \mathrm{EXP} \left(\sum_{m=1}^{\infty} \frac{N_m}{m} t^m \right), \tag{1.4}$$

where $\mathrm{EXP}(x)$ stands for the function e^x. The above function is called the **zeta function** of $\mathcal{X}$. From the definition, we know that, once this function is given, we have information on the number of $\mathbb{F}_{q^m}$-rational points on $\mathcal{X}$, for all $m \geq 1$.

Before studying this zeta function, we first look at an example.

Example 1.5.2 Let $\mathcal{L}$ be the projective line defined by $aX + bY + cZ = 0$ over $\mathbb{F}_q$ with $(a, b, c) \neq (0, 0, 0)$. We may assume that $a \neq 0$. First, we consider the number of "finite points." For each $y = \beta \in \mathbb{F}_{q^m}$, there is a unique solution for the equation $ax + b\beta + c = 0$. This implies that we get q^m "finite points" $\{[-(b\beta+c)/a, \beta, 1] : \beta \in \mathbb{F}_{q^m}\}$. Together with the "point at infinity" $[-b, a, 0]$, we obtain altogether $q^m + 1$ $\mathbb{F}_{q^m}$-rational points on $\mathcal{L}$. Hence, we obtain, where ln is the natural logarithm,

$$
\begin{aligned}
\ln\left(Z_\mathcal{L}(t)\right) &= \sum_{m=1}^{\infty} \frac{q^m + 1}{m} t^m \\
&= \sum_{m=1}^{\infty} \frac{(qt)^m}{m} + \sum_{m=1}^{\infty} \frac{1}{m} t^m \\
&= \ln\left(\frac{1}{1 - qt}\right) + \ln\left(\frac{1}{1 - t}\right) \\
&= \ln\left(\frac{1}{(1 - t)(1 - qt)}\right),
\end{aligned}
$$

i.e.,

$$
Z_\mathcal{L}(t) = \frac{1}{(1 - t)(1 - qt)}.
$$

The above example shows that the zeta function of a projective line is a simple rational function. In fact, this is true for an arbitrary projective curve, due to the following **Weil Theorem**.

Theorem 1.5.3 *Let $\mathcal{X}$ be a smooth projective curve of genus g over $\mathbb{F}_q$. Then the zeta function of $\mathcal{X}$ is a rational function of the form*

$$
Z_\mathcal{X}(t) = \frac{L(t)}{(1 - t)(1 - qt)}, \tag{1.5}
$$

where $L(t)$ is a polynomial of degree $2g$ in $\mathbb{Z}[t]$ with $L(0) = 1$. Furthermore, if we factor $L(t)$ into a product $\prod_{i=1}^{2g}(1 - w_i t) \in \mathbb{C}[t]$, then $|w_i| = \sqrt{q}$ for all $1 \leq i \leq 2g$.

The numerator $L(t)$ of the zeta function in the above theorem is called the **L-function** of $\mathcal{X}$. The proof of the Weil Theorem was a breakthrough in the history of number theory and algebraic geometry. There are several proofs for the Weil Theorem, ranging from relatively elementary ones to more advanced ones using ℓ-adic cohomology. The reader may refer to [151, Chapter V] for an elementary proof and to [75, Appendix C] for an advanced proof.

One of the nice consequences of the Weil Theorem is that a bound on the number of $\mathbb{F}_{q^m}$-rational points on a curve can be derived.

Corollary 1.5.4 *Let $\mathcal{X}$ be a smooth projective curve of genus g with L-function $L(t) = \sum_{i=0}^{2g} a_i t^i$. Write $L(t)$ into a product $\prod_{i=1}^{2g}(1 - w_i t)$ and let N_m be the number of $\mathbb{F}_{q^m}$-rational points on $\mathcal{X}$. Then*

(i) $a_{2g-i} = q^{g-i} a_i$ *for all $0 \leq i \leq 2g$, in particular, $a_{2g} = q^g$;*

(ii) $N_m = q^m + 1 - \sum_{i=1}^{2g} w_i^m$ for all $m \geq 1$, in particular, $N_1 = q + 1 - \sum_{i=1}^{2g} w_i = q + 1 + a_1$;

(iii) **(Hasse-Weil bound)** $|N_m - q^m - 1| \leq 2g\sqrt{q^m}$ for all $m \geq 1$, in particular, $N_1 \leq q + 1 + 2g\sqrt{q}$.

Proof. (i) Consider the reciprocal polynomial of $L(t)$

$$\tilde{L}(t) \overset{\text{def}}{=} t^{2g} L\left(\frac{1}{t}\right) = \prod_{i=1}^{2g}(t - w_i).$$

As all the coefficients of $L(t)$ are real numbers, the complex conjugate of a root of $L(t)$ is again a root of $L(t)$. On the other hand, the complex conjugate of w_i is q/w_i since $|w_i| = \sqrt{q}$. This implies that $\prod_{i=1}^{2g} w_i = q^g$ and $L(t) = \prod_{i=1}^{2g}(1 - (q/w_i)t) = \prod_{i=1}^{2g}((q/w_i)t - 1)$. Thus,

$$\tilde{L}(qt) \overset{\text{def}}{=} (qt)^{2g} L\left(\frac{1}{qt}\right) = \prod_{i=1}^{2g}(qt - w_i) = q^g \prod_{i=1}^{2g}\left(\frac{q}{w_i}t - 1\right) = q^g L(t).$$

Comparing the coefficients of $\tilde{L}(qt)$ with those of $q^g L(t)$ gives the desired result.

(ii) Combining (1.4) and (1.5), we have

$$Z_{\mathcal{X}}(t) = \text{EXP}\left(\sum_{m=1}^{\infty} \frac{N_m}{m} t^m\right) = \frac{\prod_{i=1}^{2g}(1 - w_i t)}{(1 - t)(1 - qt)}.$$

By taking the logarithm for both sides of the above identity, then expanding the right-hand side into a power series and finally comparing coefficients of the two power series, we obtain $N_m = q^m + 1 - \sum_{i=1}^{2g} w_i^m$ for all $m \geq 1$. As $a_1 = -\sum_{i=1}^{2g} w_i$, we get $N_1 = q + 1 + a_1$.

(iii) By (ii), we have

$$|N_m - q^m - 1| = \left|\sum_{i=1}^{2g} w_i^m\right| \leq \sum_{i=1}^{2g} |w_i^m| = 2g\sqrt{q^m}.$$

The proof is completed. $\qquad\square$

From the above corollary, we see that there are only g unknown coefficients $a_1, \ldots, a_g$ in the L-function of $\mathcal{X}$ and these g coefficients can be determined by $N_1, \ldots, N_g$. This means that, once we know the numbers of $\mathbb{F}_{q^m}$-rational points for $m = 1, \ldots, g$, we can totally determine the zeta function of $\mathcal{X}$ and hence the number of $\mathbb{F}_{q^m}$-rational points for all $m \geq 1$.

Example 1.5.5 (i) Consider the plane curve $\mathcal{E}$ over $\mathbb{F}_3$ defined by the affine equation $y^2 = x^3 - x + 1$. It is easy to check that all points (including the "point at infinity" $[0, 1, 0]$) on $\mathcal{E}$ are nonsingular. Hence, the genus of this curve is 1. By Corollary 1.5.4, to find the zeta function of this curve, it is sufficient to determine the number of $\mathbb{F}_3$-rational points. It is easy to verify that all the $\mathbb{F}_3$-rational points of this curve are $(0, 1), (0, 2), (1, 1), (1, 2), (2, 1), (2, 2)$, together with the "point at infinity," i.e., $N_1 = 7$. Hence, by Corollary 1.5.4(ii), we have $a_1 = N_1 - 3 - 1 = 3$. Therefore, the zeta function is

$$Z_{\mathcal{E}}(t) = \frac{1 + 3t + 3t^2}{(1 - t)(1 - 3t)}.$$

From this zeta function, we can obtain the number of $\mathbb{F}_{3^m}$-rational points for all $m \geq 1$. The two reciprocal roots of $3t^2 + 3t + 1$ are $(-3 + \sqrt{-3})/2$ and $(-3 - \sqrt{-3})/2$. Hence, we have the number of $\mathbb{F}_{3^m}$-rational points on $\mathcal{E}$ as

$$N_m = 3^m + 1 - \left(\frac{-3 + \sqrt{-3}}{2}\right)^m - \left(\frac{-3 - \sqrt{-3}}{2}\right)^m.$$

For instance, there are seven $\mathbb{F}_9$-rational points and 28 $\mathbb{F}_{27}$-rational points.

(ii) Consider the Hermitian curve $\mathcal{H}$ over $\mathbb{F}_{r^2}$ defined in Example 1.1.1(iv). By Example 1.4.4(iii), the genus g of this curve is $r(r - 1)/2$. For any $x = \alpha \in \mathbb{F}_{r^2}$, $\alpha^{r+1} \in \mathbb{F}_r$. Thus, there is an element $\beta \in \mathbb{F}_{r^2}$ such that $\text{Tr}(\beta) = \alpha^{r+1}$, where Tr stands for the trace map from $\mathbb{F}_{r^2}$ to $\mathbb{F}_r$. Hence, the equation $y^r + y - \alpha^{r+1} = 0$ becomes $\text{Tr}(y - \beta) = 0$. We know that the kernel of the trace map has r elements (see [95, Theorem 2.23]), i.e., there are r solutions for this equation. This implies that, for any given $\alpha \in \mathbb{F}_{r^2}$, we get r points on $\mathcal{H}$. Together with the "point at infinity" $[0, 1, 0]$, we have $1 + r \cdot r^2 = 1 + r^3$ $\mathbb{F}_{r^2}$-rational points. Therefore, it follows from Corollary 1.5.4(ii) that

$$N_1 = 1 + r^3 = r^2 + 1 - \sum_{i=1}^{2g} w_i,$$

where $\{w_i\}_{i=1}^{2g}$ stand for the reciprocal roots of the L-function of $\mathcal{H}$. From the above identity, we have $-\sum_{i=1}^{2g} w_i = r^2(r - 1) = 2gr$. As $|w_i| = r$, we must have $w_i = -r$ for all $1 \leq i \leq 2g$. Finally, we conclude that the zeta function of $\mathcal{H}$ is

$$Z_{\mathcal{H}}(t) = \frac{(1 + rt)^{r(r-1)}}{(1 - t)(1 - r^2 t)}.$$

In particular, $N_2 = (r^2)^2 + 1 - 2g(-r)^2 = r^3 + 1 = N_1$. This implies that all the $\mathbb{F}_{r^4}$-rational points are also $\mathbb{F}_{r^2}$-rational.

For an algebraic curve $\mathcal{X}$, we denote by $N(\mathcal{X})$ the number of $\mathbb{F}_q$-rational points on $\mathcal{X}$. By the Hasse-Weil bound, we know that the number of $\mathbb{F}_q$-rational points on a curve is upper bounded in terms of q and its genus. One might wonder how large this number could be, i.e., we want to determine the values of the following quantity

$$N_q(g) = \max_{\mathcal{X}} N(\mathcal{X}),$$

where $\mathcal{X}$ ranges over all projective, smooth algebraic curves of genus g over $\mathbb{F}_q$.

One can imagine that it is not easy at all to determine the exact value $N_q(g)$ for an arbitrary pair (q, g). The complete solution to this problem has been found only for $g = 0, 1, 2$ (see [47, 139, 138, 140, 141]). The reader may refer to [164] and [116, pages 120–121] for tables on the values of $N_q(g)$ for some small values of q and g.

To look at the asymptotic behavior of $N_q(g)$ when q is fixed and g tends to ∞, we can define the following asymptotic quantity

$$A(q) \stackrel{\text{def}}{=} \limsup_{g \to \infty} \frac{N_q(g)}{g}.$$

An upper bound on $A(q)$ was given by Vlǎduţ and Drinfeld [167]:

$$A(q) \leq \sqrt{q} - 1.$$

For applications, it is more important to find lower bounds for this asymptotic quantity. Ihara [78] first showed, by using modular curves, that $A(q) \geq \sqrt{q} - 1$ for any square power q. This result determines the exact value of $A(q)$ for all square powers, i.e., for any square power q, we have

$$A(q) = \sqrt{q} - 1. \tag{1.6}$$

On the other hand, no single value of $A(q)$ is known if q is a nonsquare. However, some lower bounds have been obtained so far. For instance, by using modular curves and explicit function fields, Zink [182] and Bassa-Garcia-Stichtenoth [7] showed that

$$A(q^3) \geq \frac{2(q^2 - 1)}{q + 2}. \tag{1.7}$$

Serre [141, Part II, Theorem 24] made use of class field theory to show that there is an absolute positive constant c such that

$$A(q) \geq c \cdot \log(q)$$

for all prime powers q, where, here and throughout this book, log is taken to mean $\log_2$.

In another direction, lower bounds have already been obtained for $A(q)$, for small nonsquare q such as $q = 2, 3, 5, 7, 11, 13$, etc. For instance, in [180], Xing and Yeo showed that

$$A(2) \geq 0.258.$$

Chapter 2

Introduction to Error-Correcting Codes

2.1 Introduction

Cryptography is concerned with sensitive data transmission, storage, etc., while coding theory deals with the error-prone process of data transmission across noisy channels and data storage, via clever means, so that errors that occur can be corrected. Interestingly, though their primary objectives are different, much interplay between these two areas of research has been discovered. For example, the McEliece (or, equivalently, Niederreiter) cryptosystem is a public key cryptosystem based on error-correcting codes (see [105, 112]), though it does not fall within the scope of this book. However, we will see in some of the later chapters that error-correcting codes play prominent roles in cryptographic primitives such as secret sharing schemes and authentication codes. They also constitute one of the ways in which algebraic curves enter into cryptography, through algebraic geometry codes, i.e., error-correcting codes constructed using algebraic curves. This chapter serves to provide a brief introduction to error-correcting codes. For a more comprehensive introduction to the subject, the reader may refer to [96].

2.2 Linear Codes

For an integer $q \geq 2$, a q-ary **code** C of length n is a nonempty subset of

$$A^n \stackrel{\text{def}}{=} \{(u_1, \ldots, u_n) : u_i \in A\},$$

where A is a finite set of q elements, called the **code alphabet**. An element of C is called a **codeword** and an element of A^n is just called a "word."

If the size of C is M, we say that C is a q-ary (n, M)-code. If $q = 2, 3$, or 4, we call C a binary, ternary, or quaternary code, respectively.

Example 2.2.1 (i) Consider the source encoding of four characters, "A, B, C, D", as follows:

$$A \rightarrow 00, \quad B \rightarrow 01, \quad C \rightarrow 10, \quad D \rightarrow 11.$$

Suppose the message "A," which is encoded as 00, is transmitted over a noisy channel. It may become distorted and assume that 01 is received. Then, according to this encoding rule, the receiver decodes it to "B," which means that the communication fails in this case.

In this example, the collection of the encoded messages $\{00, 01, 10, 11\}$ forms a binary $(2, 4)$-code. This code does not allow correction or detection of errors.

(ii) We now introduce some redundancy in order to detect errors. For instance, we can modify the above encoding scheme as follows

$$A \rightarrow 00\textit{0}, \quad B \rightarrow 01\textit{1}, \quad C \rightarrow 10\textit{1}, \quad D \rightarrow 11\textit{0}.$$

Then we can detect one error. For example, suppose the message "A," which is encoded as 000, is transmitted over a noisy channel. It may become distorted and assume that 010 is received. Since 010 is not one of the four codewords, we know that there must be errors. The collection of the encoded messages $\{000, 011, 101, 110\}$ forms a binary $(3, 4)$-code. It can detect one error. However, it is less efficient than the previous $(2, 4)$-code as a 3-bit message carries only 2-bit information.

The **information rate** of a q-ary (n, M)-code is defined by $\mathcal{R}(C) \overset{\text{def}}{=} (\log_q M)/n$, where $\log_q$ is the logarithm to the base q. It is easy to see that $\mathcal{R}(C) \leq 1$ for all codes as $M = |C| \leq |A^n| = q^n$. In Example 2.2.1(i), the information rate of the code is $(\log_2 4)/2 = 1$, which achieves the maximum possible. However, the code does not allow any error detection. In Example 2.2.1(ii), the information rate of the code is $(\log_2 4)/3 = 2/3$, which means that a 3-bit message carries only 2-bit information. From this example, we see that the information rate measures the transmission speed of the message, i.e., the transmission efficiency of the code.

Thus, we know that, for a fixed length n, the size of a q-ary code measures the transmission efficiency of the code. A natural question is then which parameter(s) of a code would measure its error correction or detection capability. We now introduce this parameter, namely, the minimum distance of a code.

Definition 2.2.2 Let $\mathbf{x}$ and $\mathbf{y}$ be words of length n over an alphabet A. The **(Hamming) distance** from $\mathbf{x}$ to $\mathbf{y}$, denoted by $d(\mathbf{x}, \mathbf{y})$, is defined to be the number of places at which $\mathbf{x}$ and $\mathbf{y}$ differ. If $\mathbf{x} = x_1 \cdots x_n$ and $\mathbf{y} = y_1 \cdots y_n$, then

$$d(\mathbf{x}, \mathbf{y}) = d(x_1, y_1) + \cdots + d(x_n, y_n), \tag{2.1}$$

where x_i and y_i are regarded as words of length 1, and

$$d(x_i, y_i) = \begin{cases} 1 & \text{if } x_i \neq y_i \\ 0 & \text{if } x_i = y_i. \end{cases}$$

Example 2.2.3 (i) Let $A = \{0, 1\}$ and let $\mathbf{x} = 000000$, $\mathbf{y} = 011101$, $\mathbf{z} = 111111$. Then

$$d(\mathbf{x}, \mathbf{y}) = 4, \quad d(\mathbf{y}, \mathbf{z}) = 2, \quad d(\mathbf{z}, \mathbf{x}) = 6.$$

(ii) Let $A = \{0, 1, 2\}$ and let $\mathbf{x} = 1200$, $\mathbf{y} = 1012$, $\mathbf{z} = 0011$. Then

$$d(\mathbf{x}, \mathbf{y}) = 3, \quad d(\mathbf{y}, \mathbf{z}) = 2, \quad d(\mathbf{z}, \mathbf{x}) = 4.$$

Proposition 2.2.4 *Let* $\mathbf{x}, \mathbf{y}, \mathbf{z}$ *be words of length* n *over* A. *Then we have*

(i) $0 \leq d(\mathbf{x}, \mathbf{y}) \leq n$;

(ii) $d(\mathbf{x}, \mathbf{y}) = 0$ *if and only if* $\mathbf{x} = \mathbf{y}$;

(iii) $d(\mathbf{x}, \mathbf{y}) = d(\mathbf{y}, \mathbf{x})$;

(iv) **(Triangle inequality)** $d(\mathbf{x}, \mathbf{z}) \leq d(\mathbf{x}, \mathbf{y}) + d(\mathbf{y}, \mathbf{z})$.

Proof. (i), (ii), and (iii) are obvious from the definition of the Hamming distance. It is enough to prove (iv) when $n = 1$, which we now assume.

If $\mathbf{x} = \mathbf{z}$, then (iv) is obviously true since $d(\mathbf{x}, \mathbf{z}) = 0$.

If $\mathbf{x} \neq \mathbf{z}$, then either $\mathbf{y} \neq \mathbf{x}$ or $\mathbf{y} \neq \mathbf{z}$, so (iv) is again true. $\square$

Definition 2.2.5 For a code C containing at least two words, the **(minimum) distance** of C, denoted by $d(C)$, is

$$d(C) = \min\{d(\mathbf{x}, \mathbf{y}) : \mathbf{x}, \mathbf{y} \in C, \ \mathbf{x} \neq \mathbf{y}\}.$$

A code, of length n, size M, and distance d, is referred to as an (n, M, d)-**code**. The numbers n, M, d are called the **parameters** of the code.

Example 2.2.6 (i) Let $C = \{00000, 10112, 22222\}$ be a ternary code (i.e., with code alphabet $\{0, 1, 2\}$). Then $d(C) = 4$ since

$$d(00000, 10112) = 4,$$
$$d(00000, 22222) = 5,$$
$$d(10112, 22222) = 4.$$

Hence, C is a ternary (5,3,4)-code.

(ii) Let $C = \{000000, 111111, 222222\}$ be a ternary code. Then $d(C) = 6$ since

$$d(000000, 111111) = 6,$$
$$d(000000, 222222) = 6,$$
$$d(111111, 222222) = 6.$$

Hence, C is a ternary (6,3,6)-code.

Remark 2.2.7 For a q-ary (n, M, d)-code, the minimum distance d in fact measures its error detection and correction capabilities (see [96]).

We have just introduced arbitrary codes without algebraic or combinatorial structures. If we can equip codes with certain structures, it could make both theoretical study and practical applications easier. A natural way is to equip codes with a linear structure, i.e., make a code into a linear space. In order to do so, we set the code alphabet to be the finite field $\mathbb{F}_q$ of q elements. Then $\mathbb{F}_q^n$ is an $\mathbb{F}_q$-vector space of dimension n. A **linear code** of length n over $\mathbb{F}_q$ is a subspace of $\mathbb{F}_q^n$.

From now on in this section, we focus on linear codes.

A q-ary linear code of length n and dimension k is referred to as an $[n, k]$-linear code. Furthermore, if the minimum distance is d, it is referred to as an $[n, k, d]$-linear code.

One of the advantages of linear codes is that the minimum distance can be easily determined through the Hamming weight defined below.

Definition 2.2.8 Let $\mathbf{x}$ be a word in $\mathbb{F}_q^n$. The **(Hamming) weight** of $\mathbf{x}$, denoted by $\mathrm{wt}(\mathbf{x})$, is defined to be the number of nonzero coordinates in $\mathbf{x}$, i.e.,

$$\mathrm{wt}(\mathbf{x}) = d(\mathbf{x}, \mathbf{0}),$$

where $\mathbf{0}$ is the zero word.

Remark 2.2.9 For every element x of $\mathbb{F}_q$, we can define the Hamming weight as follows:

$$\mathrm{wt}(x) = d(x, 0) = \begin{cases} 1 & \text{if } x \neq 0 \\ 0 & \text{if } x = 0. \end{cases}$$

Then, writing $\mathbf{x} \in \mathbb{F}_q^n$ as $\mathbf{x} = (x_1, x_2, \ldots, x_n)$, the Hamming weight of $\mathbf{x}$ can also be equivalently defined as

$$\mathrm{wt}(\mathbf{x}) = \mathrm{wt}(x_1) + \mathrm{wt}(x_2) + \cdots + \mathrm{wt}(x_n). \tag{2.2}$$

Lemma 2.2.10 *If* $\mathbf{x}, \mathbf{y} \in \mathbb{F}_q^n$, *then* $d(\mathbf{x}, \mathbf{y}) = \mathrm{wt}(\mathbf{x} - \mathbf{y})$.

Proof. For $x, y \in \mathbb{F}_q$, $d(x, y) = 0$ if and only if $x = y$, which is true if and only if $x - y = 0$ or, equivalently, $\mathrm{wt}(x - y) = 0$. The desired result now follows from (2.1) and (2.2). $\qquad\square$

Definition 2.2.11 Let C be a code (not necessarily linear). The **minimum (Hamming) weight** of C, denoted $\mathrm{wt}(C)$, is the smallest of the weights of the nonzero codewords of C.

Theorem 2.2.12 *Let* C *be a linear code over* $\mathbb{F}_q$. *Then* $d(C) = \mathrm{wt}(C)$.

Proof. Recall that, for any words $\mathbf{x}, \mathbf{y}$, we have $d(\mathbf{x}, \mathbf{y}) = \text{wt}(\mathbf{x} - \mathbf{y})$.

By definition, there exist $\mathbf{x}', \mathbf{y}' \in C$ such that $d(\mathbf{x}', \mathbf{y}') = d(C)$, so

$$d(C) = d(\mathbf{x}', \mathbf{y}') = \text{wt}(\mathbf{x}' - \mathbf{y}') \geq \text{wt}(C),$$

since $\mathbf{x}' - \mathbf{y}' \in C$.

Conversely, there is a $\mathbf{z} \in C \backslash \{\mathbf{0}\}$ such that $\text{wt}(C) = \text{wt}(\mathbf{z})$, so

$$\text{wt}(C) = \text{wt}(\mathbf{z}) = d(\mathbf{z}, \mathbf{0}) \geq d(C).$$

This completes the proof. □

Example 2.2.13 Consider the binary linear code

$$C = \{000000, 111000, 000111, 111111\}.$$

We see that

$$\text{wt}(111000) = 3, \quad \text{wt}(000111) = 3, \quad \text{wt}(111111) = 6.$$

Hence, $d(C) = 3$.

For a q-ary linear code C, the **dual code** $C^{\perp}$ of C is defined to be the orthogonal complement of C, i.e.,

$$C^{\perp} = \{\mathbf{x} \in \mathbb{F}_q^n : \mathbf{x} \cdot \mathbf{c} = 0 \text{ for all } \mathbf{c} \in C\},$$

where $\cdot$ denotes the usual **inner product**. The **dual minimum distance** of C, denoted by $d^{\perp}(C)$ or $d(C^{\perp})$, is defined to be the minimum distance of $C^{\perp}$. If C is a subspace of $C^{\perp}$, then C is called a **self-orthogonal** code. Furthermore, if $C = C^{\perp}$, then C is said to be **self-dual**.

Example 2.2.14 The binary code $\{0000, 1111\}$ is self-orthogonal. Its dual code is $\{0000, 1111, 1100, 0110, 0011, 1001, 0101, 1010\}$.

The binary code $\{0000, 1100, 0011, 1111\}$ is self-dual.

Note that, in the Euclidean space $\mathbb{R}^n$ (where $\mathbb{R}$ denotes the real field), the intersection of a subspace with its orthogonal complement is always 0. However, this is no longer true in $\mathbb{F}_q^n$, as we have seen in the above example.

Since a linear code C is determined by a basis, we can form a matrix by putting the codewords in a basis as rows. Such a matrix is called a **generator matrix**. If $C \subseteq \mathbb{F}_q^n$ has dimension k, then a generator matrix of C has size $k \times n$. A generator matrix of the dual code of C is called a **parity-check matrix** of C. The size of a parity-check matrix is $(n - k) \times n$. It is clear that the code C is also determined by its parity-check matrices.

Theorem 2.2.15 *For a q-ary linear code C of length n, we have*

$$\dim(C) + \dim(C^{\perp}) = n,$$

where the dimension is over $\mathbb{F}_q$.

Proof. Let G be a generator matrix of C, then the rank $r(G)$ of G is $\dim(C)$. Then $C^{\perp}$ is the solution space of the system of equations $G\mathbf{x} = \mathbf{0}$. Hence, $\dim(C^{\perp}) = n - r(G) = n - \dim(C)$. The proof is completed. $\qquad\square$

Example 2.2.16 The matrices

$$\begin{pmatrix} 1 & 1 & 1 & 0 & 0 \\ 0 & 0 & 1 & 1 & 1 \end{pmatrix}, \qquad \begin{pmatrix} 1 & 1 & 0 & 0 & 0 \\ 0 & 0 & 0 & 1 & 1 \\ 0 & 1 & 1 & 1 & 0 \end{pmatrix}$$

are generator and parity-check matrices, respectively, of the binary code

$$\{00000, 11100, 00111, 11011\}.$$

If a generator matrix (respectively, parity-check matrix) of a linear code C is given, then all the parameters of C are determined. It is easy to see that the length and dimension of C can be determined by the size of these two matrices. Furthermore, the minimum distance of C is also determined by a parity-check matrix through the following result.

Theorem 2.2.17 *Let C be a linear code and let H be a parity-check matrix for C. Then*

(i) *C has distance $\geq d$ if and only if any $d - 1$ columns of H are linearly independent;*

(ii) *C has distance $\leq d$ if and only if H has d columns that are linearly dependent.*

Proof. Let $\mathbf{v} = (v_1, \ldots, v_n) \in C$ be a codeword of weight $e > 0$. Suppose the nonzero coordinates are in the positions $i_1, \ldots, i_e$, so that $v_j = 0$ if $j \notin \{i_1, \ldots, i_e\}$. Let $\mathbf{c}_i$ $(1 \leq i \leq n)$ denote the ith column of H.

By the definition of a parity-check matrix, it is easy to see that C contains a nonzero word $\mathbf{v} = (v_1, \ldots, v_n)$ of weight e (whose nonzero coordinates are $v_{i_1}, \ldots, v_{i_e}$) if and only if

$$\mathbf{0} = \mathbf{v}H^T = v_{i_1}\mathbf{c}_{i_1}^T + \cdots + v_{i_e}\mathbf{c}_{i_e}^T,$$

which is true if and only if there are e columns of H (namely, $\mathbf{c}_{i_1}, \ldots, \mathbf{c}_{i_e}$) that are linearly dependent. Here, ^{T} denotes the transpose of a matrix or a vector.

To say that the distance of C is $\geq d$ is equivalent to saying that C does not contain any nonzero word of weight $\leq d-1$, which is in turn equivalent to saying that any $\leq d - 1$ columns of H are linearly independent. This proves (i).

Similarly, to say that the distance of C is $\leq d$ is equivalent to saying that C contains a nonzero word of weight $\leq d$, which is in turn equivalent to saying that H has $\leq d$ columns (and hence d columns) that are linearly dependent. This proves (ii). $\qquad\square$

An immediate corollary of Theorem 2.2.17 is the following result.

Corollary 2.2.18 *Let C be a linear code and let H be a parity-check matrix for C. Then the following statements are equivalent:*

(i) *C has distance d;*

(ii) *any $d-1$ columns of H are linearly independent and H has d columns that are linearly dependent.*

Example 2.2.19 Let C be a binary linear code with a parity-check matrix

$$H = \begin{pmatrix} 1 & 0 & 0 & 1 & 1 & 1 \\ 0 & 1 & 0 & 1 & 0 & 1 \\ 0 & 0 & 1 & 0 & 1 & 1 \end{pmatrix}.$$

Then its minimum distance is 3 since any two columns are linearly independent, while the first two and the fourth columns are linearly dependent.

For a binary linear code, its extended code is obtained by adding a parity-check coordinate. This idea can be generalized to codes over any finite field.

Definition 2.2.20 For any code C over $\mathbb{F}_q$, the **extended code of** C, denoted by $\overline{C}$, is defined to be

$$\overline{C} = \left\{ (c_1, \ldots, c_n, -\sum_{i=1}^{n} c_i) \; : \; (c_1, \ldots, c_n) \in C \right\}.$$

When $q = 2$, the extra coordinate $-\sum_{i=1}^{n} c_i = \sum_{i=1}^{n} c_i$ added to the codeword $(c_1, \ldots, c_n)$ is called the **parity-check** coordinate.

Theorem 2.2.21 *If C is an (n, M, d)-code over $\mathbb{F}_q$, then $\overline{C}$ is an $(n + 1, M, d')$-code over $\mathbb{F}_q$, with $d \le d' \le d + 1$. If C is linear, then so is $\overline{C}$. Moreover, when C is linear,*

$$\left(\begin{array}{c|c} H & \begin{matrix} 0 \\ \vdots \\ 0 \end{matrix} \\ \hline 1 \cdots 1 & 1 \end{array} \right)$$

is a parity-check matrix of $\overline{C}$ if H is a parity-check matrix of C.

The proof is straightforward, so it is left to the reader.

Example 2.2.22 (i) Consider the binary linear code

$$C_1 = \{0000, 1100, 0011, 1111\}.$$

It has parameters $[4, 2, 2]$. The extended code

$$\overline{C_1} = \{00000, 11000, 00110, 11110\}$$

is a binary $[5, 2, 2]$-linear code.

(ii) Consider the binary linear code $C_2 = \{0000, 0111, 0011, 0100\}$. It has parameters $[4, 2, 1]$. The extended code

$$\overline{C_2} = \{00000, 01111, 00110, 01001\}$$

is a binary $[5, 2, 2]$-linear code.

This example shows that the minimum distance $d(\overline{C})$ can achieve both $d(C)$ and $d(C) + 1$.

The Golay codes were discovered by Golay in the late 1940s. It turns out that the Golay codes are essentially unique in the sense that binary codes with the same parameters as them can be shown to be equivalent to them.[1]

Definition 2.2.23 Let G be the 12×24 matrix

$$G = (I_{12}|A),$$

where I_{12} is the 12×12 identity matrix and A is the 12×12 matrix

$$A = \begin{pmatrix} 0 & 1 & 1 & 1 & 1 & 1 & 1 & 1 & 1 & 1 & 1 & 1 \\ 1 & 1 & 1 & 0 & 1 & 1 & 1 & 0 & 0 & 0 & 1 & 0 \\ 1 & 1 & 0 & 1 & 1 & 1 & 0 & 0 & 0 & 1 & 0 & 1 \\ 1 & 0 & 1 & 1 & 1 & 0 & 0 & 0 & 1 & 0 & 1 & 1 \\ 1 & 1 & 1 & 1 & 0 & 0 & 0 & 1 & 0 & 1 & 1 & 0 \\ 1 & 1 & 1 & 0 & 0 & 0 & 1 & 0 & 1 & 1 & 0 & 1 \\ 1 & 1 & 0 & 0 & 0 & 1 & 0 & 1 & 1 & 0 & 1 & 1 \\ 1 & 0 & 0 & 0 & 1 & 0 & 1 & 1 & 0 & 1 & 1 & 1 \\ 1 & 0 & 0 & 1 & 0 & 1 & 1 & 0 & 1 & 1 & 1 & 0 \\ 1 & 0 & 1 & 0 & 1 & 1 & 0 & 1 & 1 & 1 & 0 & 0 \\ 1 & 1 & 0 & 1 & 1 & 0 & 1 & 1 & 1 & 0 & 0 & 0 \\ 1 & 0 & 1 & 1 & 0 & 1 & 1 & 1 & 0 & 0 & 0 & 1 \end{pmatrix}.$$

The binary linear code with generator matrix G is called the **extended binary Golay code** and will be denoted by G_{24}.

Let $\hat{G}$ be the 12×23 matrix

$$\hat{G} = (I_{12}|\hat{A}),$$

where I_{12} is the 12×12 identity matrix and $\hat{A}$ is the 12×11 matrix obtained from the matrix A by deleting the last column of A. The binary linear code with generator matrix $\hat{G}$ is called the **binary Golay code** and will be denoted by G_{23}.

As a generator matrix of a Golay code is explicitly given, we can easily determine its parameters.

---

[1] Two codes are said to be **equivalent** if one can be obtained from the other by permuting coordinates, multiplying nonzero scalars to coordinates, or composing operations of both types.

Theorem 2.2.24 *The extended binary Golay code G_{24} is a $[24, 12, 8]$-linear code, while the Golay code G_{23} is a $[23, 8, 7]$-linear code. Furthermore, G_{24} is self-dual.*

We leave the proof of this theorem to the reader.

Finally, we introduce a class of linear codes, Reed-Muller codes, in this section.

There are many ways to define the Reed-Muller codes. We choose an inductive definition. Note that we consider only binary Reed-Muller codes here.

Definition 2.2.25 The **(first-order) Reed-Muller codes** $\mathcal{R}(1, m)$ are binary codes defined, for all integers $m \geq 1$, recursively as follows:

(i) $\mathcal{R}(1,1) = \mathbb{F}_2^2 = \{00, 01, 10, 11\}$.

(ii) For $m \geq 1$,

$$\mathcal{R}(1, m+1) = \{(\mathbf{u}, \mathbf{u}) \; : \; \mathbf{u} \in \mathcal{R}(1, m)\} \cup \{(\mathbf{u}, \mathbf{u} + \mathbf{1}) \; : \; \mathbf{u} \in \mathcal{R}(1, m)\},$$

where $\mathbf{1}$ is the all-one vector.

The parameters of the Reed-Muller codes can be readily determined as well.

Theorem 2.2.26 *For $m \geq 1$, the Reed-Muller code $\mathcal{R}(1, m)$ is a binary $[2^m, m+1, 2^{m-1}]$-linear code, in which every codeword, except $\mathbf{0}$ and $\mathbf{1}$, has weight 2^{m-1}.*

We leave the proof of this theorem to the reader again.

2.3 Bounds

Most of the results and their proofs in this section are straightforward, so we omit the proofs. The reader may refer to [96] for the detailed proofs.

To consider codes with good parameters, we need to introduce the following quantity, one of the most important for the theoretical study of error-correcting codes.

Definition 2.3.1 For a given code alphabet A of size q (with $q > 1$) and given values of n and d, let $A_q(n, d)$ denote the largest possible size M for which there exists an (n, M, d)-code over A. Thus

$$A_q(n, d) = \max\{M : \text{there exists an } (n, M, d)\text{-code over } A\}.$$

Any (n, M, d)-code C that has the maximum size, that is, for which $M = A_q(n, d)$, is called an **optimal code**.

It is clear that $A_q(n, d)$ depends only on the size of A, n, and d. It is independent of the choice of A (except for its size).

The numbers $A_q(n, d)$ play a central role in coding theory and much effort has been made in determining their values. In fact, the problem of determining the values of $A_q(n, d)$ is sometimes known as the **main coding theory problem**.

Instead of considering all codes, we may restrict ourselves to linear codes and obtain the following similar definition:

Definition 2.3.2 For a given prime power q and given values of n and d, let $B_q(n, d)$ denote the largest possible size q^k for which there exists an $[n, k, d]$-code over $\mathbb{F}_q$. Thus

$$B_q(n, d) = \max\{q^k : \text{there exists an } [n, k, d]\text{-code over } \mathbb{F}_q\}.$$

While it is in general rather difficult to determine the exact values of $A_q(n, d)$ and $B_q(n, d)$, there are some properties that afford easy proofs.

Theorem 2.3.3 ([96, Theorem 5.1.7]) *Let $q \geq 2$ be a prime power. Then*

(i) $B_q(n, d) \leq A_q(n, d) \leq q^n$ *for all $1 \leq d \leq n$;*

(ii) $B_q(n, 1) = A_q(n, 1) = q^n$;

(iii) $B_q(n, n) = A_q(n, n) = q$.

Note that all the results on $A_q(n, d)$ in Theorem 2.3.3 are still true for any positive integers $q > 1$.

In the case of binary codes, there are additional elementary results on $A_2(n, d)$ and $B_2(n, d)$.

Theorem 2.3.4 ([96, Theorem 5.1.11]) *Suppose d is odd.*

(i) *Then a binary (n, M, d)-code exists if and only if a binary $(n+1, M, d+1)$-code exists. Therefore, if d is odd, $A_2(n+1, d+1) = A_2(n, d)$.*

(ii) *Similarly, a binary $[n, k, d]$-linear code exists if and only if a binary $[n+1, k, d+1]$-linear code exists, so $B_2(n+1, d+1) = B_2(n, d)$.*

Remark 2.3.5 The last statement in Theorem 2.3.4(i) is equivalent to: if d is even, then $A_2(n, d) = A_2(n-1, d-1)$. There is also an analogue for (ii).

While the determination of the exact values of $A_q(n, d)$ and $B_q(n, d)$ may be difficult, several well-known bounds, both upper and lower ones, do exist. We discuss some of them in the rest of this section.

A list of lower bounds and, in some cases, exact values for $A_2(n, d)$ may be found at the following web page maintained by Simon Litsyn of Tel Aviv University:

http://www.eng.tau.ac.il/~litsyn/tableand/index.html.

The following website, maintained by Markus Grassl, contains tables which give the best known bounds (upper and lower) on the distance d for q-ary linear codes ($q \leq 9$) of given length and dimension:

http://www.codetables.de.

We discuss five well-known bounds: the sphere-covering bound, the Gilbert-Varshamov bound, the sphere-packing bound, the Singleton bound, and the Plotkin bound.

To study these bounds, we have to consider a sphere of a given radius and its volume.

Definition 2.3.6 Let A be an alphabet of size q, where $q > 1$. For any vector $\mathbf{u} \in A^n$ and any integer $r \geq 0$, the **sphere** of radius r and center $\mathbf{u}$, denoted $S_A(\mathbf{u}, r)$, is the set $\{\mathbf{v} \in A^n : d(\mathbf{u}, \mathbf{v}) \leq r\}$.

Definition 2.3.7 For a given integer $q > 1$, a positive integer n and an integer $r \geq 0$, define $V_q^n(r)$ to be

$$V_q^n(r) = \begin{cases} \binom{n}{0} + \binom{n}{1}(q-1) + \binom{n}{2}(q-1)^2 + \cdots + \binom{n}{r}(q-1)^r & \text{if } 0 \leq r \leq n \\ q^n & \text{if } n \leq r. \end{cases}$$

Lemma 2.3.8 ([96, Lemma 5.2.3]) *For all integers $r \geq 0$, a sphere of radius r in A^n contains exactly $V_q^n(r)$ vectors, where A is an alphabet of size $q > 1$.*

We are now ready to state the sphere-covering bound.

Theorem 2.3.9 (Sphere-covering bound)([96, Theorem 5.2.4]) *For an integer $q > 1$ and integers n, d such that $1 \leq d \leq n$, we have*

$$\frac{q^n}{\sum_{i=0}^{d-1} \binom{n}{i}(q-1)^i} = \frac{q^n}{V_q^n(d-1)} \leq A_q(n, d).$$

The Gilbert-Varshamov bound is a lower bound for $B_q(n, d)$ (i.e., for linear codes) known since the 1950s. There is also an asymptotic version of the Gilbert-Varshamov bound (see Section 2.5), which concerns infinite sequences of codes whose lengths tend to infinity. For a long time, the asymptotic Gilbert-Varshamov bound was the best lower bound known to be attainable by an infinite family of linear codes, so it became a kind of benchmark for judging the "goodness" of an infinite sequence of linear codes. Between 1977 and 1982, Goppa constructed algebraic geometry codes using algebraic curves over finite fields with many rational points. A major breakthrough in coding theory was achieved shortly after these discoveries when it was shown that there are sequences of algebraic geometry codes that perform better than the asymptotic Gilbert-Varshamov bound for certain sufficiently large q (see Theorem 2.5.3).

Theorem 2.3.10 (Gilbert-Varshamov bound)([96, Theorem 5.2.6]) *Let* n, k, *and* d *be integers satisfying* $2 \leq d \leq n$ *and* $1 \leq k \leq n$. *If*

$$V_q^{n-1}(d-2) = \sum_{i=0}^{d-2} \binom{n-1}{i}(q-1)^i < q^{n-k}, \tag{2.3}$$

then there exists an $[n, k]$-*linear code over* $\mathbb{F}_q$ *with minimum distance at least* d.

Corollary 2.3.11 ([96, Corollary 5.2.7]) *For a prime power* $q > 1$ *and integers* n, d *such that* $2 \leq d \leq n$, *we have*

$$B_q(n, d) \geq q^{n - \lceil \log_q(V_q^{n-1}(d-2)+1) \rceil} \geq \frac{q^{n-1}}{V_q^{n-1}(d-2)},$$

where, for a real number x, $\lceil x \rceil$ *is the smallest integer greater than or equal to* x.

The first upper bound for $A_q(n, d)$ that we discuss is the Hamming bound, also known as the sphere-packing bound.

Theorem 2.3.12 (Hamming or sphere-packing bound)([96, Theorem 5.3.1]) *For an integer* $q > 1$ *and integers* n, d *such that* $1 \leq d \leq n$, *we have*

$$A_q(n, d) \leq \frac{q^n}{\sum_{i=0}^{\lfloor (d-1)/2 \rfloor} \binom{n}{i}(q-1)^i} = \frac{q^n}{V_q^n(\lfloor (d-1)/2 \rfloor)},$$

where, for a real number x, $\lfloor x \rfloor$ *is the largest integer less than or equal to* x.

It is natural to wonder if there exist any codes attaining the sphere-packing bound.

Definition 2.3.13 A q-ary code that attains the Hamming (or sphere-packing) bound, i.e., one which has $q^n \big/ \left(\sum_{i=0}^{\lfloor (d-1)/2 \rfloor} \binom{n}{i}(q-1)^i \right)$ codewords, is called a **perfect code**.

Some of the earliest known codes, such as the Hamming codes and the Golay codes, are perfect codes.

Hamming codes were discovered by Richard W. Hamming. They form an important class of codes; they have interesting properties and are easy to encode and decode.

Let $q \geq 2$ be any prime power. Note that any nonzero vector $\mathbf{v} \in \mathbb{F}_q^m$ generates a subspace $<\mathbf{v}>$ of dimension 1. Furthermore, for $\mathbf{v}, \mathbf{w} \in \mathbb{F}_q^m \backslash \{\mathbf{0}\}$, $<\mathbf{v}> = <\mathbf{w}>$ if and only if there is a nonzero scalar $\lambda \in \mathbb{F}_q \backslash \{0\}$ such that $\mathbf{v} = \lambda \mathbf{w}$. Therefore, there are exactly $(q^m - 1)/(q - 1)$ distinct subspaces of dimension 1 in $\mathbb{F}_q^m$.

Definition 2.3.14 Let $m \geq 2$. A q-ary linear code, whose parity-check matrix H has the property that the columns of H are made up of precisely one nonzero vector from each vector subspace of dimension 1 of $\mathbb{F}_q^m$, is called a q-**ary Hamming code**, often denoted as $\mathrm{Ham}(m, q)$.

Proposition 2.3.15 (Parameters of the q-ary Hamming codes)([96, Proposition 5.3.15]) *The Hamming code* $\mathrm{Ham}(m, q)$ *is a* $[(q^m-1)/(q-1), (q^m-1)/(q-1) - m, 3]$-*code.*

Definition 2.3.16 The dual of the q-ary Hamming code $\mathrm{Ham}(m, q)$ is called a q-ary **simplex code**. It is sometimes denoted by $S(m, q)$.

The next bound, the Singleton bound, is an interesting one that is related to the well-known Reed-Solomon codes.

Theorem 2.3.17 (Singleton bound)([96, Theorem 5.4.1]) *For any integer* $q > 1$, *any positive integer* n *and any integer* d *such that* $1 \leq d \leq n$, *we have*

$$A_q(n, d) \leq q^{n-d+1}.$$

In particular, when q *is a prime power, the parameters* $[n, k, d]$ *of any linear code over* $\mathbb{F}_q$ *satisfy*

$$k + d \leq n + 1.$$

Remark 2.3.18 The following is an easy direct proof for the inequality $k + d \leq n + 1$ in the case of an $[n, k, d]$-linear code C:

Given any parity-check matrix H for C, the row rank, and hence the rank, of H is, by definition, $n - k$. Therefore, any $n - k + 1$ columns of H form a linearly dependent set. By Theorem 2.2.17(ii), $d \leq n - k + 1$.

Definition 2.3.19 A linear code with parameters $[n, k, d]$ such that $k + d = n + 1$ is called a **maximum distance separable (MDS) code**.

Remark 2.3.20 An alternative way to state the Singleton bound is: for any q-ary (n, M, d)-code C, we have

$$\mathcal{R}(C) + \delta(C) \leq 1,$$

where $\mathcal{R}(C)$ is the information rate given by $(\log_q M)/n$ and $\delta(C)$ is the **relative minimum distance** $(d(C) - 1)/n$. A linear code C is MDS if and only if $\mathcal{R}(C) + \delta(C) = 1$.

One of the interesting properties of MDS codes is the following.

Theorem 2.3.21 ([96, Theorem 5.4.5]) *Let* C *be a linear code over* $\mathbb{F}_q$ *with parameters* $[n, k, d]$. *Let* G, H *be a generator matrix and a parity-check matrix, respectively, for* C. *Then, the following statements are equivalent:*

(i) *C is an MDS code;*

(ii) *every set of $n - k$ columns of H is linearly independent;*

(iii) *every set of k columns of G is linearly independent;*

(iv) $C^\perp$ *is an MDS code.*

Definition 2.3.22 An MDS code C over $\mathbb{F}_q$ is **trivial** if and only if C satisfies one of the following:

(i) $C = \mathbb{F}_q^n$;

(ii) C is equivalent to the code generated by $\mathbf{1} = (1, \ldots, 1)$; or

(iii) C is equivalent to the dual of the code generated by $\mathbf{1}$.

Otherwise, C is said to be **nontrivial**.

An interesting family of examples of MDS codes is given by the (generalized) Reed-Solomon codes.

Definition 2.3.23 Let $n \leq q$. Let $\boldsymbol{\alpha} = (\alpha_1, \alpha_2, \ldots, \alpha_n)$, where α_i $(1 \leq i \leq n)$ are distinct elements of $\mathbb{F}_q$. Let $\mathbf{v} = (v_1, v_2, \ldots, v_n)$, where $v_i \in \mathbb{F}_q^*$ for all $1 \leq i \leq n$. For $k \leq n$, the **generalized Reed-Solomon code** $GRS_k(\boldsymbol{\alpha}, \mathbf{v})$ is defined to be

$$\{(v_1 f(\alpha_1), v_2 f(\alpha_2), \ldots, v_n f(\alpha_n)) \ : \ f(x) \in \mathbb{F}_q[x] \text{ and } \deg(f(x)) < k\}.$$

The elements $\alpha_1, \alpha_2, \ldots, \alpha_n$ are called the **code locators** of $GRS_k(\boldsymbol{\alpha}, \mathbf{v})$.

In the literature, the generalized Reed-Solomon code $GRS_k(\boldsymbol{\alpha}, \mathbf{1})$ is often referred to as a **Reed-Solomon code**.

It is easy to verify that generalized Reed-Solomon codes are linear.

Theorem 2.3.24 *The generalized Reed-Solomon code $GRS_k(\boldsymbol{\alpha}, \mathbf{v})$ has parameters $[n, k, n - k + 1]$, so it is an MDS code.*

Proof. It is obvious that $GRS_k(\boldsymbol{\alpha}, \mathbf{v})$ has length n. It is clear that a nonzero polynomial gives a nonzero codeword since a nonzero polynomial of degree at most $k - 1$ has at most $k - 1$ zeros. Thus, the dimension of the code is the same as that of the space of polynomials of degree $< k$, i.e., the dimension is k. It remains to show that the minimum distance of $GRS_k(\boldsymbol{\alpha}, \mathbf{v})$ is $n - k + 1$.

To do this, we count the maximum number of zeros in a nonzero codeword. Suppose $f(x)$ is not identically zero. Since $\deg(f(x)) < k$, the polynomial $f(x)$ can only have at most $k - 1$ zeros, i.e., the codeword $(v_1 f(\alpha_1), v_2 f(\alpha_2), \ldots, v_n f(\alpha_n))$ has at most $k - 1$ zeros among its coordinates. In other words, its weight is at least $n - k + 1$, so the minimum distance d of $GRS_k(\boldsymbol{\alpha}, \mathbf{v})$ satisfies $d \geq n - k + 1$. However, the Singleton bound shows that $d \leq n - k + 1$, so $d = n - k + 1$. Hence, $GRS_k(\boldsymbol{\alpha}, \mathbf{v})$ is MDS. $\square$

It is interesting to note that the dual of a generalized Reed-Solomon code is again a generalized Reed-Solomon code.

Theorem 2.3.25 *The dual of the generalized Reed-Solomon code $GRS_k(\boldsymbol{\alpha}, \mathbf{v})$ over $\mathbb{F}_q$ of length n is $GRS_{n-k}(\boldsymbol{\alpha}, \mathbf{v}')$ for some $\mathbf{v}' \in (\mathbb{F}_q^*)^n$.*

Proof. First, let $k = n - 1$. From Theorems 2.3.21 and 2.2.15, the dual of $GRS_{n-1}(\boldsymbol{\alpha}, \mathbf{v})$ is an MDS code of dimension 1, so it has parameters $[n, 1, n]$. In particular, its basis consists of a vector $\mathbf{v}' = (v_1', \ldots, v_n')$, where $v_i' \in \mathbb{F}_q^*$ for all $1 \leq i \leq n$. Clearly, this dual code is $GRS_1(\boldsymbol{\alpha}, \mathbf{v}')$.

It follows, in particular, that, for all $f(x) \in \mathbb{F}_q[x]$ of degree $< n - 1$, we have

$$v_1 v_1' f(\alpha_1) + \cdots + v_n v_n' f(\alpha_n) = 0, \tag{2.4}$$

where $\mathbf{v} = (v_1, \ldots, v_n)$.

Now, for arbitrary k, we claim that $GRS_k(\boldsymbol{\alpha}, \mathbf{v})^\perp = GRS_{n-k}(\boldsymbol{\alpha}, \mathbf{v}')$.

A typical codeword in $GRS_k(\boldsymbol{\alpha}, \mathbf{v})$ is $(v_1 f(\alpha_1), \ldots, v_n f(\alpha_n))$, where $f(x) \in \mathbb{F}_q[x]$ with degree $\leq k - 1$, while a typical codeword in $GRS_{n-k}(\boldsymbol{\alpha}, \mathbf{v}')$ has the form $(v_1' g(\alpha_1), \ldots, v_n' g(\alpha_n))$, with $g(x) \in \mathbb{F}_q[x]$ of degree $\leq n - k - 1$. Since $\deg(f(x)g(x)) \leq n - 2 < n - 1$, we have

$$(v_1 f(\alpha_1), \ldots, v_n f(\alpha_n)) \cdot (v_1' g(\alpha_1), \ldots, v_n' g(\alpha_n)) = 0$$

from (2.4).

Therefore, $GRS_{n-k}(\boldsymbol{\alpha}, \mathbf{v}') \subseteq GRS_k(\boldsymbol{\alpha}, \mathbf{v})^\perp$. Comparing the dimensions of both codes, the theorem follows. $\square$

Now we introduce the last bound for this section.

Theorem 2.3.26 (Plotkin bound)([96, Theorem 5.5.2]) *For any integer $q > 1$, any positive integer n and any integer d such that $rn < d$, where $r = 1 - q^{-1}$, we have*

$$A_q(n, d) \leq \left\lfloor \frac{d}{d - rn} \right\rfloor.$$

Remark 2.3.27 The Plotkin bound holds for codes for which d is large relative to n. It is often better than many of the other upper bounds, although it is only applicable to a smaller range of values of d.

2.4 Algebraic Geometry Codes

Algebraic geometry codes play important new roles in both coding theory and cryptography, the latter of which is the theme of this book. We introduce them in this section.

First, we look at some examples.

Example 2.4.1 (i) Let $\mathcal{L}$ be the plane line over $\mathbb{F}_q$ defined by $ax+by+c = 0$ with $a \neq 0$. Denote by ∞ the point $[-b, a, 0]$. Let $P_i = (-(b\alpha_i + c)/a, \alpha_i)$ $(i = 1, \ldots, n$ with $n \leq q)$ be n distinct $\mathbb{F}_q$-rational points. Consider the code

$$\{(f(P_1), \ldots, f(P_n)) : f \in \mathcal{L}(k\infty)\}.$$

By Example 1.4.1(i), the set $\mathcal{L}(k\infty)$ consists of all univariate polynomials $f(y)$ of degree at most k. Thus, the codeword $(f(P_1), \ldots, f(P_n))$ is the same as $(f(\alpha_1), f(\alpha_2), \ldots, f(\alpha_n))$. Hence, when we describe the above code in the polynomial language, it is none other than

$$\{(f(\alpha_1), \ldots, f(\alpha_n)) : f \in \mathbb{F}_q[y], \ \deg(f) \leq k\}.$$

This is precisely a generalized Reed-Solomon code discussed in the previous section.

(ii) Let $\mathcal{H}$ be the Hermitian curve defined by $y^r + y = x^{r+1}$ over $\mathbb{F}_{r^2}$. Denote by ∞ the point $[0, 1, 0]$. Let $P_i = (\alpha_i, \beta_i)$ $(i = 1, \ldots, n$ with $n \leq r^3)$ be n distinct $\mathbb{F}_{r^2}$-rational points. Consider the code over $\mathbb{F}_{r^2}$ defined by

$$\{(f(P_1), \ldots, f(P_n)) : f \in \mathcal{L}(m\infty)\}.$$

Assume that $n > m \geq 2g - 1 = r(r-1) - 1$ with g being the genus of $\mathcal{H}$. Then, by the Riemann-Roch Theorem, we have $\dim \mathcal{L}(m\infty) = m + 1 - g$. On the other hand, by the fact that y is algebraic over the field $\mathbb{F}_q(x)$ of degree r, we know that the set $\{x^i y^j : 0 \leq ri + (r+1)j \leq m, i \geq 0$ and $0 \leq j \leq r-1\}$ is linearly independent over $\mathbb{F}_{r^2}$ and it is contained in $\mathcal{L}(m\infty)$. By counting the number of elements in this set, we conclude that it is a basis of $\mathcal{L}(m\infty)$. Therefore, the code is

$$\left\{ \begin{array}{l} (f(\alpha_1, \beta_1), \ldots, f(\alpha_n, \beta_n)) : \\[2mm] \qquad f(x, y) = \displaystyle\sum_{0 \leq ri+(r+1)j \leq m, i \geq 0, \ 0 \leq j \leq r-1} a_{ij} x^i y^j \in \mathbb{F}_{r^2}[x, y] \end{array} \right\}.$$

We will investigate the parameters of this code later.

For $\mathcal{X}$ a smooth curve over $\mathbb{F}_q$ and a divisor $D = \sum_{P \in \mathcal{X}} m_P P$, the **support** of D, denoted by $\mathrm{Supp}(D)$, is defined to be the set $\{P : m_P \neq 0\}$. Thus, for any $f \in \mathcal{L}(D)$ and any point $Q \notin \mathrm{Supp}(D)$, Q is not a pole of f and we can evaluate f at the point Q. An algebraic geometry code, introduced by Goppa, is defined as follows.

Definition 2.4.2 Let $\mathcal{X}$ be a projective smooth curve over $\mathbb{F}_q$ and let D be a divisor on $\mathcal{X}$. Assume that $\mathcal{P} \stackrel{\text{def}}{=} \{P_1, \ldots, P_n\}$ is a set of n distinct $\mathbb{F}_q$-rational points on $\mathcal{X}$ such that $\mathcal{P} \cap \mathrm{Supp}(D) = \emptyset$. Then the code

$$C(D; \mathcal{P}) \stackrel{\text{def}}{=} \{(f(P_1), \ldots, f(P_n)) : f \in \mathcal{L}(D)\}$$

is called an **algebraic geometry code**.

Clearly, the two examples in Example 2.4.1 are special cases of algebraic geometry codes. In particular, generalized Reed-Solomon codes are actually algebraic geometry codes.

By the Riemann-Roch Theorem, we can estimate the parameters of an algebraic geometry code.

Theorem 2.4.3 *Let X be a smooth projective curve over $\mathbb{F}_q$ of genus g and let D be a divisor on X over $\mathbb{F}_q$ of degree m with $m < n$. Assume that $\mathcal{P} \stackrel{\text{def}}{=} \{P_1, \ldots, P_n\}$ is a set of n distinct $\mathbb{F}_q$-rational points on X such that $\mathcal{P} \cap \text{Supp}(D) = \emptyset$. Then the code $C(D; \mathcal{P})$ has parameters $[n, k, d]$ with*

$$k \geq m + 1 - g, \quad d \geq n - m,$$

and hence $k + d \geq n + 1 - g$. Furthermore, if $m \geq 2g - 1$, then $k = m + 1 - g$.

Proof. Let f be a nonzero element of $\mathcal{L}(D)$. We consider the Hamming weight of $\mathbf{c}_f = (f(P_1), \ldots, f(P_n))$. Let I be the set $\{1 \leq i \leq n : f(P_i) = 0\}$. Then $f \in \mathcal{L}(D - \sum_{i \in I} P_i)$. Hence, by Lemma 1.4.6(iii), we have $\deg(D - \sum_{i \in I} P_i) \geq 0$ since f is a nonzero element. Therefore, we get $\text{wt}(\mathbf{c}_f) = n - |I| \geq n - \deg(D) = n - m > 0$. From this, we conclude that (i) the minimum distance of $C(D; \mathcal{P})$ is at least $n - m$; (ii) the dimension of $C(D; \mathcal{P})$ is the same as that of $\mathcal{L}(D)$. The desired result on the dimension now follows from the Riemann-Roch Theorem. $\square$

Remark 2.4.4 The dual code $C(D; \mathcal{P})^\perp$ of an algebraic geometry code $C(D; \mathcal{P})$ can be described in terms of the residues of certain differentials [151, 163]. In this book, we also refer to the dual code of an algebraic geometry code as a **residual code**. Under the condition $\deg(D) = m \geq 2g - 1$, the parameters of the residual code $C(D; \mathcal{P})^\perp$ for the code $C(D; \mathcal{P})$ of Theorem 2.4.3 are $[n, n - \ell(D) = n - m + g - 1, d \geq m - 2g + 2]$ (see [151, Theorem 2.2.7] and [163, Theorem 3.1.43]).

Example 2.4.5 (i) Let $\mathcal{L}$ be a projective line defined over $\mathbb{F}_q$. Then it has genus $g = 0$. Thus, an algebraic geometry code $C(D; \mathcal{P})$ from $\mathcal{L}$ has parameters $[n, k \geq m + 1, d \geq n - m]$. By the Singleton bound, we must have $k = m + 1$ and $d = n - m$. This means that $C(D; \mathcal{P})$ is an MDS code. Note that the length of the code $C(D; \mathcal{P})$ is at most $q + 1$ as $\mathcal{L}$ has just $q + 1$ points. If $D = m\infty$, then $C(D; \mathcal{P})$ coincides with Example 2.4.1(i).

(ii) Let $\mathcal{E}$ be a curve of genus $g = 1$ over $\mathbb{F}_q$. Then an algebraic geometry code $C(D; \mathcal{P})$ from $\mathcal{E}$ has parameters satisfying $n \leq k + d \leq n + 1$. The first inequality follows from Theorem 2.4.3, while the second inequality follows from the Singleton bound. Note that, in this case, the length of an algebraic geometry code from a curve of genus 1 is at most $q + 1 + \lfloor 2\sqrt{q} \rfloor$ by the Hasse-Weil bound.

Now let $\mathcal{E}$ be the elliptic curve $y^2 + y = x^3 + x$ over $\mathbb{F}_2$ defined in Example 1.1.1(i). Let $\mathcal{P}$ consist of all the four "finite points" $P_1 = (0,0)$, $P_2 = (0,1)$, $P_3 = (1,0)$, $P_4 = (1,1)$ and let $D = m\infty$.

For $m = 2$, we have $\mathcal{L}(2\infty) = \{0, 1, x, 1+x\}$. Thus,

$$C(D; \mathcal{P}) = \{0000, 1111, 0011, 1100\}.$$

This is a binary $[4, 2, 2]$-code and it meets the lower bound of parameters in Theorem 2.4.3.

For $m = 3$, we have $\mathcal{L}(3\infty) = \{0, 1, x, 1+x, y, y+1, y+x, y+x+1\}$. Thus,

$$C(D; \mathcal{P}) = \{0000, 1111, 0011, 1100, 0101, 1010, 0110, 1001\}.$$

This is a binary $[4, 3, 2]$-MDS code. The distance of this code is better than the lower bound $n - m = 1$ given in Theorem 2.4.3.

(iii) Consider an algebraic geometry code $C(D; \mathcal{P})$ from the Hermitian curve $\mathcal{H}$ defined by $y^r + y = x^{r+1}$ over $\mathbb{F}_{r^2}$. Then $k + d \geq n + 1 - g = n + 1 - r(r-1)/2$. If $D = m\infty$, then $C(D; \mathcal{P})$ coincides with Example 2.4.1(ii).

Remark 2.4.6 From Example 2.4.5(ii), we know that the lower bound on the parameters of algebraic geometry codes given in Theorem 2.4.3 could be improved. There has indeed been a lot of effort on improving the parameters of algebraic geometry codes, but we are not going to investigate these improvements in this chapter. Instead, we revisit this topic in Chapter 6, when we consider the construction of frameproof codes.

2.5 Asymptotic Behavior of Codes

An important question in coding theory is the construction of good codes with large length. In other words, we want to construct asymptotically good codes. We first consider some examples.

Example 2.5.1 (i) Consider the q-ary Hamming code $\mathrm{Ham}(m, q)$ defined in Definition 2.3.14. It has length $n_m = (q^m - 1)/(q - 1)$, dimension $k_m = (q^m - 1)/(q - 1) - m$, and distance $d_m = 3$ (see Proposition 2.3.15). Thus, we get

$$\lim_{m \to \infty} \frac{k_m}{n_m} = 1, \qquad \lim_{m \to \infty} \frac{d_m - 1}{n_m} = 0.$$

(ii) Consider the binary Reed-Muller code $\mathcal{R}(1, m)$ defined in Definition 2.2.25. It has length $n_m = 2^m$, dimension $k_m = m + 1$, and distance $d_m = 2^{m-1}$ (see Theorem 2.2.26). Thus, we obtain

$$\lim_{m \to \infty} \frac{k_m}{n_m} = 0, \quad \lim_{m \to \infty} \frac{d_m - 1}{n_m} = \frac{1}{2}.$$

The above examples show that either the information rate or the relative minimum distance tends to zero when the length tends to infinity. Now the question is whether we can find a sequence of codes with length tending to infinity such that both the information rate and the relative minimum distance tend to positive numbers. To further study the asymptotic behavior of codes, we need to define some asymptotic quantities.

For a code C over $\mathbb{F}_q$, we denote by $n(C), M(C)$, and $d(C)$ the length, the size, and the minimum distance of C, respectively. Let U_q be the set of ordered pairs $(\delta, R) \in \mathbb{R}^2$ for which there exists a family $\{C_i\}_{i=1}^\infty$ of codes over $\mathbb{F}_q$ with $n(C_i) \to \infty$ and

$$\delta = \lim_{i \to \infty} \frac{d(C_i) - 1}{n(C_i)}, \quad R = \lim_{i \to \infty} \frac{\log_q M(C_i)}{n(C_i)}.$$

The following description of U_q can be found in [163, Section 1.3.1].

Lemma 2.5.2 *There exists a continuous function $\alpha_q(\delta)$, for $\delta \in [0, 1]$, such that*

$$U_q = \{(\delta, R) \in \mathbb{R}^2 : 0 \le R \le \alpha_q(\delta), \ 0 \le \delta \le 1\}.$$

Moreover, $\alpha_q(0) = 1, \alpha_q(\delta) = 0$ for $\delta \in [(q-1)/q, 1]$, and $\alpha_q(\delta)$ decreases on the interval $[0, (q-1)/q]$.

One of the major problems in coding theory is to determine the domain U_q, or equivalently, the function $\alpha_q(\delta)$. However, it is not an easy task to determine the function $\alpha_q(\delta)$. In fact, one knows very little about this function. Nowadays, researchers attempt to find lower and upper bounds for this function instead. The first benchmark bound for $\alpha_q(\delta)$ is the Gilbert-Varshamov bound given in the following theorem.

Theorem 2.5.3 *One has*

$$\alpha_q(\delta) \ge 1 - H_q(\delta) \tag{2.5}$$

for $\delta \in (0, 1 - 1/q)$, where $H_q(\delta)$ is the q-ary entropy function

$$H_q(\delta) \overset{\text{def}}{=} \delta \log_q(q - 1) - \delta \log_q \delta - (1 - \delta) \log_q(1 - \delta).$$

Proof. For every $n \ge 2$, put $d_n \overset{\text{def}}{=} \lfloor \delta n \rfloor$. Then, by the definition of $A_q(n, d)$, there exists a q-ary $(n, A_q(n, d_n))$-code C_n. It is clear that $\lim_{n \to \infty} \frac{d(C_n) - 1}{n(C_n)} =$

δ. By the sphere-covering bound (see Theorem 2.3.9), one has

$$\alpha_q(\delta) \geq \lim_{n \to \infty} \frac{\log_q M(C_n)}{n(C_n)}$$

$$= \lim_{n \to \infty} \frac{\log_q A_q(n, d_n)}{n(C_n)}$$

$$\geq 1 - \lim_{n \to \infty} \frac{\log_q \left(\sum_{i=0}^{d-1} \binom{n}{i}(q-1)^i \right)}{n}$$

$$= 1 - H_q(\delta),$$

where the last equality follows from Stirling's formula. □

From the proof of the above theorem, we know that the asymptotic Gilbert-Varshamov bound is not constructive since the sphere-covering bound is not constructive. In other words, the proof does not lead to any polynomial-time construction of codes attaining the asymptotic Gilbert-Varshamov bound. The Gilbert-Varshamov bound had remained the best known bound for about 30 years before algebraic geometry codes were introduced by Goppa. Next, we derive the asymptotic bounds from the algebraic geometry codes introduced in Section 2.4.

Theorem 2.5.4 *One has*

$$\alpha_q(\delta) \geq 1 - \delta - \frac{1}{A(q)} \tag{2.6}$$

for $\delta \in (0, 1 - 1/A(q))$, where $A(q)$ is the asymptotic quantity defined in Section 1.5.

Proof. Let $\{\mathcal{X}/\mathbb{F}_q\}$ be a family of curves such that $g(\mathcal{X}) \to \infty$ and $N(\mathcal{X})/g(\mathcal{X}) \to A(q)$, where $g(\mathcal{X})$ denotes the genus of $\mathcal{X}$. Choose an $\mathbb{F}_q$-rational point P_0 and $n = N(\mathcal{X}) - 1$ other distinct $\mathbb{F}_q$-rational points $P_1, \ldots, P_n$. Choose a family $\{m\}$ of positive integers such that $m/n \to 1 - \delta$. The algebraic geometry code $C(D; \mathcal{P})$ defined in Section 2.4 (with $D = mP_0$ and $\mathcal{P} = \{P_1, \ldots, P_n\}$) is an $[n, \geq m - g(\mathcal{X}) + 1, \geq n - m]$-linear code over $\mathbb{F}_q$. It has a subcode that is a q-ary $[n, m - g(\mathcal{X}) + 1, n - m]$-linear code. It is easy to see that, for $k = m - g(\mathcal{X}) + 1$, we have $k/n \to 1 - \delta - 1/A(q)$. The desired result follows. □

Remark 2.5.5 Through the above proof and the construction of algebraic geometry codes in Section 2.4, we know that the bound (2.6) is constructive provided the sequences of curves attaining $A(q)$ and the associated Riemann-Roch spaces can be constructed in polynomial time.

Example 2.5.6 Consider the family of curves $\{\mathcal{X}_i\}_{i \geq 1}$ over $\mathbb{F}_{q^2}$ in [65, 66], where the function field of $\mathcal{X}_i$ is $\mathbb{F}_{q^2}(x_1, \ldots, x_i)$, with $x_1, x_2, \ldots$ defined by

$$x_{i+1}^q + x_{i+1} = \frac{x_i^q}{x_i^{q-1} + 1}.$$

This family of curves is explicit. Moreover, the associated Riemann-Roch space $\mathcal{L}(mP)$, for an $\mathbb{F}_q$-rational point P on $\mathcal{X}_i$, can be constructed in polynomial time (see [143]). This family of curves satisfies

$$g(\mathcal{X}_i) = \begin{cases} (q^{i/2} - 1)^2 & \text{if } i \text{ is even} \\ (q^{(i+1)/2} - 1)(q^{(i-1)/2} - 1) & \text{if } i \text{ is odd} \end{cases}$$

and

$$N(\mathcal{X}_i) \geq q^{i-1}(q^2 - q) + 1.$$

Thus, $N(\mathcal{X}_i)/g(\mathcal{X}_i) \to q - 1$.

Hence, there is a family of algebraic geometry codes over $\mathbb{F}_{q^2}$ which can be explicitly constructed and which attains the bound in Theorem 2.5.4.

Chapter 3

Elliptic Curves and Their Applications to Cryptography

With the advent of public key cryptography in the 1970s, the role of mathematics in real-world applications entered a new and exciting phase. Indeed, many well-established mathematical problems were thrust into the limelight with many computational problems taking on an added significance. Essentially, at the heart of a public key scheme is a pair of public/private keys such that we can be reasonably assured that the private key (kept hidden from all but the originator) cannot be computed from the public key (accessible to all) using the available resources.

To achieve such a property, "computationally hard" mathematical problems are exploited. Some notable examples include the **integer factorization problem** and the **discrete logarithm problem**, which form the basis of the earliest public key encryption schemes, namely, the RSA and ElGamal encryption schemes, respectively. In particular, these schemes work with the ring of integers modulo n (for some large composite integers n) or the multiplicative subgroups of finite fields.

In view of the rapid growth in information technology and its penetration into our daily lives, it is inevitable that security issues become an increasing concern. As proofs of the computational infeasibility of the above two problems remain elusive, different hard mathematical problems continue to be sought. In addition, encryption schemes which can be adapted to small devices and new scenarios are required. For instance, it may be more practical for the public key to reflect the identity of the users. Such emerging trends and requirements fuel the continual evolution in the use of mathematical tools in cryptography.

In Chapter 1, we briefly introduced elliptic curves as examples of algebraic plane curves. In fact, elliptic curves have been a well-studied mathematical object, fascinating many renowned mathematicians, as these curves offer a rich and insightful structure. It is therefore not surprising that, over the past few decades, researchers have discovered that elliptic curves, particularly those defined over finite fields, lend themselves as ideal candidates to address some of the key issues in cryptography. The main areas include the following:

- many discrete logarithm-based schemes can be adapted to elliptic curves which are suitable for small devices;

- pairings defined on elliptic curves give rise to new applications such as identity-based schemes and tripartite key exchange;

- elliptic curves can be used to factor integers, and the algorithm is relatively efficient for integers with moderately large prime factors.

This chapter is devoted to an exposition on the role of elliptic curves in the area of public key cryptography. As this topic has attracted much active research leading to a large repository of interesting results, a thorough treatment is unfortunately an impossible endeavor. As such, we only seek to provide the reader with an overall flavor by introducing important notions and results on elliptic curves as well as the main applications in cryptography. We encourage the reader to consult the extensive literature on this subject to delve deeper into any aspect of particular interest [12, 19, 20, 42, 48, 53, 60, 74, 77].

3.1 Basic Introduction

In Example 1.1.1(i), we gave an example of an elliptic curve over the binary field $\mathbb{F}_2$. For fields with odd characteristic, we mentioned in Example 1.1.1(ii) that a hyperelliptic curve $y^2 = f(x)$, where $f(x)$ is a square-free polynomial over $\mathbb{F}_q$, is an elliptic curve when the degree of $f(x)$ is 3. Since such curves are algebraic plane curves, all the results of Chapter 1 apply.

We begin this section by revisiting examples from Chapter 1 and recalling some of the results on algebraic curves when applied to such curves. We then present the general definitions of elliptic curves over finite fields and study the divisor class groups of such curves. By establishing a one-to-one correspondence between the divisor class group and the points on the curve, we define a group law governing the points on the curve.

In Example 1.1.1(i), we considered the affine curve defined over $\mathbb{F}_2$:

$$\mathcal{E} \; : \; y^2 + y = x^3 + x.$$

Its projective form is $Y^2Z + YZ^2 = X^3 + XZ^2$. This is an example of an elliptic curve. We recall some of the properties of $\mathcal{E}$, given in Chapter 1, below:

- The elliptic curve $\mathcal{E}$ is smooth (Example 1.1.10(iii)).

- The elliptic curve $\mathcal{E}$ has five rational points: $O = [0, 1, 0]$, $[0, 0, 1]$, $[0, 1, 1]$, $[1, 0, 1]$, $[1, 1, 1]$ (Example 1.1.1(i)). The point O is known as the "point at infinity," while the other four points are finite points.

- The genus of $\mathcal{E}$ is 1 (Theorem 1.4.3).

- The Riemann-Roch space $\mathcal{L}(kO)$ has dimension k for any positive integer k (Example 1.4.1(ii)). A basis for this space is $\{x^i y^j \;:\; i \geq 0,\ 0 \leq j \leq 1,\ 2i + 3j \leq k\}$.

- By the Hasse-Weil bound (Corollary 1.5.4(iii)), $\mathcal{E}$ has at most $2 + 1 + 2\sqrt{2}$ $\mathbb{F}_2$-rational points, so $\mathcal{E}$ has the largest possible number of $\mathbb{F}_2$-rational points.

- The L-function of $\mathcal{E}$ is $L(t) = 1 + 2t + 2t^2$ (Theorem 1.5.3). Since $L(t) = (1 - (-1 + i)t)(1 - (-1 - i)t)$, for any positive integer m, $\mathcal{E}$ has $2^m + 1 - ((-1 + i)^m + (-1 - i)^m)$ $\mathbb{F}_{2^m}$-rational points.

For odd characteristic p, we see from Example 1.1.1(ii) that an elliptic curve $\mathcal{E}$ can take the form

$$\mathcal{E} : y^2 = f(x),$$

where $f(x)$ is a square-free polynomial of degree 3. An example is the curve $y^2 = x^3 - x + 1$ over $\mathbb{F}_3$ given in Example 1.5.5(i). Once again, $\mathcal{E}$ has genus 1 and properties similar to the above ones hold as well.

These examples may suggest that an elliptic curve is an algebraic curve of genus 1. This is indeed true. We now turn to the general definition of an elliptic curve.

Definition 3.1.1 An **elliptic curve** $\mathcal{E}$ over $\mathbb{F}_q$ is a projective plane curve defined by

$$\mathcal{E} \;:\; Y^2 Z + A_1 XYZ + A_3 YZ^2 = X^3 + A_2 X^2 Z + A_4 XZ^2 + A_6 Z^3, \quad (3.1)$$

where $A_1, A_2, A_3, A_4, A_6 \in \mathbb{F}_q$. This equation is known as the generalized **projective Weierstrass equation** for $\mathcal{E}$.

Remark 3.1.2   (i) Note that $\mathcal{E}$ will, in general, refer to a subset of $\mathbf{P}^2(\overline{\mathbb{F}}_q)$. The $\mathbb{F}_q$-rational points of $\mathcal{E}$ will be denoted by $\mathcal{E}(\mathbb{F}_q)$.

(ii) Observe that, by letting $Z = 0$, the point $O = [0, 1, 0]$ lies on $\mathcal{E}$ for any finite field. By convention, O is referred to as the "point at infinity."

(iii) If $Z \neq 0$, then we can dehomogenize the generalized projective Weierstrass equation to obtain the generalized **affine Weierstrass equation**:

$$y^2 + h(x)y = s(x), \quad (3.2)$$

where $h(x) = A_1 x + A_3$ and $s(x) = x^3 + A_2 x^2 + A_4 x + A_6$. In other words, an elliptic curve $\mathcal{E}$ can also be defined as the set of points comprising O and all (x, y) satisfying $w(x, y) = 0$, where $w(x, y) = y^2 + h(x)y - s(x)$. Henceforth, we will use this affine definition to describe the elliptic curve $\mathcal{E}$.

The next lemma gives a map that transforms a point on $\mathcal{E}$ to another point on $\mathcal{E}$.

Lemma 3.1.3 *Let $\mathcal{E}$ be an elliptic curve over $\mathbb{F}_q$ defined by $y^2 + h(x)y = s(x)$. For each finite point $P = (x, y)$ on $\mathcal{E}$, we have that $\chi_{\mathcal{E}}(P) \overset{\text{def}}{=} (x, -y-h(x)) \in \mathcal{E}$ as well. In particular, if $h(x) = 0$, then $\chi_{\mathcal{E}}(x, y) = (x, -y) \in \mathcal{E}$. Furthermore, $\chi_{\mathcal{E}}$ is an involution, i.e., $\chi_{\mathcal{E}} \circ \chi_{\mathcal{E}}$ is the identity map on $\mathcal{E}$.*

Proof. Let x and y be such that $y^2 + h(x)y = s(x)$. Then $(-y - h(x))^2 + h(x)(-y - h(x)) = y^2 + 2h(x)y + h(x)^2 - h(x)y - h(x)^2 = y^2 + h(x)y = s(x)$, so $(x, -y - h(x)) \in \mathcal{E}$. Clearly, $\chi_{\mathcal{E}} \circ \chi_{\mathcal{E}}(x, y) = \chi_{\mathcal{E}}(x, -y - h(x)) = (x, y)$. $\square$

For purposes of applications to cryptography, we are only interested in smooth elliptic curves. Recall from Chapter 1 that $\mathcal{E}$ is smooth if $\mathcal{E}$ is nonsingular at all its points. Let $w(x, y) = y^2 + h(x)y - s(x)$. Then $\mathcal{E}$ is singular at a point P if and only if $\frac{\partial w}{\partial x}(P) = \frac{\partial w}{\partial y}(P) = 0$. By considering projective coordinates and the homogenized form of w, it is clear that $\mathcal{E}$ is nonsingular at O. A criterion for $\mathcal{E}$ to be nonsingular, in terms of the coefficients A_i's, is given in the next proposition. The reader may refer to [144, Chapter III, Proposition 1.4] for the proof.

For an elliptic curve $\mathcal{E}$ given by the generalized affine Weierstrass equation in (3.2), the **discriminant** of $\mathcal{E}$ is given by

$$\triangle = -B_2^2 B_8 - 8B_4^3 - 27B_6^2 + 9B_2 B_4 B_6, \tag{3.3}$$

where

$$\begin{aligned}
B_2 &= A_1^2 + 4A_2 \\
B_4 &= 2A_4 + A_1 A_3 \\
B_6 &= A_3^2 + 4A_6 \\
B_8 &= A_1^2 A_6 + 4A_2 A_6 - A_1 A_3 A_4 + A_2 A_3^2 - A_4^2.
\end{aligned}$$

Proposition 3.1.4 *Let $\mathcal{E}$ be an elliptic curve given by the generalized affine Weierstrass equation in (3.2). Then $\mathcal{E}$ is smooth if and only if $\triangle \neq 0$.*

The generalized Weierstrass equation can be somewhat too complicated especially when the characteristic of $\mathbb{F}_q$ is not 2 or 3. Indeed, for such characteristics, the generalized Weierstrass equation can be transformed, by a suitable change of variables, into the following **Weierstrass normal form**:

$$W(x, y) \overset{\text{def}}{=} y^2 - x^3 - Ax - B = 0, \tag{3.4}$$

with $A, B \in \mathbb{F}_q$. In this case, the discriminant of the elliptic curve becomes

$$\triangle = -16(4A^3 + 27B^2). \tag{3.5}$$

We refer the reader to [144, Chapter III, Section 1] for the details of this transformation.

From now on, for the sake of simplicity, we assume that $p \neq 2, 3$, where p is the characteristic of $\mathbb{F}_q$, so that $\mathcal{E}$ can be defined by the equation (3.4).

Remark 3.1.5 (i) When q is even, an elliptic curve $\mathcal{E}$ over $\mathbb{F}_q$ can be transformed into one of the following forms: $y^2 + y = s(x)$, $\deg(s(x)) = 3$, or $y^2 + y = x + 1/(ax + b)$, $a, b \in \mathbb{F}_q$, $a \neq 0$ (see [151, Proposition 6.1.2]).

(ii) When $p = 3$, an elliptic curve $\mathcal{E}$ over $\mathbb{F}_q$ can be transformed into the following form: $y^2 = s(x)$, $\deg(s(x)) = 3$ and $s(x)$ is square-free (see [151, Proposition 6.1.2]).

(iii) Apart from the Weierstrass form, other coordinate systems for elliptic curves, such as the Edwards coordinates, have also been proposed with the view to speed up some computational problems. More details on the Edwards coordinates can be found at the URL: http://cr.yp.to/newelliptic/newelliptic.html.

Using $W(x, y)$ and Proposition 3.1.4, the smoothness of $\mathcal{E}$ can be checked more easily. In fact, it can be directly verified that $\mathcal{E}$ is smooth if and only if $4A^3 + 27B^2 \neq 0$. Let $s(x) = x^3 + Ax + B$ so that $y^2 = s(x)$. Recall that $\mathcal{E}$ is smooth means that $s(x)$ has no multiple root. In other words, over $\overline{\mathbb{F}}_q$, $s(x)$ can be factored as $s(x) = (x - a_1)(x - a_2)(x - a_3)$, where a_1, a_2, a_3 are distinct.

At this juncture, it is useful for us to restate the definition of elliptic curve that we will work with for the remainder of this chapter.

Definition 3.1.6 Let $p \geq 5$ be a prime and let $q = p^r$. Suppose that we have $A, B \in \mathbb{F}_q$ such that $4A^3 + 27B^2 \neq 0$. Then a (nonsingular) **elliptic curve** $\mathcal{E}$ over $\mathbb{F}_q$ is the set of points comprising the point at infinity O as well as the pairs (x, y) such that $y^2 = x^3 + Ax + B$. We also let $W(x, y) = y^2 - x^3 - Ax - B$ and $s(x) = x^3 + Ax + B$.

In Chapter 1, the notions of the function field and discrete valuation for an algebraic curve were introduced. Evidently, these concepts apply to elliptic curves as well. We now demonstrate these ideas and prove some of the related results when restricted to elliptic curves.

We define the **coordinate ring** of $\mathcal{E}$ to be $\mathbb{F}_q[\mathcal{E}] = \mathbb{F}_q[x, y]/(W(x, y))$. The **function field** of $\mathcal{E}$, denoted by $\mathbb{F}_q(\mathcal{E}) = \mathbb{F}_q(x, y)$, is defined as the quotient field of $\mathbb{F}_q[\mathcal{E}]$. The elements of $\mathbb{F}_q[\mathcal{E}]$ are known as polynomials on $\mathcal{E}$ while the elements of $\mathbb{F}_q(\mathcal{E})$ are known as rational functions on $\mathcal{E}$. In particular, two rational functions f_1/g_1 and f_2/g_2 are equivalent if $f_1 g_2 - f_2 g_1 = V(x, y)W(x, y)$ for some $V(x, y) \in \mathbb{F}_q[x, y]$. Clearly, every rational function maps a point of $\mathcal{E}$ to an element of $\mathbb{F}_q \cup \{\infty\}$.

Write $s(x) = (x - a_1)(x - a_2)(x - a_3)$. For $i = 1, 2, 3$, let P_i denote the point $(a_i, 0)$ on $\mathcal{E}$.

Example 3.1.7 (i) Let $P = (a, b) \in \mathcal{E}$, where $a \neq a_i$, $i = 1, 2, 3$. Then $(x - a)(P) = 0$ and $(x - a)(\chi_{\mathcal{E}}(P)) = 0$. Hence, P and $\chi_{\mathcal{E}}(P)$ are zeros of $x - a$. Moreover, $x - a$ has no other zero.

(ii) For $i = 1, 2, 3$, we have $y(P_i) = 0$. Moreover, y has no other zero.

(iii) Consider the function x/y. Using the projective coordinates $O = [0, 1, 0]$, we have $(x/y)(O) = (xz/yz)(O) = 0$.

We know from Chapter 1 that the discrete valuation of every rational function at any rational point on $\mathcal{E}$ can be defined. The next theorem gives us the discrete valuations explicitly. We refer the reader to [37, Theorem 4.3] for the proof.

Theorem 3.1.8 *Let $\mathcal{E}$ be an elliptic curve given by $y^2 = (x - a_1)(x - a_2)(x - a_3)$, where a_1, a_2, a_3 are distinct, let O be the point at infinity, and let $P_i = (a_i, 0)$ $(1 \le i \le 3)$ be points on $\mathcal{E}$.*

(i) *Let $f \in \mathbb{F}_q(\mathcal{E})$ and let $P = (a, b) \in \mathcal{E}$ be such that $P \notin \{P_1, P_2, P_3, O\}$. Write $f = (x - a)^d g$, where $g(P) \neq 0, \infty$. Then $\nu_P(f) = d$.*

(ii) *Let $P = P_i$, for $i = 1, 2, 3$. For $f \in \mathbb{F}_q(\mathcal{E})$, write $f = y^d g$, where $g(P) \neq 0, \infty$. Then $\nu_P(f) = d$.*

(iii) *For $f \in \mathbb{F}_q(\mathcal{E})$, write $f = (x/y)^d g$, where $g(O) \neq 0, \infty$. Then $\nu_O(f) = d$. In particular, $\nu_O(x) = -2$ and $\nu_O(y) = -3$.*

Example 3.1.9 We consider the elliptic curve $y^2 = x^3 - 4x$ defined over $\mathbb{F}_{11}$.

(i) Since $P = (-1, 5)$ and $Q = (-1, 6)$ lie on $\mathcal{E}$, $\nu_P(x + 1) = \nu_Q(x + 1) = 1$. We also have $\nu_O(x+1) = -2$. For all other points R on $\mathcal{E}$, $\nu_R(x+1) = 0$. Hence, $\sum_{R \in \mathcal{E}} \nu_R(x + 1) = 0$.

(ii) If $P \in \{(0, 0), (2, 0), (-2, 0)\}$, then $\nu_P(y) = 1$. Furthermore, $\nu_O(y) = -3$. For all other points R on $\mathcal{E}$, $\nu_R(y) = 0$. Hence, $\sum_{R \in \mathcal{E}} \nu_R(y) = 0$.

(iii) Since $x = y^2/(x + 2)(x - 2)$, $\nu_{(0,0)}(x) = 2$.

(iv) By Definition 1.3.2(iv), $\nu_{(0,0)}(2y + x) = 1$.

Remark 3.1.10 By Lemma 1.4.5, counting multiplicities, the number of zeros of a function of $\mathcal{E}$ must be equal to the number of its poles. Recall that a plane line is defined by $\ell(x, y) = \alpha y + \beta x + \gamma = 0$, where $\alpha, \beta, \gamma \in \mathbb{F}_q$ and $(\alpha, \beta) \neq (0, 0)$. Clearly, O is the only pole of $\ell(x, y)$ and $\nu_O(\ell) = -3$ or -2 according to $\alpha \neq 0$ or $\alpha = 0$, respectively. This shows that the line $\ell(x, y) = 0$ can have either two or three zeros (counting multiplicities) on $\mathcal{E}$. This fact will be used to define a group law on $\mathcal{E}$ in Subsection 3.1.2.

We recall the notions of a divisor and the Riemann-Roch space from Chapter 1. A divisor of $\mathcal{E}$ is a formal sum of points on $\mathcal{E}$. In other words, a divisor of $\mathcal{E}$ can be written in the form $\sum_{P \in \mathcal{E}} m_P P$, where $m_P \in \mathbb{Z}$ for all $P \in \mathcal{E}$, and $m_P = 0$ for all but finitely many points $P \in \mathcal{E}$. For a positive integer k, the Riemann-Roch space $\mathcal{L}(kO)$ is defined as the space

$\mathcal{L}(kO) = \{f \in \mathbb{F}_q(\mathcal{E}) : \nu_O(f) \geq -k, \nu_P(f) \geq 0 \text{ for all } P \in \mathcal{E} \setminus \{O\}\} \cup \{0\}$.
The Riemann-Roch Theorem (Theorem 1.4.2) tells us that $\mathcal{L}(kO)$ is a finite-dimensional vector space over $\mathbb{F}_q$ with dimension $k + 1 - 1 = k$. In fact, a basis for $\mathcal{L}(kO)$ is $\{x^i y^j : i \geq 0, \ 0 \leq j \leq 1, \ 2i + 3j \leq k\}$.

According to Theorem 1.4.3, the genus of any elliptic curve is exactly 1. An interesting question is whether elliptic curves give all the possible algebraic curves of genus 1. The next theorem answers this question. The reader may refer to [144, Chapter III, Proposition 3.1] for details of the proof.

Theorem 3.1.11 *Let $\mathcal{E}$ be a smooth absolutely irreducible algebraic curve with genus 1. If $\mathcal{E}$ contains at least one $\mathbb{F}_q$-rational point O, then $\mathcal{E}$ is an elliptic curve. In other words, there exist $x, y \in \mathbb{F}_q(\mathcal{E})$ satisfying a generalized Weierstrass equation $w(x, y) = 0$.*

3.1.1 The Divisor Class Group of an Elliptic Curve

In this subsection, we investigate the divisor class group of degree zero of $\mathcal{E}$, that is, the group $\mathrm{Pic}^0(\mathcal{E}) \overset{\text{def}}{=} \mathrm{Div}^0(\mathcal{E})/\mathrm{Princ}(\mathcal{E})$ (see Section 1.4). In fact, $\mathrm{Pic}^0(\mathcal{E})$ comprises all divisor classes of degree zero of $\mathcal{E}$. In particular, we show that there exists a one-to-one correspondence between $\mathrm{Pic}^0(\mathcal{E})$ and the points on $\mathcal{E}$.

We begin by showing that the divisor class group of degree zero contains at least as many elements as $\mathcal{E}$.

Lemma 3.1.12 *Let P and Q be two distinct points on $\mathcal{E}$. Then there does not exist any function f such that $\mathrm{div}(f) = P - Q$.*

Proof. We prove by contradiction. Suppose that there exists some function f such that $\mathrm{div}(f) = P - Q$. Then $f \in \mathcal{L}(Q)$ (since $\mathrm{div}(f) + Q = P \geq 0$). By the Riemann-Roch Theorem, $\ell(Q) = 1$ and we know that $\mathcal{L}(Q)$ contains $\mathbb{F}_q$. It follows that $\mathcal{L}(Q) = \mathbb{F}_q$ and f is a constant. This is a contradiction since P and Q are distinct. □

For two divisors D and G of the same degree, by $D \cong G$ we mean that D is equivalent to G. For a divisor D of degree $\mathrm{Div}^0(\mathcal{E})$, we denote by $[D]$ the equivalence class in $\mathrm{Pic}^0(\mathcal{E})$ containing D.

Corollary 3.1.13 (i) *For two points P and Q of $\mathcal{E}$, $P \cong Q$ as divisors if and only if $P = Q$ as points of $\mathcal{E}$.*

(ii) *The divisor classes $[P - O]$ are all distinct, for distinct $P \in \mathcal{E}$. In particular, $\mathrm{Pic}^0(\mathcal{E})$ must be at least as large as $\mathcal{E}$.*

Let f be a polynomial on $\mathcal{E}$. Then it is easy to see that $\nu_O(f) \leq -2$, so f must have at least two zeros. Lemma 3.1.12 shows that, in fact, any nonconstant rational function on $\mathcal{E}$ must have at least two zeros and two poles. Indeed, if there is some function f with exactly one zero P and one pole Q, then $\text{div}(f) = P - Q$, which implies that the divisors P and Q are equivalent, contradicting Corollary 3.1.13(i).

Theorem 3.1.14 *Let D be a divisor of degree 0. Then there exists a unique point $P \in \mathcal{E}$ such that $D \cong P - O$.*

Proof. Let $D' = O + D$. Then $\deg(D') = 1$. By the Riemann-Roch Theorem, $\mathcal{L}(D')$ has dimension 1. In particular, $\mathcal{L}(D') \neq \{0\}$. Let f be a nonzero function in $\mathcal{L}(D')$. We have $\text{div}(f) + D + O \geq 0$. Since $\deg(\text{div}(f)) = 0$, there must exist some P on $\mathcal{E}$ such that $\text{div}(f) = P - D - O$. Consequently, $D \cong P - O$. The uniqueness follows from Corollary 3.1.13. □

Theorem 3.1.14 shows that every equivalence class in $\text{Pic}^0(\mathcal{E})$ contains $P - O$ for a unique $P \in \mathcal{E}$. In other words, $\text{Pic}^0(\mathcal{E})$ comprises precisely all the classes $[P - O]$, for $P \in \mathcal{E}$. Let τ and σ be the maps

$$\tau : \text{Pic}^0(\mathcal{E}) \to \mathcal{E}$$
$$[P - O] \mapsto P$$

and

$$\sigma : \mathcal{E} \to \text{Pic}^0(\mathcal{E})$$
$$P \mapsto [P - O].$$

Clearly, σ and τ are inverses of each other with $\sigma(O) = [0]$, the 0 divisor class. Consequently, there is a one-to-one correspondence between the sets $\text{Pic}^0(\mathcal{E})$ and $\mathcal{E}$.

For each $D \in \text{Div}^0(\mathcal{E})$, while Theorem 3.1.14 shows the existence of a unique $P \in \mathcal{E}$ such that D is equivalent to $P - O$, it does not give an explicit way of finding this P. Part (ii) in the following example illustrates how this can be done.

Example 3.1.15 Consider the elliptic curve $\mathcal{E} : y^2 = x^3 + 4x + 4$ over $\mathbb{F}_7$.

(i) Consider the line defined by $\ell_1(x, y) = y + x + 2 = 0$. To find the zeros of $\ell_1(x, y)$ on $\mathcal{E}$, we substitute $y = -x - 2$ into the equation of $\mathcal{E}$, yielding $(x + 2)^2 = x^3 + 4x + 4$, or equivalently, $x^3 - x^2 = 0$ or $x^2(x - 1) = 0$. Hence, $x = 0$ is a double root while $x = 1$ is a simple root. For $x = 0$, we have $y = 5$, and for $x = 1$, we have $y = 4$. Consequently, $(0, 5)$ is a double zero of $\ell_1(x, y)$ while $(1, 4)$ is a simple zero of $\ell_1(x, y)$. Using the divisor notation, $\text{div}(\ell_1(x, y)) = 2(0, 5) + (1, 4) - 3O$.

(ii) Consider the points $P = (0, 2)$ and $Q = (1, 4)$ on $\mathcal{E}$. We now want to find a point R on $\mathcal{E}$ such that $R - O$ is equivalent to $P + Q - 2O$ as divisors. To this end, consider the line defined by $\ell_2(x, y) = y - 2 - \frac{4-2}{1-0} \cdot x = y - 2x - 2 = 0$. One checks easily that P and Q are zeros of $\ell_2(x, y)$. To find the third zero of $\ell_2(x, y)$, we substitute $y = 2x + 2$

into the equation of $\mathcal{E}$ so that $(2x + 2)^2 = x^3 + 4x + 4$. Solving for x, we find that $x = 0, 1$, or 3. Consequently, $S = (3, 1)$ is the third zero of $\ell_2(x, y)$ and we conclude that $\ell_2(x, y)$ has three simple zeros. Therefore, $\operatorname{div}(\ell_2(x, y)) = P + Q + S - 3O$.

Next, consider $\operatorname{div}(x - 3) = (3, 1) + (3, -1) - 2O$. Hence, $\operatorname{div}(\ell_2(x, y)/(x - 3)) = P + Q - (3, -1) - O$ or $(P - O) + (Q - O) - ((3, -1) - O) = \operatorname{div}(\ell_2(x, y)/(x - 3))$. In other words, $D = P + Q - 2O$ is equivalent to $(3, -1) - O$ as divisors.

Lemma 3.1.16 *Let* $P = (a, b) \in \mathcal{E}$. *Recall that* $\chi_\mathcal{E}(P) = (a, -b)$. *We have* $-[P - O] = [\chi_\mathcal{E}(P) - O]$.

Proof. Note that $x - a$ has a zero at P, a zero at $\chi_\mathcal{E}(P)$, and a double pole at O. When $b = 0$, then $P = (a, 0)$ is a double zero of $x - a$. Hence, $\operatorname{div}(x - a) = P + \chi_\mathcal{E}(P) - 2O$. Since $[P - O] + [\chi_\mathcal{E}(P) - O] = [P + \chi_\mathcal{E}(P) - 2O] = [0]$, it follows that $[\chi_\mathcal{E}(P) - O] = -[P - O]$. $\qquad\square$

3.1.2 The Group Law on Elliptic Curves

In Subsection 3.1.1, we have identified points on an elliptic curve $\mathcal{E}$ with divisor classes in $\operatorname{Pic}^0(\mathcal{E})$ via the bijective map σ. Since $\operatorname{Pic}^0(\mathcal{E})$ is an abelian group, this suggests that we can define a group structure on $\mathcal{E}$ as well by turning σ into a group homomorphism. More specifically, let P and Q be two points on $\mathcal{E}$. Let $R \in \mathcal{E}$ be such that $[P - O] + [Q - O] = [R - O]$. This is possible since $\operatorname{Pic}^0(\mathcal{E})$ is a group and R exists by Theorem 3.1.14. We can then define $P \oplus Q = R$. Indeed, $R = \tau(\sigma(P) + \sigma(Q))$.

Example 3.1.17 Refer to the elliptic curve of Example 3.1.15. Then $(0, 2) \oplus (1, 4) = (3, -1)$.

Theorem 3.1.18 *An elliptic curve* $\mathcal{E}$, *under the operation* $\oplus$ *defined above, is an abelian group.*

Proof. First, we show that O acts as our identity. This is true since, for any point $P \in \mathcal{E}$, $O \oplus P = \tau([0] + [P - O]) = \tau([P - O]) = P$. Next, it is clear that $\ominus P = \chi_\mathcal{E}(P)$, where $\ominus P$ denotes the unique point on $\mathcal{E}$ such that $P \oplus (\ominus P) = O$, according to Lemma 3.1.16. Finally, the associativity and commutativity of $\oplus$ follow from the respective corresponding properties for the addition in $\operatorname{Pic}^0(\mathcal{E})$. $\qquad\square$

It is possible to obtain explicit expressions for the coordinates of $R = P \oplus Q$ in terms of the coordinates of P and Q.

To this end, let P and Q be two points on $\mathcal{E}$. Suppose that $P = (a, b)$ and $Q = (c, d)$. If $a = c$ and $b = -d$, then $Q = \ominus P$ so that $P \oplus Q = O$ (see proof of Theorem 3.1.18). Therefore, we assume that we do not have $a = c$ and $b = -d$. The main idea is to find R via the zeros on lines as follows.

- Let $\ell = 0$ be the line on which P and Q lie.

- Since $P \neq \ominus Q$, ℓ must be of the form $\ell = y - Mx - C$, i.e., the line is given by $y - Mx - C = 0$.

- Since $\nu_O(\ell) = -3$, it follows that ℓ has three zeros.

- Let P, Q, and $S = (g', h')$ be these zeros.

- Then $\operatorname{div}(\ell) = P + Q + S - 3O$.

- Further, with $v = x - g'$, we have $\operatorname{div}(v) = S + (g', -h') - 2O$.

- Hence, $\operatorname{div}(\ell/v) = P + Q - (g', -h') - O = (P-O) + (Q-O) - ((g', -h') - O)$.

- Consequently, we have $R = P \oplus Q = \chi_{\mathcal{E}}(S) = (g', -h')$.

We are now ready to give the coordinates of $P \oplus Q$.

Theorem 3.1.19 *Let* $P = (a, b)$ *and* $Q = (c, d)$ *be two points on* $\mathcal{E}$ *such that* $P \neq \ominus Q$. *Define* M *as*

$$M = \begin{cases} \frac{d-b}{c-a} & \text{if } a \neq c \\ \frac{3a^2 + A}{2b} & \text{if } a = c. \end{cases}$$

Furthermore, let $C = b - Ma$. *Then the point* $P \oplus Q$ *is given by* $R = (g, h)$, *where*

$$g = M^2 - a - c \quad \text{and} \quad h = -Mg - C.$$

Proof. Let $\ell = 0$ be the line passing through P and Q. It is clear that ℓ is given by $\ell = y - Mx - C$, where M and C are as defined. Let S be the third zero of ℓ. Write $S = (g', h')$. Now, we substitute $y = Mx + C$ into the equation for $\mathcal{E}$ to obtain

$$(Mx + C)^2 = x^3 + Ax + B.$$

Expanding this produces $f(x) = x^3 - M^2 x^2 + (A - 2CM)x + B - C^2 = 0$. Now, a, c and g' must be roots of this equation since they are the x-coordinates of P, Q and S, which satisfy both the equation of the elliptic curve and $\ell = 0$. Thus, $f(x) = (x - a)(x - c)(x - g')$ and comparing the coefficients of the x^2 terms gives $a + c + g' = M^2$. Hence, $g' = M^2 - a - c$. It is now clear that $h' = Mg' + C$. Finally, according to the procedure of computing $P \oplus Q$, we obtain $g = g'$ and $h = -h' = -Mg - C$. $\square$

The formula for $P \oplus Q$ in Theorem 3.1.19 is known as the **addition formula**. In the case $P = Q$, the formula given in Theorem 3.1.19 to compute $[2]P$ is often called the **duplication formula**.

Remark 3.1.20 It may be instructive to consider an elliptic curve over the real field $\mathbb{R}$ which can be represented graphically on the xy-plane. Here, O can be interpreted as the point that lies on all vertical lines. For each point P on $\mathcal{E}$, $\chi_{\mathcal{E}}(P)$ is in fact the reflection of P in the x-axis. Given two points P and Q on $\mathcal{E}$, we can draw a line $\ell = 0$ passing through P and Q. In the case $P = Q$, then $\ell = 0$ is the tangent line to $\mathcal{E}$ at P. It can be easily seen that $\ell = 0$ cuts $\mathcal{E}$ at one other point, which we denote by S. Finally, reflecting S in the x-axis gives $R = P \oplus Q$.

Note that, in some elementary textbooks on elliptic curves, the group law on $\mathcal{E}$ is often defined via the geometric or algebraic approach. However, proving the associativity can be very messy in this case. As such, in this chapter, we have opted to introduce the group law via the divisor class group, where all the group properties are easily proved.

Summing up all the discussion we have had thus far yields the following result:

Theorem 3.1.21 *The set of points on $\mathcal{E}$ forms an abelian group that is isomorphic to the divisor class group of degree zero via the isomorphism σ such that $\sigma(P) = [P - O]$.*

In particular, this isomorphism enables us to derive a criterion to decide when a divisor is principal, as the next proposition demonstrates.

Proposition 3.1.22 *A divisor $D = \sum_{P \in \mathcal{E}} m_P P$, where $m_P \in \mathbb{Z}$ for all P, is principal if and only if $\deg(D) = \sum_{P \in \mathcal{E}} m_P = 0$ and $\bigoplus_{P \in \mathcal{E}} m_P P = O$, where $\oplus$ refers to the sum of points on $\mathcal{E}$, $-P$ means $\ominus P$, and $mP = P \oplus \cdots \oplus P$ (m times).*

Proof. First, suppose that $\bigoplus_{P \in \mathcal{E}} m_P P = O$ and $\deg(D) = 0$. Then $[\sum_{P \in \mathcal{E}} m_P P - \sum_{P \in \mathcal{E}} m_P O] = [O - O]$, so $\sum_{P \in \mathcal{E}} m_P P - \sum_{P \in \mathcal{E}} m_P O$ is principal. Since $\deg(D) = 0$, $\sum_{P \in \mathcal{E}} m_P = 0$ so that $\sum_{P \in \mathcal{E}} m_P O$ is the zero divisor. Consequently, $D = D - \sum_{P \in \mathcal{E}} m_P O$ is principal.

Conversely, suppose that D is principal. It is clear that $\deg(D) = 0$. Let $S = \bigoplus_{P \in \mathcal{E}} m_P P$. If $S \neq O$, then $[\sum_{P \in \mathcal{E}} m_P P - \sum_{P \in \mathcal{E}} m_P O] = [S - O]$, which implies that D is equivalent to $S - O$, which is nonprincipal (see Lemma 3.1.12). We conclude that $S = O$. □

From now on, we simply denote the addition operation in $\mathcal{E}$ by $+$, i.e., $P \oplus Q = P + Q$. Similarly, $\ominus P = -P$. For any positive integer m, we use $[m]P$ to denote $P + P + \cdots + P$ (m times) and $[-m]P$ to denote $-P - P - \cdots - P$ (m times). We note that, despite the similarity in notation with the use of $+$ in divisors, the meaning of $+$ should be clear from the context.

Corollary 3.1.23 *For a positive integer m, let P be a point on $\mathcal{E}$ so that $[m]P = O$. Then there exists a function $f_{m,P}$, unique up to a constant multiple, such that $\operatorname{div}(f_{m,P}) = mP - mO$.*

Proof. The existence of $f_{m,P}$ follows immediately from Proposition 3.1.22. Now suppose that g is another function such that $\mathrm{div}(g) = mP - mO$. Then $\mathrm{div}(f_{m,P}/g)$ is the zero divisor. Consequently, $f_{m,P}/g$ must be a constant. $\square$

Example 3.1.24 Suppose that P is a point on an elliptic curve $\mathcal{E}$ such that $[2]P = O$. This means that $P = -P$. Since the characteristic of the field is not 2, it follows that the y-coordinate of P must be 0. Conversely, if a is a root of $x^3 + Ax + B = 0$, then $P = (a, 0)$ is such that $[2]P = O$. We have seen earlier that $\mathrm{div}(x - a) = 2P - 2O$ (cf. proof of Lemma 3.1.16).

Remark 3.1.25 The function $f_{m,P}$ in Corollary 3.1.23 is known as a **Miller function** at P. More discussion on the Miller functions is found in Subsection 3.4.2.

3.2 Maps between Elliptic Curves

We have so far defined an elliptic curve to be a set of points satisfying the Weierstrass normal form $y^2 = x^3 + Ax + B$, where $A, B \in \mathbb{F}_q$ are such that $4A^3 + 27B^2 \neq 0$. Note that there are altogether at most $q^2 - 1$ such curves. How are these curves related? Are there any similarities or differences between the curves? Suppose that we choose a certain curve for our applications to cryptography. Will the security of our cryptographic scheme be compromised in any way if we transform to another curve?

These questions motivate us to study various maps between elliptic curves in this section. In particular, we discuss morphisms, isomorphisms, and isogenies between curves. Moreover, we introduce and study the Frobenius map and the multiplication map in some detail.

3.2.1 Morphisms of Elliptic Curves

We begin with the most general maps, namely, morphisms.

Definition 3.2.1 Let $\mathcal{E}_1$ and $\mathcal{E}_2$ be two elliptic curves defined over $\mathbb{F}_q$ given by the Weierstrass equations: $\mathcal{E}_1 : y^2 = x^3 + A_1 x + B_1$ and $\mathcal{E}_2 : y^2 = x^3 + A_2 x + B_2$. A **morphism** ψ between $\mathcal{E}_1$ and $\mathcal{E}_2$ is a pair of rational functions on $\mathcal{E}_1$, namely, $\psi = (h_x, h_y)$, where $h_x, h_y \in \mathbb{F}_q(\mathcal{E}_1)$ such that $h_y^2 = h_x^3 + A_2 h_x + B_2$.

Let $\psi = (h_x, h_y)$ be a morphism between $\mathcal{E}_1$ and $\mathcal{E}_2$. For a point $P = (a, b)$ on $\mathcal{E}_1$, we have $\psi(P) = (h_x(P), h_y(P)) = (h_x((a, b)), h_y((a, b)))$. Note that the representation of ψ may not be unique. In particular, $\psi = (h_x, h_y)$ and $\psi' = (h_x', h_y')$ are equivalent if the following conditions hold:

- $h_x \equiv h'_x \pmod{y^2 - x^3 - A_1 x - B_1}$;

- $h_y \equiv h'_y \pmod{y^2 - x^3 - A_1 x - B_1}$;

- $h_y^2 = h_x^3 + A_2 h_x + B_2$;

- $h_y'^2 = h_x'^3 + A_2 h'_x + B_2$.

It is clear that the set of equivalent morphisms forms an equivalence class. Let $\psi = (h_x, h_y)$ be a morphism from $\mathcal{E}_1$ to $\mathcal{E}_2$. Observe that $h_y(P)$ is not defined if and only if $h_x(P)$ is not defined. In this case, we set $\psi(P) = O$. Note that, in all cases, the value of $\psi(P)$ does not depend on the particular choice of the functions h_x and h_y in the representation of ψ.

Example 3.2.2 (i) Let $\mathcal{E}_1$ and $\mathcal{E}_2$ be any two elliptic curves over a field $\mathbb{F}_q$. Fix a point $P \in \mathcal{E}_2$. The map $\psi = P$ is the **constant morphism** which sends all $Q \in \mathcal{E}_1$ to $P \in \mathcal{E}_2$.

(ii) For any elliptic curve $\mathcal{E}$, the map $\mathcal{E} \to \mathcal{E}$ defined by $P \mapsto P$, for every $P \in \mathcal{E}$, is called the **identity morphism**.

(iii) For an elliptic curve $\mathcal{E}$, the map $\chi_\mathcal{E}$ that sends P to $-P$ is clearly a morphism from $\mathcal{E}$ to itself (see Lemma 3.1.3). It is called the **negation morphism**.

(iv) Let $\mathcal{E}_1$ and $\mathcal{E}_2$ be elliptic curves over $\mathbb{F}_{11}$ defined by $\mathcal{E}_1 : y^2 = x^3 + x$ and $\mathcal{E}_2 : y^2 = x^3 + 5x$. Define $\psi : \mathcal{E}_1 \to \mathcal{E}_2$ such that $\psi((x, y)) = (4x, 8y)$. Now, we have $(8y)^2 = 9y^2 = 9(x^3 + x) = (4x)^3 + 5(4x)$. Thus, ψ is a morphism from $\mathcal{E}_1$ to $\mathcal{E}_2$ with $\psi(O) = O$.

In the next example, we look at the translation map of an elliptic curve.

Example 3.2.3 Let $\mathcal{E}_1 = \mathcal{E}_2 = \mathcal{E}$ be given by the Weierstrass equation: $y^2 = x^3 + Ax + B$. Fix a finite point $P = (a, b) \in \mathcal{E}$. We define the **translation morphism** at P by $\tau_P(Q) = Q + P$. We now try to derive h_x and h_y. By the addition formula, we have

$$
\begin{aligned}
h_x &= ((y - b)/(x - a))^2 - x - a \\
&= \frac{(y-b)^2 - (x+a)(x-a)^2}{(x-a)^2} \\
&= \frac{x^3 + Ax + B - 2by + a^3 + Aa + B - (x^3 + a^3 - a^2 x - ax^2)}{(x-a)^2} \\
&= \frac{(x+a)(A+ax) - 2by + 2B}{(x-a)^2}.
\end{aligned}
$$

Similarly, we have

$$
\begin{aligned}
h_y &= ((y - b)/(x - a)) \cdot (a - h_x) - b \\
&= \frac{(y-b)(a(x-a)^2 - (x+a)(A+ax) + 2by - 2B) - b(x-a)^3}{(x-a)^3}.
\end{aligned}
$$

As expected, τ_P is not defined for $x = a$ and $y = -b$. In the case where $x = a$ and $y = b(\neq -b)$, we can perform the following computations.

Since $P = (a, b) \in \mathcal{E}$, we have $b^2 = a^3 + Aa + B$. Subtracting this from $y^2 = x^3 + Ax + B$ produces $y^2 - b^2 = x^3 - a^3 + A(x-a) = (x-a)(x^2 + ax + a^2 + A)$. Hence, $\frac{y-b}{x-a} = \frac{x^2 + ax + a^2 + A}{y+b}$. Substituting this into the addition formula yields h'_x and h'_y, which can be evaluated at points for which $x = a$.

As expected, $\tau_P(-P) = O$. Thus, τ_P is a morphism on $\mathcal{E}$ and is called the translation map at P. In fact, we can show similarly that the addition formula is a group homomorphism from $\mathcal{E} \times \mathcal{E}$ to $\mathcal{E}$.

Given three elliptic curves $\mathcal{E}_1, \mathcal{E}_2, \mathcal{E}_3$ defined over $\mathbb{F}_q$ and morphisms $\psi_1 : \mathcal{E}_1 \to \mathcal{E}_2$ and $\psi_2 : \mathcal{E}_2 \to \mathcal{E}_3$, we define the **composition** $\psi_2 \circ \psi_1 : \mathcal{E}_1 \to \mathcal{E}_3$ by $\psi_2 \circ \psi_1(P) = \psi_2(\psi_1(P))$ for all $P \in \mathcal{E}_1$. For example, for any two points P_1 and P_2 on an elliptic curve $\mathcal{E}$, $\tau_{P_1} \circ \tau_{P_2} = \tau_{P_1 + P_2}$.

Next, we state two standard results on morphisms between curves. Their proofs can be found in [144, Chapter II, Theorem 2.3 and Proposition 2.6(b)].

Lemma 3.2.4 *Let $\mathcal{E}_1$ and $\mathcal{E}_2$ be two elliptic curves defined over $\mathbb{F}_q$, and let ψ be a morphism from $\mathcal{E}_1$ to $\mathcal{E}_2$.*

(i) *The morphism ψ is either the constant morphism or it is surjective.*

(ii) *Let Q be a point on $\mathcal{E}_2$. Then the set $\{P \in \mathcal{E}_1 : \psi(P) = Q\}$ is finite.*

Definition 3.2.5 *Let $\mathcal{E}_1$ and $\mathcal{E}_2$ be two elliptic curves and let ψ be a morphism from $\mathcal{E}_1$ to $\mathcal{E}_2$. Then ψ is an **isomorphism** if there exists a morphism ψ' from $\mathcal{E}_2$ to $\mathcal{E}_1$ such that $\psi \circ \psi'$ is the identity morphism on $\mathcal{E}_2$ and $\psi' \circ \psi$ is the identity on $\mathcal{E}_1$. In the case where ψ is an isomorphism, we say that $\mathcal{E}_1$ and $\mathcal{E}_2$ are **isomorphic curves**. If $\mathcal{E}_1 = \mathcal{E}_2$, then we say that ψ is an **automorphism**.*

Example 3.2.6 *Let $\mathcal{E}$ be an elliptic curve and P an arbitrary point on $\mathcal{E}$. Then the translation map τ_P at P is an automorphism on $\mathcal{E}$ with inverse τ_{-P}. Furthermore, the negation morphism $\chi_{\mathcal{E}}$ is an automorphism too with $\chi_{\mathcal{E}}$ itself as the inverse.*

Let $\mathcal{E}_1$ and $\mathcal{E}_2$ be two elliptic curves defined over $\mathbb{F}_q$ and let $\mathbb{F}_q(\mathcal{E}_1)$ and $\mathbb{F}_q(\mathcal{E}_2)$ be their respective function fields. Let ψ be a morphism from $\mathcal{E}_1$ to $\mathcal{E}_2$. For all functions $f \in \mathbb{F}_q(\mathcal{E}_2)$, we can compose f with ψ to get $f \circ \psi(P) = f(\psi(P))$ for all $P \in \mathcal{E}_1$. Thus, $f \circ \psi$ is an element of $\mathbb{F}_q(\mathcal{E}_1)$. It follows that ψ induces a map ψ^* from $\mathbb{F}_q(\mathcal{E}_2)$ to $\mathbb{F}_q(\mathcal{E}_1)$. In fact, ψ^* is a homomorphism of fields. Moreover, if ψ is nonconstant, then ψ^* is an injective homomorphism from $\mathbb{F}_q(\mathcal{E}_2)$ to $\mathbb{F}_q(\mathcal{E}_1)$. From this, we see that $\mathbb{F}_q(\mathcal{E}_2)$ can be identified with a subfield of $\mathbb{F}_q(\mathcal{E}_1)$ via ψ^*. We define the **degree** of ψ to be the degree of this field extension $\deg(\psi) = [\mathbb{F}_q(\mathcal{E}_1) : \psi^*(\mathbb{F}_q(\mathcal{E}_2))]$. Furthermore, ψ is said to be **separable** (respectively, **inseparable**) if the field extension $\mathbb{F}_q(\mathcal{E}_1)/\psi^*(\mathbb{F}_q(\mathcal{E}_2))$ is separable (respectively, (purely) inseparable). We denote the separable and inseparable degrees of the extension $\mathbb{F}_q(\mathcal{E}_1)/\psi^*(\mathbb{F}_q(\mathcal{E}_2))$ by $\deg_s(\psi)$ and $\deg_i(\psi)$, respectively. Note that, if $\psi = (h_x, h_y)$, then $\psi^*(\mathbb{F}_q(\mathcal{E}_2)) = \mathbb{F}_q(h_x, h_y)$, which is clearly a subfield of $\mathbb{F}_q(\mathcal{E}_1) = \mathbb{F}_q(x, y)$.

Another way to define an isomorphism of elliptic curves is given by the next proposition. We leave the proof to the reader.

Proposition 3.2.7 *A morphism ψ of elliptic curves is an isomorphism if and only if ψ^* is an isomorphism of fields. Hence, ψ is an isomorphism if and only if $\deg(\psi) = 1$.*

Remark 3.2.8 Let $\mathcal{E}_1$ and $\mathcal{E}_2$ be two elliptic curves defined over $\mathbb{F}_q$. According to Lemma 3.2.4(ii), for a point $Q \in \mathcal{E}_2$, the set $\psi^{-1}(Q) \overset{\text{def}}{=} \{P \in \mathcal{E}_1 : \psi(P) = Q\}$ is finite. Indeed, $\psi^{-1}(Q)$ can have at most $\deg(\psi)$ elements. Moreover, $|\psi^{-1}(Q)| = \deg(\psi)$ for all but finitely many points $Q \in \mathcal{E}_2$.

We now turn to an important class of morphisms between elliptic curves known as isogenies. An **isogeny** σ between two elliptic curves $\mathcal{E}_1$ and $\mathcal{E}_2$ is a morphism that sends $O \in \mathcal{E}_1$ to $O \in \mathcal{E}_2$. As we have noted, a morphism is either a constant or is surjective. Hence, an isogeny is either the zero map or a nonconstant morphism with finite kernel. Two elliptic curves are known as **isogenous** if there is an isogeny σ between them. For instance, the morphisms in Example 3.2.2(ii) and (iii) are isogenies.

Let λ be any morphism between $\mathcal{E}_1$ and $\mathcal{E}_2$. Suppose that P is a point on $\mathcal{E}_1$ with $\lambda(P) = O$. Then $\sigma = \lambda \circ \tau_P$ sends O to O, so σ is again an isogeny. Equivalently, $\lambda = \sigma \circ \tau_{-P}$, so any morphism is the composition of an isogeny with a translation map.

Another interesting property of an isogeny is the following.

Theorem 3.2.9 *Let σ be an isogeny from $\mathcal{E}_1$ to $\mathcal{E}_2$. Then σ is a group homomorphism.*

Once again, we leave the proof of Theorem 3.2.9 for the interested reader to check out from books such as [144, Chapter III, Theorem 4.8].

From Theorem 3.2.9, it follows that a nonconstant isogeny is a group homomorphism with finite kernel. Let σ be an isogeny from $\mathcal{E}_1$ to $\mathcal{E}_2$. Then its kernel is a subgroup of $\mathcal{E}_1$. Further, for each $Q \in \mathcal{E}_2$, $|\sigma^{-1}(Q)|$ is constant. Fix an $R \in \mathcal{E}_1$ so that $\sigma(R) = Q$. Then $\sigma^{-1}(Q) = \{P + R : P \in \ker(\sigma)\}$. In fact, we have the following result, which we state without proof (see [144, Chapter II, Proposition 2.6(b)]).

Theorem 3.2.10 *Let σ be an isogeny from $\mathcal{E}_1$ to $\mathcal{E}_2$. Then, for each $Q \in \mathcal{E}_2$, we have $|\sigma^{-1}(Q)| = \deg_s(\sigma)$, where $\deg_s$ refers to the degree of separability of σ. In particular, $|\ker(\sigma)| = \deg_s(\sigma)$.*

Recall that we consider only elliptic curves over finite fields of characteristic different from 2 and 3. Suppose that we have two elliptic curves $\mathcal{E}_1 : y^2 = x^3 + A_1 x + B_1$ and $\mathcal{E}_2 : y^2 = x^3 + A_2 x + B_2$. We consider now the question of isomorphism between $\mathcal{E}_1$ and $\mathcal{E}_2$, and how the coefficients A_1, B_1, A_2, B_2 are related when $\mathcal{E}_1$ and $\mathcal{E}_2$ are isomorphic. More precisely, we provide two criteria to determine if two curves are isomorphic. This then allows us to calculate the number of isomorphism classes of elliptic curves over a finite field $\mathbb{F}_q$ (two elliptic curves belong to the same isomorphism class if and only if they are isomorphic).

We omit the proof of the following theorem. The reader may refer to [144, pages 49–50] for a discussion.

Theorem 3.2.11 *Two elliptic curves $\mathcal{E}_1 : y^2 = x^3 + A_1 x + B_1$ and $\mathcal{E}_2 : y^2 = x^3 + A_2 x + B_2$ are isomorphic over $\mathbb{F}_q$ if and only if there exists a nonzero constant $u \in \mathbb{F}_q^*$ such that $A_2 = A_1 u^4$ and $B_2 = B_1 u^6$. Moreover, the isomorphism is given by $\psi : \mathcal{E}_1 \to \mathcal{E}_2$ such that $\psi(x) = u^2 x$ and $\psi(y) = u^3 y$.*

Remark 3.2.12 This result shows us that an isomorphism of two elliptic curves over $\mathbb{F}_q$ is determined by just one parameter $u \in \mathbb{F}_q^*$. Note that it is possible for two curves to be nonisomorphic over $\mathbb{F}_q$ but to become isomorphic over an extension field $\mathbb{F}_{q^k}$ for some k. In this case, $u \in \mathbb{F}_{q^k} \setminus \mathbb{F}_q$.

Recall that we have defined the discriminant of an elliptic curve $\mathcal{E}$ under the generalized Weierstrass form in (3.3). Using the Weierstrass normal form, it can be simplified to $\triangle = -16(4A^3 + 27B^2)$ (see (3.5)). We only consider smooth curves (see Definition 3.1.6), so $\triangle \neq 0$. In this case, we also define the *j*-**invariant** of $\mathcal{E}$ by $j(\mathcal{E}) = -1728(4A)^3/\triangle$.

The next theorem shows us that isomorphic curves (over $\overline{\mathbb{F}}_q$) have the same *j*-invariants.

Theorem 3.2.13 *Let $\mathcal{E}_1$ and $\mathcal{E}_2$ be two elliptic curves defined over $\mathbb{F}_q$. Then $\mathcal{E}_1$ and $\mathcal{E}_2$ are isomorphic (over $\overline{\mathbb{F}}_q$) if and only if $j(\mathcal{E}_1) = j(\mathcal{E}_2)$.*

Proof. Suppose that $\mathcal{E}_1$ and $\mathcal{E}_2$ are isomorphic (over $\overline{\mathbb{F}}_q$). It follows from Theorem 3.2.11 that there exists some nonzero constant $u \in \overline{\mathbb{F}}_q$ such that $A_2 = A_1 u^4$ and $B_2 = B_1 u^6$. Therefore, $j(\mathcal{E}_2) = 1728(4u^4 A_1)^3/(16(4(u^4 A_1)^3 + 27(B_1 u^6)^2)) = j(\mathcal{E}_1)$.

Conversely, suppose that $j(\mathcal{E}_1) = j(\mathcal{E}_2)$. This yields $A_1^3 B_2^2 = A_2^3 B_1^2$. We wish to look for a nonzero $u \in \overline{\mathbb{F}}_q$ so that $x' = u^2 x$ and $y' = u^3 y$, where x, y define the equation of $\mathcal{E}_1$ and x', y' define the equation of $\mathcal{E}_2$. If $A_1 = 0$, then B_1 cannot be 0 since $\mathcal{E}_1$ is nonsingular. Hence, $A_2 = 0$ and $B_2 \neq 0$. This gives $j = 0$ and we choose $u = (B_2/B_1)^{1/6} \neq 0$. Next, suppose that $B_1 = 0$. As before, we see that $B_2 = 0$ and $A_1, A_2 \neq 0$. Further, $j = 1728$. We let $u = (A_2/A_1)^{1/4}$ in this case. Finally, suppose that $A_1 B_1 \neq 0$. In this case, we set $u = (A_2/A_1)^{1/4} = (B_2/B_1)^{1/6}$. $\qquad\square$

Remark 3.2.14 Note that $\mathcal{E}_1$ and $\mathcal{E}_2$ having identical *j*-invariants shows that they are isomorphic over the algebraic closure $\overline{\mathbb{F}}_q$ of $\mathbb{F}_q$, but not necessarily over $\mathbb{F}_q$. To show that $\mathcal{E}_1$ and $\mathcal{E}_2$ are indeed isomorphic over $\mathbb{F}_q$, we need to find a u in $\mathbb{F}_q$ as in Theorem 3.2.11.

Example 3.2.15 Let $\mathcal{E}_1$ and $\mathcal{E}_2$ be two elliptic curves defined over $\mathbb{F}_7$ given by the Weierstrass equations: $\mathcal{E}_1 : y^2 = x^3 + x$ and $\mathcal{E}_2 : y^2 = x^3 - x$. Since $j(\mathcal{E}_1) = j(\mathcal{E}_2) = 1728$, both $\mathcal{E}_1$ and $\mathcal{E}_2$ are isomorphic over $\overline{\mathbb{F}}_q$. Note that, in this case, we need a u with $u^4 = -1$. Thus, $\mathcal{E}_1$ and $\mathcal{E}_2$ are not isomorphic over $\mathbb{F}_7$, but since $4|((7^2 - 1)/2)$, there exists a $u \in \mathbb{F}_{49}^*$ such that $u^4 = -1$, so $\mathcal{E}_1$ and $\mathcal{E}_2$ are isomorphic over $\mathbb{F}_{49}$.

Definition 3.2.16 We say that two elliptic curves $\mathcal{E}_1$ and $\mathcal{E}_2$ are **twists** of each other if they are isomorphic over $\overline{\mathbb{F}}_q$. Furthermore, they are called **quadratic twists** if they are not isomorphic over $\mathbb{F}_q$ but are isomorphic over $\mathbb{F}_{q^2}$.

Fix a nonsquare $v \in \mathbb{F}_q$ and a $u \in \mathbb{F}_{q^2}$ such that $u^2 = v$. Let $\mathcal{E}$ be an elliptic curve defined over $\mathbb{F}_q$ with Weierstrass equation $\mathcal{E} : y^2 = x^3 + Ax + B$. Then the curve defined by $\mathcal{E}' : y'^2 = x'^3 + A'x' + B'$, with $y' = yu^3, x' = xu^2, A' = Au^4, B' = Bu^6$, is a quadratic twist of $\mathcal{E}$. This is clear since $\mathcal{E}$ and $\mathcal{E}'$ are not isomorphic over $\mathbb{F}_q$ but are isomorphic over $\mathbb{F}_q(u)$. Furthermore, it can be shown that, up to isomorphism over $\mathbb{F}_q$, this is the only quadratic twist of $\mathcal{E}$.

Theorem 3.2.17 *Let $\mathcal{E}$ be an elliptic curve with $\mathcal{E}'$ as its quadratic twist. Suppose that N and N' denote the numbers of $\mathbb{F}_q$-rational points on $\mathcal{E}$ and $\mathcal{E}'$, respectively. Then $N + N' = 2q + 2$.*

Proof. Let $\mathcal{E} : y^2 = x^3 + Ax + B$ and $\mathcal{E}' : y'^2 = x'^3 + A'x' + B'$, where $A' = Au^4$, $B' = Bu^6$, $y' = yu^3$, $x' = xu^2$ for some $u \in \mathbb{F}_{q^2} \backslash \mathbb{F}_q$. First, we have an identity element in $\mathcal{E}$ and $\mathcal{E}'$ each, thereby constituting two distinct points. We now count the number of finite points on $\mathcal{E}(\mathbb{F}_q) \cup \mathcal{E}'(\mathbb{F}_q)$.

Suppose that $v = u^2 \in \mathbb{F}_q$, and since $u \notin \mathbb{F}_q$, v must be a nonsquare in $\mathbb{F}_q$. Let $x \in \mathbb{F}_q$. Then $x' = xu^2 = xv \in \mathbb{F}_q$. Since $x'^3 + A'x' + B' = u^6(x^3 + Ax + B)$, and $u^6 = v^3$, which is a nonsquare in $\mathbb{F}_q$, it follows that $x^3 + Ax + B$ is a square in $\mathbb{F}_q$ if and only if $x'^3 + A'x' + B'$ is a nonsquare in $\mathbb{F}_q$. This implies that, for each $x \in \mathbb{F}_q$ for which $x^3 + Ax + B \neq 0$, we can either find two distinct values of y such that $(x, y) \in \mathcal{E}(\mathbb{F}_q)$ or there are two distinct values of $y' \in \mathbb{F}_q$ such that $(x', y') \in \mathcal{E}'(\mathbb{F}_q)$. Furthermore, if $x \in \mathbb{F}_q$ and $x^3 + Ax + B = 0$, then $x' \in \mathbb{F}_q$ and $x'^3 + A'x' + B' = 0$, so that $(x, 0)$ and $(x', 0)$ are in $\mathcal{E}(\mathbb{F}_q)$ and $\mathcal{E}'(\mathbb{F}_q)$, respectively. We conclude that each $x \in \mathbb{F}_q$ gives rise to two distinct points in $\mathcal{E}(\mathbb{F}_q) \cup \mathcal{E}'(\mathbb{F}_q)$. It follows that there are $2q$ finite points in $\mathcal{E}(\mathbb{F}_q) \cup \mathcal{E}'(\mathbb{F}_q)$ and, hence, a total of $2q + 2$ points.

Next, we suppose that $B = 0$ and $u^2 \notin \mathbb{F}_q$. Since we need $u^4 \in \mathbb{F}_q$, this is only possible if $4|(q + 1)$, so $q \equiv 3 \pmod 4$. Now, for each A and $x \in \mathbb{F}_q^*$, $x^3 + Ax = x(x^2 + A)$, so $x^3 + Ax$ is a square in $\mathbb{F}_q$ if and only if $(-x)((-x)^2 + A)$ is a nonsquare. This means that $\mathcal{E}(\mathbb{F}_q)$ has $2(q - 1)/2 + 1 + 1 = q + 1$ points. Thus, if $\mathcal{E}'$ is a quadratic twist of $\mathcal{E}$, it has Weierstrass equation $y'^2 = x'^3 + A'x'$, so $|\mathcal{E}(\mathbb{F}_q)| + |\mathcal{E}'(\mathbb{F}_q)| = q + 1 + q + 1 = 2q + 2$, as required.

Finally, we consider the case in which $A = 0$. Suppose that $u^2 \notin \mathbb{F}_q$ and $u^6 \in \mathbb{F}_q$. This is only possible if $6|(q + 1)$, i.e., $q \equiv 5 \pmod 6$. Fix a $B \in \mathbb{F}_q^*$.

Since $3 \nmid (q-1)$, the map $\alpha \mapsto \alpha^3$, for $\alpha \in \mathbb{F}_q$, is one-to-one, so there is exactly one x such that $x^3 + B = y^2$, for every $y \in \mathbb{F}_q$. Consequently, $\mathcal{E}(\mathbb{F}_q)$ has $q+1$ elements and by the same argument as above, $|\mathcal{E}(\mathbb{F}_q)| + |\mathcal{E}'(\mathbb{F}_q)| = 2q + 2$.

In conclusion, we have shown that $\mathcal{E}$ and $\mathcal{E}'$ have a total of $2q + 2$ $\mathbb{F}_q$-rational points in all cases. $\qquad\Box$

Before we end this subsection, we use Theorem 3.2.11 to count the number of nonisomorphic elliptic curves over a finite field $\mathbb{F}_q$. We illustrate the technique for prime fields $\mathbb{F}_p$.

The elliptic curves are of the form $y^2 = x^3 + Ax + B$, with $A, B \in \mathbb{F}_p$. Hence, there are altogether p^2 possible curves over $\mathbb{F}_p$ given by such equations, but such a curve is singular if and only if $4A^3 + 27B^2 = 0$. This occurs if and only if there exists some constant C for which $A = -3C^2$ and $B = 2C^3$. In particular, each C uniquely determines the pair (A, B) and *vice versa*. Consequently, there are p singular curves and $p^2 - p$ (nonsingular) elliptic curves.

Fix an elliptic curve $\mathcal{E} : y^2 = x^3 + Ax + B$. We have seen that another curve $\mathcal{E}' : y^2 = x^3 + A'x^3 + B'$ is isomorphic to $\mathcal{E}$ if and only if there exists a nonzero constant u such that $A' = Au^4$ and $B' = Bu^6$. The number of different elliptic curves isomorphic to $\mathcal{E}$ over $\mathbb{F}_p$ is now $(p-1)/|\mathrm{Aut}(\mathcal{E})|$, where $\mathrm{Aut}(\mathcal{E})$ refers to the group of automorphisms of $\mathcal{E}$. Summing over all representatives of elliptic curves in each isomorphism class yields $\sum_{\mathcal{E}} (p-1)/|\mathrm{Aut}(\mathcal{E})| = p(p-1)$. Thus, $\sum_{\mathcal{E}} (1/|\mathrm{Aut}(\mathcal{E})|) = p$. We now look at $\mathrm{Aut}(\mathcal{E})$. Recall that an automorphism of $\mathcal{E}$ is an isomorphism from $\mathcal{E}$ to $\mathcal{E}$. In other words, it comprises u such that $A = Au^4$ and $B = Bu^6$. If $B = 0$, then $|\mathrm{Aut}(\mathcal{E})| = \gcd(4, p-1)$. If $A = 0$, then $|\mathrm{Aut}(\mathcal{E})| = \gcd(6, p-1)$. Otherwise, $|\mathrm{Aut}(\mathcal{E})| = 2$.

It follows that, in the case $p \equiv -1 \pmod{12}$, $|\mathrm{Aut}(\mathcal{E})| = 2$ for all $\mathcal{E}$, so that the number of nonisomorphic elliptic curve classes over $\mathbb{F}_p$ is exactly $2p$. To see this, let N be the total number of nonisomorphic elliptic curve classes over $\mathbb{F}_p$. Since $\gcd(6, p-1) = \gcd(4, p-1) = 2$, we have $N/2 = p$ or $N = 2p$. The other cases can be similarly estimated or computed.

In particular, the number of nonisomorphic elliptic curves classes is $2p+6$, $2p + 2$, $2p + 4$, $2p$ for $p \equiv 1, 5, 7, 11 \pmod{12}$, respectively.

3.2.2 The Frobenius and Multiplication Morphisms

We now discuss two important isogenies, namely, the **Frobenius map** and the **multiplication map**. These two isogenies play an important role in proving many important results on elliptic curves as we shall see in the remainder of the chapter.

Throughout this subsection, we let $q = p^k$ for some prime p ($p \neq 2, 3$) and positive integer k. Fix an elliptic curve over $\mathbb{F}_q$ as $\mathcal{E} : y^2 = x^3 + Ax + B$. We define the qth-power **Frobenius morphism** ϕ_q by $\phi_q((x, y)) = (x^q, y^q)$ for every finite point $(x, y) \in \mathcal{E}$ and $\phi_q(O) = O$.

We first show that $\phi_q((x, y))$ lies on $\mathcal{E}$. Indeed, raising the Weierstrass normal form of $\mathcal{E}$ to the qth power yields $y^{2q} = x^{3q} + Ax^q + B$ or $(y^q)^2 = (x^q)^3 + Ax^q + B$. Thus, $(x^q, y^q) \in \mathcal{E}$. Since $\phi_q(O) = O$, ϕ_q is an isogeny. Given any two elliptic curves $\mathcal{E}_1, \mathcal{E}_2$ and any isogeny $\sigma : \mathcal{E}_1 \to \mathcal{E}_2$, it is easy to see that $\sigma \circ \phi_q = \phi_q \circ \sigma$, where ϕ_q on the left-hand side is on $\mathcal{E}_1$ while that on the right-hand side is on $\mathcal{E}_2$.

Let e be a positive integer. It is clear that the composition $\phi_q \circ \cdots \circ \phi_q$ (e times) $= \phi_{q^e}$. Since any element z in $\mathbb{F}_{q^e}$ satisfies $z^{q^e} = z$, the following lemma is immediate.

Lemma 3.2.18 *Let P be a point on $\mathcal{E}$. Then P is an $\mathbb{F}_{q^e}$-rational point of $\mathcal{E}$ if and only if $\phi_{q^e}(P) = P$.*

In the next proposition, we state some important properties for ϕ_q. The reader may refer to [144, Chapter V, Theorem 3.1] for the proofs.

Proposition 3.2.19 (i) *The Frobenius morphism ϕ_q is injective.*

(ii) *The Frobenius morphism ϕ_q is purely inseparable of degree q.*

Lemma 3.2.18 enables us to prove a very nice result on isogenous curves.

Theorem 3.2.20 *Let $\mathcal{E}_1$ and $\mathcal{E}_2$ be two isogenous elliptic curves over $\mathbb{F}_q$. Then they have the same number of $\mathbb{F}_q$-rational points, i.e., $|\mathcal{E}_1(\mathbb{F}_q)| = |\mathcal{E}_2(\mathbb{F}_q)|$.*

Proof. Let σ be an isogeny from $\mathcal{E}_1$ to $\mathcal{E}_2$. Let ϕ_q denote the qth-power Frobenius morphism. By Lemma 3.2.18, $\phi_q(P) = P$ for all $P \in \mathcal{E}_1(\mathbb{F}_q)$ and for all $P \in \mathcal{E}_2(\mathbb{F}_q)$. In other words, $\mathcal{E}_1(\mathbb{F}_q) = \ker(1 - \phi_q)$ on $\mathcal{E}_1$ and $\mathcal{E}_2(\mathbb{F}_q) = \ker(1 - \phi_q)$ on $\mathcal{E}_2$. For a fixed $Q \in \mathcal{E}_2$ and for all $P \in \mathcal{E}_1$ with $\sigma(P) = Q$, we have $Q \in \mathcal{E}_2(\mathbb{F}_q)$ if and only if $P \in \ker((1 - \phi_q) \circ \sigma)$.

Now, $|\mathcal{E}_2(\mathbb{F}_q)| = |\ker((1 - \phi_q) \circ \sigma)| / |\ker(\sigma)| = |\ker(\sigma \circ (1 - \phi_q))| / |\ker(\sigma)|$. Note that we have used the facts that ϕ_q commutes with σ and that $1 - \phi_q$ operates on $\mathcal{E}_2$ and $\mathcal{E}_1$, respectively, in the above equality. Since isogenies are group homomorphisms (see Theorem 3.2.9), we obtain $|\mathcal{E}_2(\mathbb{F}_q)| = |\ker(1 - \phi_q)|$, where $1 - \phi_q$ is on $\mathcal{E}_1$, and this is equal to $|\mathcal{E}_1(\mathbb{F}_q)|$ as required. $\square$

In fact, the converse of the statement in Theorem 3.2.20 is also true, that is, two elliptic curves having the same number of $\mathbb{F}_q$-rational points are isogenous. This is known as Tate's isogeny theorem and was first proposed and proved by John Tate in 1966 (see [161]).

To end this subsection, we discuss yet another useful isogeny.

Let m be an integer and let $\mathcal{E}$ be an elliptic curve. The map $[m] : P \mapsto [m]P$ is a map from $\mathcal{E}$ to $\mathcal{E}$ defined by $[m]P = P + P + \cdots + P$ (m times). The map $[m]$ is called a **multiplication morphism**. Observe that, when $m = -1$, we obtain the negation morphism $\chi_\mathcal{E}$. By first showing that $[2]$ is a morphism and using induction, we can show that $[m]$ is a morphism for any integer m. Since $[m](O) = O$, the morphism $[m]$ is in fact an isogeny.

Here are some basic properties of the map $[m]$.

First, it is clear that, for any two integers m and n, we have $[m] \circ [n] = [mn]$. Next, we show that $[m]$ is not a constant for all nonzero integers m. Clearly, the map $[2]$ is not a constant since there are only four points P on $\mathcal{E}$ for which $[2]P = O$. Now suppose that, for some integer m, the map $[m]$ is a constant map. If m is even, we can write $m = 2n$. Since $[2]$ is not a constant map, it follows immediately that $[n]$ is a constant map. We can thus assume that m is odd. Pick a point $P \neq O$ such that $[2]P = O$. Then $[m]P = [m-1]P + P = P \neq O$. Consequently, $[m]$ is not the zero map and cannot be a constant. In fact, we have the following result (cf. [144, Chapter III, Corollary 5.5 and Theorem 6.2]).

Proposition 3.2.21 *Let m be an integer.*

(i) *The map $[m]$ is separable if and only if $p \nmid m$.*

(ii) *The degree of $[m]$ satisfies $\deg([m]) = m^2$.*

The points P on $\mathcal{E}$ for which $[m]P = O$ are called m-**torsion points**. We now proceed to study more about these points.

3.3   The Group $\mathcal{E}(\mathbb{F}_q)$ and Its Torsion Subgroups

In this section, our aim is to investigate two particular subgroups of $\mathcal{E}$, namely, the m-torsion subgroup $\mathcal{E}[m]$ and the group $\mathcal{E}(\mathbb{F}_q)$ of $\mathbb{F}_q$-rational points of $\mathcal{E}$. In addition, we present a useful class of elliptic curves known as supersingular elliptic curves.

3.3.1   The Torsion Group $\mathcal{E}[m]$

For an integer $m \neq 0$, let $\mathcal{E}[m]$ denote the kernel of the multiplication map $[m]$. All points of $\mathcal{E}[m]$ are called m-torsion points. Note that, since $[m]P = O$ if and only if $[-m]P = O$, it suffices to consider $m > 0$. Furthermore, we consider all points $P \in \mathcal{E}(\overline{\mathbb{F}}_q)$ rather than only points in $\mathcal{E}(\mathbb{F}_q)$. In other words, $\mathcal{E}[m]$ may not be a subgroup of $\mathcal{E}(\mathbb{F}_q)$.

Since $[m]$ is an isogeny, $\mathcal{E}[m]$ is finite. We begin by finding the number of m-torsion points.

Let m be an integer with $\gcd(m, p) = 1$, where p is the characteristic of $\mathbb{F}_q$. By Proposition 3.2.21, $[m]$ is separable (since $p \nmid m$). In addition, $|\mathcal{E}[m]| = |\ker([m])| = \deg([m]) = m^2$ (by Theorem 3.2.10).

What is the group structure of $\mathcal{E}[m]$? We first take a look at an example.

Example 3.3.1 Let $m = 2$. Suppose that $[2]P = O$. Write $P = (a, b)$. Since we have $P = -P$, it follows that $b = 0$, and hence, a is a root of $x^3 + Ax + B$. This yields three possible values of a. As $2O = O$, $\mathcal{E}[2]$ has four points, and as a group, $\mathcal{E}[2] \cong \mathbb{Z}_2 \times \mathbb{Z}_2$.

For a positive integer m, we first observe that $\mathcal{E}[m]$ is not cyclic, for otherwise $\mathcal{E}[m]$ will contain an element of order m^2.

Now, suppose that ℓ is a prime that divides m. Since $\mathcal{E}[\ell]$ is not cyclic, we can only have $\mathcal{E}[\ell] = \mathbb{Z}_\ell \times \mathbb{Z}_\ell$. Since $\mathcal{E}[m]$ is abelian, it follows from the Fundamental Theorem of Finite Abelian Groups that $\mathcal{E}[m] \cong \mathbb{Z}_{m_1} \times \mathbb{Z}_{m_2} \times \cdots \times \mathbb{Z}_{m_k}$ with $m_1 | m_2 | \cdots | m_k$. Since $\mathcal{E}[\ell] \subseteq \mathcal{E}[m]$ and $\mathcal{E}[\ell] \cong \mathbb{Z}_\ell \times \mathbb{Z}_\ell$, we must have $k = 2$, i.e., $\mathcal{E}[m] \cong \mathbb{Z}_{m_1} \times \mathbb{Z}_{m_2}$, where $m_1 | m_2$ and $m_1 m_2 = m^2$. Consequently, $m_1 = m_2 = m$, so that $\mathcal{E}[m] \cong \mathbb{Z}_m \times \mathbb{Z}_m$.

Next, we consider the case where m is a power of p.

Lemma 3.3.2 *For all positive integers e, $\mathcal{E}[p^e]$ is either $\{O\}$ or isomorphic to $\mathbb{Z}_{p^e}$.*

Proof. By Proposition 3.2.21(i), the map $[p^e]$ is inseparable. Therefore, it follows from Theorem 3.2.10 that $|\mathcal{E}[p^e]| = \deg_s([p^e]) < \deg([p^e]) = p^{2e}$. In particular, for $e = 1$, $|\mathcal{E}[p]| = 1$ or p. Thus, the result holds for $e = 1$.

Suppose that $\mathcal{E}[p] = \{O\}$. Then, it is clear that, for all $e > 1$, $\mathcal{E}[p^e] = \{O\}$ (for otherwise, there would be a point of order p).

Since $\mathcal{E}[p] \subseteq \mathcal{E}[p^e]$, $\mathcal{E}[p^e]$ must be cyclic; otherwise, $\mathcal{E}[p^e]$ contains $\mathbb{Z}_p \times \mathbb{Z}_p$, which then means $\mathcal{E}[p] \supseteq \mathbb{Z}_p \times \mathbb{Z}_p$, a contradiction. Now consider the group morphism θ: $\mathcal{E}[p^e] \to \mathcal{E}[p^{e-1}]$ by sending P to $[p]P$, for $e \geq 2$. It is clear that the kernel of θ is $\mathcal{E}[p]$ and it is surjective (see Lemma 3.2.4). Hence, $|\mathcal{E}[p^e]| = p|\mathcal{E}[p^{e-1}]|$. By induction, we obtain the desired result. $\square$

Definition 3.3.3 An elliptic curve $\mathcal{E}$ is called a **supersingular elliptic curve** if $\mathcal{E}[p] = \{O\}$. Otherwise, $\mathcal{E}$ is an **ordinary elliptic curve**.

Example 3.3.4 From the proof of Theorem 3.2.17, if $q \equiv 3 \pmod 4$, then the curve $\mathcal{E} : y^2 = x^3 + x$ has $q + 1$ $\mathbb{F}_q$-rational points. Similarly, for all odd integers d, we have $q^d \equiv 3 \pmod 4$ and, thus, $\mathcal{E}(\mathbb{F}_{q^d})$ has $q^d + 1$ points. Suppose that $\mathcal{E}[p] \cong \mathbb{Z}_p$, where p is the characteristic of $\mathbb{F}_q$. In other words, $\mathcal{E}$ has a point P of order p. Suppose there is some integer d such that $P \in \mathcal{E}(\mathbb{F}_{q^d})$. If d is odd, since $\mathcal{E}[p]$ is a subgroup of $\mathcal{E}(\mathbb{F}_{q^d})$, it follows that we must have $p|(q^d + 1)$, which is impossible. If $d \equiv 2 \pmod 4$, then we have $p|(1 + q^{d/2})^2$, whereas, if $d \equiv 0 \pmod 4$, then we have $p|(q^{d/2^r} - 1)^{2^r}$, where 2^r exactly divides d. Hence, we conclude that $\mathcal{E}[p] = \{O\}$ and $\mathcal{E}$ is supersingular.

Our preceding analysis leads us to the following result on the structure of $\mathcal{E}[m]$ for any arbitrary positive integer m.

Theorem 3.3.5 *Let $\mathcal{E}$ be an elliptic curve over $\mathbb{F}_q$ of characteristic p. Write $m = p^e m'$ for some m' with $\gcd(p, m') = 1$. Then $\mathcal{E}[m] \cong \mathbb{Z}_{m'} \times \mathbb{Z}_{m'}$ or $\mathbb{Z}_{m'} \times \mathbb{Z}_m$.*

Remark 3.3.6 Let m be a positive integer such that $\gcd(m,p) = 1$. By Theorem 3.3.5, if $\mathcal{E}$ is an elliptic curve over $\mathbb{F}_q$ of characteristic p, then $\mathcal{E}[m] \cong \mathbb{Z}_m \times \mathbb{Z}_m$. Hence, there exist $P_1, P_2 \in \mathcal{E}[m]$ such that all points $P \in \mathcal{E}[m]$ can be represented in the form $P = aP_1 + bP_2$, for $a,b \in \mathbb{Z}_m$. Suppose that σ is an isogeny on $\mathcal{E}$. Since $[m]\sigma(P) = \sigma([m]P) = \sigma(O) = O$ for all points $P \in \mathcal{E}[m]$ (note that it is easy to verify that $[m]$ commutes with every isogeny), it means that σ, when restricted to $\mathcal{E}[m]$, is a homomorphism from $\mathcal{E}[m]$ to $\mathcal{E}[m]$. In particular, σ can be represented by a 2×2 matrix $[\sigma]_m$ on $\mathcal{E}[m]$ with entries in $\mathbb{Z}_m$.

3.3.2   The Group $\mathcal{E}(\mathbb{F}_q)$

In Corollary 1.5.4, we have discussed the Hasse-Weil bound for the number of $\mathbb{F}_q$-rational points on an algebraic curve of genus g. Applying to elliptic curves (which are curves of genus 1 according to Theorem 3.1.11), we obtain Theorem 3.3.7.

Theorem 3.3.7 (Hasse's Theorem) *Let $N(q)$ denote the number of $\mathbb{F}_q$-rational points on $\mathcal{E}$. Then $|N(q) - q - 1| \le 2\sqrt{q}$.*

Remark 3.3.8 In fact, there is an independent proof for Hasse's Theorem for elliptic curves via the use of the Frobenius morphism ϕ_q. The main ideas of the proof can be summarized as follows.

- Recall that $N(q) = |\ker(1 - \phi_q)| = \deg_s(1 - \phi_q) = \deg(1 - \phi_q)$ since $1 - \phi_q$ is separable (cf. [144, Chapter III, Corollary 5.5]).

- For two endomorphisms ψ_1 and ψ_2 on $\mathcal{E}$, $\langle \psi_1, \psi_2 \rangle \stackrel{\text{def}}{=} \deg(\psi_1 + \psi_2) - \deg(\psi_1) - \deg(\psi_2)$ is bilinear and satisfies the Cauchy-Schwarz inequality:

$$|\langle \psi_1, \psi_2 \rangle| \le 2\sqrt{\deg(\psi_1)}\sqrt{\deg(\psi_2)}.$$

- Hence, $|N(q) - q - 1| = |\deg(1 - \phi_q) - \deg(\phi_q) - \deg(1)| = |\langle 1, -\phi_q \rangle| \le 2\sqrt{\deg(1)}\sqrt{\deg(-\phi_q)} = 2\sqrt{q}$.

For the rest of this section, we let

$$a(q) \stackrel{\text{def}}{=} q + 1 - N(q).$$

Thus far, we have the following information on $a(q)$:

- By Hasse's Theorem, $a(q)^2 \le 4q$.

- Recall that the L-function of $\mathcal{E}$ is given by $L(t) = 1 - a(q)t + qt^2$. Hence, if we write $L(t) = (1 - \alpha t)(1 - \beta t)$, it follows that $a(q) = \alpha + \beta$.

Consider the Frobenius morphism ϕ_q on $\mathcal{E}$. Let m be relatively prime to p. From Remark 3.3.6, we know that there exists a matrix $[\phi_q]_m$ that represents the action of ϕ_q on $\mathcal{E}[m]$. The following result will be useful. We refer the reader to [172, Proposition 4.11] for a proof.

Theorem 3.3.9 *Let* $\mathrm{Tr}([\phi_q]_m)$ *and* $\det([\phi_q]_m)$ *denote the trace and determinant of the matrix* $[\phi_q]_m$, *respectively. Then, we have the following:*

(i) $a(q) \equiv \mathrm{Tr}([\phi_q]_m) \pmod{m}$;

(ii) $q \equiv \det([\phi_q]_m) \pmod{m}$;

(iii) $\phi_{q^2} - [a(q)] \circ \phi_q + [q]$ *is the zero morphism on* $\mathcal{E}$.

In view of Theorem 3.3.9(iii), and since $\phi_{q^2} = \phi_q^2$, the polynomial $t^2 - a(q)t + q$ is sometimes called the **characteristic polynomial** of the Frobenius morphism ϕ_q.

In the next theorem, we classify the group orders of all the non-isogenous classes of elliptic curves over a fixed finite field $\mathbb{F}_q$. This classification is done via the endomorphism ring of $\mathcal{E}$. The reader may refer to [127] and [173] for the proof.

Theorem 3.3.10 *The set of isogeny classes of elliptic curves over* $\mathbb{F}_q$, *of characteristic* p, *is in a natural bijection with the set of integers* $a(q)$ *satisfying* $|a(q)| \leq 2\sqrt{q}$ *and any of the following conditions:*

(i) $\gcd(q, a(q)) = 1$;

(ii) q *is a square and* $a(q) = \pm 2\sqrt{q}$;

(iii) q *is a square,* $p \not\equiv 1 \pmod 3$ *and* $a(q) = \pm\sqrt{q}$;

(iv) q *is not a square,* $p = 2$ *or* 3, *and* $a(q) = \pm\sqrt{pq}$;

(v) q *is not a square and* $a(q) = 0$, *or,* q *is a square,* $p \not\equiv 1 \pmod 4$ *and* $a(q) = 0$.

The following corollary is now a direct consequence.

Corollary 3.3.11 *Suppose that* q *is a prime. Then there exists an elliptic curve over* $\mathbb{F}_q$ *with* $N(q)$ $\mathbb{F}_q$-*rational points for all integers* $N(q)$ *satisfying* $q + 1 - 2\sqrt{q} \leq N(q) \leq q + 1 + 2\sqrt{q}$.

Corollary 3.3.12 *Suppose that* $q = p^u$, *where* p *is a prime. Then the maximum number of* $\mathbb{F}_q$-*rational points that an elliptic curve defined over* $\mathbb{F}_q$ *can have satisfies*

$$N_{\max}(q) = \begin{cases} q + \lfloor 2\sqrt{q} \rfloor & \text{if } u \geq 3 \text{ is odd and } p | \lfloor 2\sqrt{q} \rfloor \\ q + 1 + \lfloor 2\sqrt{q} \rfloor & \text{otherwise.} \end{cases}$$

Moreover, there exists an elliptic curve for which this maximum is attained.

Proof. This is a direct consequence of Theorem 3.3.10. If q is a square, then Theorem 3.3.10(ii) shows that $N_{\max}(q) = q + 1 + 2\sqrt{q}$. If q is not a square and $p \nmid \lfloor 2\sqrt{q} \rfloor$, then, by Theorem 3.3.10(i), $N_{\max}(q) = q + 1 + \lfloor 2\sqrt{q} \rfloor$. Finally, if q is not a square and $p | \lfloor 2\sqrt{q} \rfloor$, we discuss this case by considering $u = 1$ or $u \geq 3$ separately. If $u = 1$, then $p = q$. In this case, we must have $p = 2$ or 3 and the case of Theorem 3.3.10(iv) achieves the maximum, i.e., $N_{\max}(q) = q + 1 + \sqrt{pq} = 2q + 1 = q + 1 + \lfloor 2\sqrt{q} \rfloor$. If $u \geq 3$, then the case of Theorem 3.3.10(i) achieves the maximum, i.e., $N_{\max}(q) = q + 1 + (\lfloor 2\sqrt{q} \rfloor - 1) = q + \lfloor 2\sqrt{q} \rfloor$ is the number of $\mathbb{F}_q$-rational points on some elliptic curve. $\square$

For brevity, let N be the number of $\mathbb{F}_q$-rational points on $\mathcal{E}$. Then $[N]P = O$ for all $P \in \mathcal{E}(\mathbb{F}_q)$. This means that $\mathcal{E}(\mathbb{F}_q) \subseteq \mathcal{E}[N]$. Now, the structure of $\mathcal{E}[N]$ is known, namely, $\mathcal{E}[N] \cong \mathbb{Z}_N \times \mathbb{Z}_N$ (unless $p|N$, in which case $\mathcal{E}[N] \cong \mathbb{Z}_{N'} \times \mathbb{Z}_{N'}$ or $\mathbb{Z}_{N'} \times \mathbb{Z}_N$, where N' is the largest divisor of N that is coprime to p). For convenience, assume that $p \nmid N$. We conclude that $\mathcal{E}(\mathbb{F}_q) \cong \mathbb{Z}_\ell \times \mathbb{Z}_m$, where $\ell | m$ and $\ell m = N$. Further, by considering the action of ϕ_q on $\mathcal{E}$, we can obtain more information on ℓ as the next lemma shows.

Lemma 3.3.13 *Suppose that* $\gcd(N, p) = 1$ *and* $\mathcal{E}(\mathbb{F}_q) \cong \mathbb{Z}_\ell \times \mathbb{Z}_m$. *Then we have that* $\ell | (a(q) - 2)$ *and* $\ell | (q - 1)$.

Proof. Since $\ell | m$, we have $\ell^2 | N$. In particular, $\mathcal{E}[\ell] \subseteq \mathcal{E}(\mathbb{F}_q)$. Consider the Frobenius morphism ϕ_q on $\mathcal{E}[\ell]$. Since every ℓ-torsion point is $\mathbb{F}_q$-rational, it follows that ϕ_q is the identity on $\mathcal{E}[\ell]$. Hence, $[\phi_q]_\ell = \begin{pmatrix} 1 & 0 \\ 0 & 1 \end{pmatrix}$. By Theorem 3.3.9, $a(q) \equiv \mathrm{Tr}([\phi_q]_\ell) \equiv 2 \pmod{\ell}$ and $q \equiv \det([\phi_q]_\ell) \equiv 1 \pmod{\ell}$. This completes the proof. $\square$

Note that, when $\ell = 1$, $\mathcal{E}(\mathbb{F}_q)$ is cyclic. We now describe the values of ℓ and m for each of the five cases in Theorem 3.3.10 (see [127]).

Theorem 3.3.14   (i) *If* $\gcd(a(q), q) = 1$, *then* $\mathcal{E}(\mathbb{F}_q)$ *can take all possible structures* $\mathbb{Z}_\ell \times \mathbb{Z}_m$, *where* $\ell m = N$ *and* $\ell | \gcd(q - 1, m)$.

(ii) *Suppose that* q *is a square and* $a(q) = \pm 2\sqrt{q}$. *Then* $\ell = m = \sqrt{q} \mp 1$.

(iii) *Suppose that* q *is a square,* $p \not\equiv 1 \pmod 3$ *and* $a(q) = \pm\sqrt{q}$. *Then* $\ell = 1$.

(iv) *Suppose that* $p = 2$ *or* 3, q *is not a square and* $a(q) = \pm\sqrt{pq}$. *Then* $\ell = 1$.

(v) *Suppose that* $a(q) = 0$ *and either one of the following conditions holds:* q *is not a square and* $p \not\equiv 3 \pmod 4$, *or* q *is a square and* $p \not\equiv 1 \pmod 4$. *Then* $\ell = 1$.

(vi) *Suppose that* q *is not a square,* $p \equiv 3 \pmod 4$ *and* $a(q) = 0$. *Then* $\ell = 1$ *or* 2 *and both possibilities may occur.*

We remark that Theorem 3.3.14 does not tell us how to construct an elliptic curve with a given structure, but rather, it only guarantees the existence of such a curve.

3.3.3 Supersingular Elliptic Curves

Supersingular elliptic curves were introduced in Definition 3.3.3 as elliptic curves which have no point of order p. In this subsection, we present another characterizing property of supersingular elliptic curves and provide some examples of these curves.

As before, we let $a(q) = q + 1 - N(q)$.

Theorem 3.3.15 *An elliptic curve $\mathcal{E}$ over $\mathbb{F}_q$, of characteristic p, is supersingular if and only if $a(q) \equiv 0 \pmod{p}$.*

Proof. For any integer $k \geq 1$, let $a(q^k) = q^k + 1 - N(q^k)$, where $N(q^k)$ denotes the number of $\mathbb{F}_{q^k}$-rational points on $\mathcal{E}$. Let $L(t) = 1 - a(q)t + qt^2 = (1 - \alpha t)(1 - \beta t)$ be the L-function of $\mathcal{E}$. Then $a(q^k) = \alpha^k + \beta^k$ by Corollary 1.5.4(ii). Observe that $\alpha^2 = a(q)\alpha - q$ and $\beta^2 = a(q)\beta - q$. Hence, for any $k \geq 1$, $\alpha^{k+2} = a(q)\alpha^{k+1} - q\alpha^k$ and $\beta^{k+2} = a(q)\beta^{k+1} - q\beta^k$. Adding these two equations, we obtain

$$a(q^{k+2}) = a(q)a(q^{k+1}) - qa(q^k). \tag{3.6}$$

Now, suppose that $a(q) \equiv 0 \pmod{p}$. Then, it is obvious that

$$a(q^2) = a(q)^2 - 2q \equiv 0 \pmod{p}, \tag{3.7}$$

while the relation in (3.6) immediately implies that $a(q^k) \equiv 0 \pmod{p}$ for all integers $k \geq 2$. Hence, $N(q^k) \equiv q^k + 1 \equiv 1 \pmod{p}$. Since $p \nmid N(q^k)$, $\mathcal{E}(\mathbb{F}_{q^k})$ cannot contain any point of order p, for all $k \geq 1$, and so, $\mathcal{E}[p] = \{O\}$. Consequently, $\mathcal{E}$ is supersingular.

Next, suppose that $a(q) \equiv a \pmod{p}$ for some $1 \leq a \leq p - 1$. It follows from the relations in (3.6) and (3.7) that $a(q^k) \equiv a(q)^k \equiv a^k \pmod{p}$, for all $k \geq 1$. Let h be the order of a mod p. Then $N(q^h) \equiv q^h + 1 - a^h \equiv q^h + 1 - 1 \equiv 0 \pmod{p}$. Thus, $\mathcal{E}[p] \neq \{O\}$, so $\mathcal{E}$ is not supersingular. $\square$

Remark 3.3.16 From Theorem 3.3.15, we observe that all the elliptic curves of Theorem 3.3.10, apart from those in (i), are supersingular.

Corollary 3.3.17 *If $q = p$ is prime, then all supersingular elliptic curves over $\mathbb{F}_p$ have $p + 1$ points.*

Proof. This is an immediate consequence of Theorem 3.3.15 and Hasse's Theorem (Theorem 3.3.7). $\square$

Example 3.3.18   (i) Let $q \equiv 3 \pmod 4$. We have seen in the proof of
Theorem 3.2.17 that all the curves given by $y^2 = x^3 + Ax$, where $A \neq 0$,
have $q + 1$ $\mathbb{F}_q$-rational points. Thus, these curves are supersingular by
Theorem 3.3.15.

(ii) Let $q \equiv 2 \pmod 3$. Consider the elliptic curve $\mathcal{E} : y^2 = x^3 + B$, with
$B \neq 0$, defined over $\mathbb{F}_q$. Once again, the proof of Theorem 3.2.17 shows
that $\mathcal{E}$ has $q + 1$ $\mathbb{F}_q$-rational points and is thus supersingular.

Remark 3.3.19   (i) In fact, it can be proved that the curves with equation
$y^2 = x^3 + Ax$ are supersingular if and only if $q \equiv 3 \pmod 4$, and the
curves with equation $y^2 = x^3 + B$ are supersingular if and only if $q \equiv 2$
$\pmod 3$. For proofs of the converse, the reader may refer to [144, pages
143–144, Examples 4.4 and 4.5] and [172, Proposition 4.35].

(ii) It may be appealing to use supersingular curves for applications to cryptography as computations can be carried out more efficiently with such
curves. However, such curves have serious drawbacks with respect to the
discrete logarithm problem. This will be discussed in greater detail in
the ensuing sections.

3.4   Computational Considerations on Elliptic Curves

We proceed next to briefly discuss some computational issues involving elliptic curves. Such considerations are especially important when elliptic curves
are used in cryptographic applications.

We continue to restrict our discussion to elliptic curves given by equations
of the form $y^2 = x^3 + Ax + B$.

3.4.1   Finding Multiples of a Point

Given a point $P = (a, b)$ on $\mathcal{E}$, we know that $-P = (a, -b)$. The duplication
formula provides an easy way to compute $[2]P$. Given a positive integer m, a
naive approach to compute $[m]P$ is to carry out $m-1$ additions $P+P+\cdots+P$.
However, it is straightforward to adapt the square-and-multiply approach for
exponentiation in finite fields to compute $[m]P$. This is described as follows:

(i) Write m in its binary form, i.e., $m = m_0 + 2m_1 + 2^2 m_2 + \cdots + 2^t m_t$,
with $m_i \in \{0, 1\}$ and $m_t = 1$.

(ii) Let $P_0 = P$.

(iii) For $i = 1, \ldots, t$, apply the duplication formula to compute $P_i \overset{\text{def}}{=} [2^i]P = [2]P_{i-1}$ recursively.

(iv) Then, $[m]P = \sum_{i=0}^{t}[m_i]P_i$.

Remark 3.4.1 In the above algorithm, $t = \lfloor \log m \rfloor$ (we recall that, throughout this book, log is taken to mean $\log_2$). It requires t additions to compute $P_1, \ldots, P_t$. In addition, it requires at most another t additions to sum up the appropriate P_i's to obtain $[m]P$. Hence, to compute $[m]P$ requires at most $2t \leq 2 \log m$ additions.

Example 3.4.2 Consider the elliptic curve $\mathcal{E}: y^2 = x^3 + x + 1$ over $\mathbb{F}_{37}$. It is easily verified that $P = (8, 22)$ lies on $\mathcal{E}$. The following table gives the value of $[2^i]P$, for $0 \leq i \leq 5$.

TABLE 3.1: Values of $[2^i]P$ for $0 \leq i \leq 5$.

i	0	1	2	3	4	5
$[2^i]P$	$(8, 22)$	$(14, 24)$	$(21, 25)$	$(28, 22)$	$(9, 6)$	$(9, 31)$

Hence, $[13]P = [8]P + [4]P + P = P_3 + P_2 + P_0 = (28, 22) + (21, 25) + (8, 22) = (6, 1)$. Further, $[-19]P = [19](-P) = [16](-P) + [2](-P) + (-P) = -P_4 - P_1 - P_0 = (9, 31) + (14, 13) + (8, 15) = (33, 28)$.

Since $-P$ is easily obtained from P for an elliptic curve, we can reduce the number of additions required in the preceding procedure by allowing for negative signs in the expansion of m. We illustrate this process with an example.

Example 3.4.3 Consider the curve $y^2 = x^3 + x + 1$ over $\mathbb{F}_{37}$ as in Example 3.4.2. Once again, consider the point $P = (8, 22)$.

(i) Suppose that we want to compute $[31]P$. Now, $31 = 1 + 2 + 4 + 8 + 16 = 32 - 1$. Thus, $[31]P = [32]P - P = P_5 - P_0 = (9, 31) + (8, 15) = (17, 26)$. Only one addition is required in this case, whereas four additions are needed in the previous approach.

(ii) Suppose that we want to compute $[23]P$. We have $23 = 1 + 2 + 4 + 16 = (1 + 2 + 4) + 16 = (-1 + 8) + 16 = -1 + 32 - 8$. Thus, $[23]P = -P - [8]P + [32]P = -P_0 - P_3 + P_5 = (8, 15) + (28, 15) + (9, 31) = (30, 24)$.

In general, if $m_i m_{i+1} \ldots m_j$ is a string of 1's in the sequence $m_0 m_1 \ldots m_t$, then $m_i 2^i + \cdots + m_j 2^j = 2^i + \cdots + 2^j = 2^{j+1} - 2^i$. In this way, it is easily seen that the number of additions required is at most $3 \log m / 2$.

There is an alternative and often more efficient method to compute multiples of $\mathbb{F}_{q^k}$-rational points on supersingular elliptic curves defined over $\mathbb{F}_q$. Suppose that $\mathcal{E}$ is an elliptic curve with $q + 1$ $\mathbb{F}_q$-rational points. According to Theorem 3.3.15, $\mathcal{E}$ is supersingular. By Theorem 3.3.9(iii), we have $\phi_{q^2} + [q] = 0$. In particular, for all points $P = (x, y) \in \mathcal{E}$, we have $[q]P = -\phi_{q^2}(P)$, i.e., $[q]P = (x^{q^2}, -y^{q^2})$. Computing (x^{q^2}, y^{q^2}) is often much easier than computing $[q](x, y)$. The procedure can now be described as follows.

(i) Write $m = m_0 + m_1 q + \cdots + m_t q^t$ with $m_0, \ldots, m_t \in \{0, 1, \ldots, q-1\}$ and $m_t \neq 0$.

(ii) For $i = 1, \ldots, t$, we have $[q^i](x, y) = (x^{q^{2i}}, (-1)^i y^{q^{2i}})$.

(iii) For $k = 0, 1, \ldots, q-1$, compute $P_k = [k](x, y) = (x_k, y_k)$.

(iv) Then $[m]P = \sum_{i=0}^{t} [q^i] P_{m_i} = \sum_{i=0}^{t} (x_{m_i}^{q^{2i}}, (-1)^i y_{m_i}^{q^{2i}})$.

3.4.2   Computing the Miller Functions

For an integer m, let $P \in \mathcal{E}[m]$ be an m-torsion point. Recall that a Miller function is a function $f_{m,P}$ such that $\mathrm{div}(f_{m,P}) = mP - mO$ (Remark 3.1.25). The Miller functions will be the key component in the definitions of the Weil and Tate pairings, to be defined in Section 3.5.

The **Miller algorithm** in Figure 3.1 seeks to output a function f, given inputs of an integer m and a point P on $\mathcal{E}$, such that $\mathrm{div}(f) = mP - [m]P - (m-1)O$. In particular, if $P \in \mathcal{E}[m]$, then $\mathrm{div}(f) = mP - mO$.

For any two points T_1 and T_2 on $\mathcal{E}$, let f_{T_1, T_2} be the function such that $\mathrm{div}(f_{T_1, T_2}) = T_1 + T_2 - (T_1 \oplus T_2) - O$, where $(T_1 \oplus T_2)$ is the point on $\mathcal{E}$ that is the sum of T_1 and T_2 under the group law. (Such a function exists by Proposition 3.1.22.) Represent m in its binary form: $m = m_0 + 2m_1 + 2^2 m_2 + \cdots + 2^t m_t$, where $m_i \in \{0, 1\}$ and $m_t = 1$. The Miller algorithm resembles the double-and-add algorithm, but operates in a left-to-right manner.

FIGURE 3.1: The Miller algorithm.

Input: an integer m and a point P on $\mathcal{E}$
Output: a function f such that $\mathrm{div}(f) = mP - [m]P - (m-1)O$

1. $f \leftarrow 1$ and $T \leftarrow P$

2. **for** $i = t-1$ **down to** 0 **do**

3. $f \leftarrow f^2 \cdot f_{T,T}$ and $T \leftarrow [2]T$

4.   **if** $m_i = 1$ **then**

5. $f \leftarrow f \cdot f_{T,P}$ and $T \leftarrow T + P$

6. **return** f

Example 3.4.4 Here, we illustrate the algorithm with a simple example. Take $m = 13 = 1 + 0 \cdot 2 + 1 \cdot 4 + 1 \cdot 8$, so that $t = 3$.

1. Initialize: $T \leftarrow P, f \leftarrow 1$.

2. $t = 2$: $f \leftarrow f^2 \cdot f_{T,T} = f_{P,P}$; $T \leftarrow [2]T = [2]P$;
   $(\operatorname{div}(f) = P + P - [2]P - O = 2P - [2]P - O)$.

   - $m_2 = 1$: $f \leftarrow f \cdot f_{[2]P,P}$; $T \leftarrow T + P = [3]P$;
     $(\operatorname{div}(f) = 2P - [2]P - O + [2]P + P - [3]P - O = 3P - [3]P - 2O)$.

3. $t = 1$: $f \leftarrow f^2 \cdot f_{[3]P,[3]P}$; $T \leftarrow [2]T = [6]P$;
   $(\operatorname{div}(f) = 2(3P - [3]P - 2O) + 2[3]P - [6]P - O = 6P - [6]P - 5O)$.

   - $m_1 = 0$.

4. $t = 0$: $f \leftarrow f^2 \cdot f_{[6]P,[6]P}$; $T \leftarrow [2]T = [12]P$;
   $(\operatorname{div}(f) = 2(6P - [6]P - 5O) + 2[6]P - [12]P - O = 12P - [12]P - 11O)$.

   - $m_0 = 1$: $f \leftarrow f \cdot f_{[12]P,P}$; $T \leftarrow [13]P$;
     $(\operatorname{div}(f) = 12P - [12]P - 11O + [12]P + P - [13]P - O = 13P - [13]P - 12O)$.

5. Output f.

Theorem 3.4.5 *The Miller algorithm outputs a function f, given inputs of an integer m and a point P on $\mathcal{E}$, such that $\operatorname{div}(f) = mP - [m]P - (m-1)O$.*

Proof. For $j = 1, \ldots, t$, let f_j be the function obtained at the jth step (i.e., $i = t - j$). For $j = 1$, if $m_{t-1} = 0$, we have $f_1 = f_{P,P}$, so $\operatorname{div}(f_1) = 2P - [2]P - O$, whereas if $m_{t-1} = 1$, we have $f_1 = f_{P,P} \cdot f_{[2]P,P}$, so $\operatorname{div}(f_1) = 3P - [3]P - 2O$. For $j = 1, \ldots, t$, let $k_j = \sum_{k=0}^{j} 2^k m_{t-j+k}$. We prove by induction that $\operatorname{div}(f_j) = k_j P - [k_j]P - (k_j - 1)O$. Now, $f_{j+1} = f_j^2 \cdot f_{[k_j]P,[k_j]P}$ if $m_{t-j} = 0$, and is $f_j^2 \cdot f_{[k_j]P,[k_j]P} \cdot f_{[2k_j]P,P}$ if $m_{t-j} = 1$. In the former case,

$$
\begin{aligned}
\operatorname{div}(f_{j+1}) &= 2\operatorname{div}(f_j) + \operatorname{div}(f_{[k_j]P,[k_j]P}) \\
&= 2k_j P - 2[k_j]P - 2(k_j - 1)O + 2[k_j]P - [2k_j]P - O \\
&= 2k_j P - [2k_j]P - (2k_j - 1)O \\
&= k_{j+1}P - [k_{j+1}]P - (k_{j+1} - 1)O.
\end{aligned}
$$

If $m_{t-j} = 1$, it follows that

$$
\begin{aligned}
\operatorname{div}(f_{j+1}) &= 2\operatorname{div}(f_j) + \operatorname{div}(f_{[k_j]P,[k_j]P}) + \operatorname{div}(f_{[2k_j]P,P}) \\
&= 2k_j P - [2k_j]P - (2k_j - 1)O + [2k_j]P + P - [(2k_j + 1)]P - O \\
&= (2k_j + 1)P - [(2k_j + 1)]P - 2k_j O \\
&= k_{j+1}P - [k_{j+1}]P - (k_{j+1} - 1)O.
\end{aligned}
$$

Hence, the result follows by induction since $k_t = m$. $\qquad\square$

3.4.3 Finding the Order of $\mathcal{E}(\mathbb{F}_q)$

Throughout this subsection, $N = N(q)$ denotes the number of $\mathbb{F}_q$-rational points on an elliptic curve $\mathcal{E}$.

Finding N is essential in most cryptographic applications. The Hasse-Weil bound gives us $q + 1 - 2\sqrt{q} \le N \le q + 1 + 2\sqrt{q}$. In this subsection, we briefly describe some of the existing methods that can be used to compute N.

The most naive approach is to list all the points on $\mathcal{E}$ and count them. This is, however, not necessary as it is sufficient to determine whether an element in the field is a square.

Let $\mathcal{E} : y^2 = x^3 + Ax + B$ be the usual Weierstrass normal form of $\mathcal{E}$.

Lemma 3.4.6 *Let $\chi : \mathbb{F}_q \to \{0, 1, -1\}$ be defined as:*

$$\chi(a) = \begin{cases} 0 & \text{if } a = 0 \\ 1 & \text{if } a \text{ is a square in } \mathbb{F}_q^* \\ -1 & \text{otherwise.} \end{cases}$$

Then $N = q + 1 + \sum_{a \in \mathbb{F}_q} \chi(a^3 + Aa + B)$.

Proof. Observe that, if $a^3 + Aa + B$ is a square in $\mathbb{F}_q^*$, then there exist two points in $\mathcal{E}(\mathbb{F}_q)$ with a as the x-coordinate. If $a^3 + Aa + B = 0$, then $(a, 0)$ is the only point in $\mathcal{E}(\mathbb{F}_q)$ with a as the x-coordinate. Finally, if $a^3 + Aa + B$ is not a square, then $\mathcal{E}(\mathbb{F}_q)$ has no point with x-coordinate equal to a. Summing up, it follows that $\mathcal{E}(\mathbb{F}_q)$ has $\chi(a^3 + Aa + B) + 1$ points with a as the x-coordinate. Together with O, the number of points in $\mathcal{E}(\mathbb{F}_q)$ must be $N = 1 + \sum_{a \in \mathbb{F}_q}(\chi(a^3 + Aa + B) + 1) = q + 1 + \sum_{a \in \mathbb{F}_q} \chi(a^3 + Aa + B)$. $\square$

The function χ can be effectively computed, because in fact, for $z \in \mathbb{F}_q$, we have $\chi(z) = z^{(q-1)/2}$.

Example 3.4.7 Let $q \equiv 3 \pmod 4$. Consider the curve $\mathcal{E} : y^2 = x^3 + x$. We now use Lemma 3.4.6 to show that $\mathcal{E}$ has $q + 1$ $\mathbb{F}_q$-rational points. Since -1 is not a square in $\mathbb{F}_q^*$, for $x \in \mathbb{F}_q^*$, $x^3 + x$ is a square if and only if $-(x^3 + x) = (-x)^3 + (-x)$ is not a square. Hence, $\sum_{a \in \mathbb{F}_q} \chi(a^3 + a) = 0$. Consequently, $N = q + 1$ as expected.

Write $a = a(q) = q + 1 - N$.

By Corollary 1.5.4, the number of $\mathbb{F}_{q^k}$-rational points on $\mathcal{E}$ can be computed once N is known. More precisely, let $L(t) = 1 - at + qt^2 = (1 - \alpha t)(1 - \beta t)$ be the L-function of $\mathcal{E}$. Then $N_k = |\mathcal{E}(\mathbb{F}_{q^k})| = q + 1 - (\alpha^k + \beta^k)$, for all positive integers k. Letting $a_k = \alpha^k + \beta^k$, the proof of Theorem 3.3.15 shows that a_k satisfies the relation $a_{k+2} = aa_{k+1} - qa_k$ (see (3.6)), for $k \ge 1$, and $a_2 = a^2 - 2q$ (see (3.7)). In this way, the values of N_k can be easily obtained.

Example 3.4.8 Let q be a square and suppose that $\mathcal{E}$ has $q + 1 + 2\sqrt{q}$ $\mathbb{F}_q$-rational points. Then $L(t) = (1 + \sqrt{q}t)^2$, so that $\alpha = \beta = -\sqrt{q}$. Hence, $a_k = (-1)^k 2q^{k/2}$ and $N_k = q^k + 1 - (-1)^k 2q^{k/2}$. Therefore, we conclude that $\mathcal{E}$ has the maximum number of $\mathbb{F}_{q^k}$-rational points whenever k is odd.

By definition, the order k of a point P is the smallest integer such that $[k]P = O$. Since $k|N$, it may sometimes be possible for us to deduce the value of N from the Hasse-Weil bound if some such k is known. We illustrate this technique with the next example.

Example 3.4.9 Consider the elliptic curve $\mathcal{E} : y^2 = x^3 + x + 2$ defined over $\mathbb{F}_{97}$. It can be directly verified that $P = (61, 8)$ lies on $\mathcal{E}$. It can be further checked that P has order 26. By the Hasse-Weil bound, N must lie in the interval $[97 + 1 - 2\sqrt{97}, 97 + 1 + 2\sqrt{97}]$ or $[79, 117]$. From the fact that $26|N$, we can conclude that $N = 104$, since 104 is the only multiple of 26 in this interval.

Remark 3.4.10 We may use the "baby-step giant-step" procedure (which requires around $O(q^{1/4})$ computations) to compute the order of a point $P \in \mathcal{E}(\mathbb{F}_q)$. The reader may refer to [172] for more details on this method. Indeed, knowledge of the orders for a few points is often sufficient to help us determine N.

All the methods we have described so far are computationally inefficient to determine the order N of the group $\mathcal{E}(\mathbb{F}_q)$. In fact, there is a deterministic polynomial-time algorithm to determine N, or more precisely, the value of $a = a(q)$. This algorithm was first proposed by Schoof, but later improved by Atkin and Elkies. In essence, it tries to find the values of $a(q) \bmod \ell$ for a large number of primes ℓ, where the product of these primes is greater than $4\sqrt{q}$. This is done with the aid of certain polynomials known as division polynomials as well as the characteristic polynomial $t^2 - a(q)t + q$ of the Frobenius morphism. Indeed, Theorem 3.3.9(iii) yields $\phi_{q^2} + [q] = [a(q)] \circ \phi_q$. For all points in $\mathcal{E}[\ell]$, it follows that $\phi_{q^2}(P) + [q \bmod \ell]P = [a(q) \bmod \ell] \circ \phi_q(P)$. Working with division polynomials then enables us to determine $a(q) \bmod \ell$. Finally, the exact value of $a(q)$ is computed via the Chinese Remainder Theorem. This algorithm has the fastest running time of $O(\log^4 q)$ with fast arithmetic. For more on this method and on division polynomials, see [172].

3.4.4 The Discrete Logarithm Problem on an Elliptic Curve

For a point $P \in \mathcal{E}$ and an integer m, we have described computationally efficient methods to compute $[m]P$. Conversely, given a point $Q = [m]P$, can we determine m efficiently?

This problem is the elliptic curve version of the **discrete logarithm problem**.

Evidently, the discrete logarithm problem can be defined for any abelian group, for instance, the multiplicative group of the nonzero elements of any finite field. Indeed, the discrete logarithm problem underpins the security of some well-known cryptographic schemes, including the ElGamal encryption scheme and the digital signature standard.

In general, the discrete logarithm problem in an arbitrary group can be solved via methods such as the ρ or the "baby-step giant-step" algorithms. However, these algorithms have a complexity of around the square root of the size of the group. We briefly describe the ρ algorithm for elliptic curve groups.

Once again, let P and Q be two points on $\mathcal{E}$ such that $Q = [m]P$. Suppose further that P has order N.

(i) Pick around s pairs of values $(a_1, b_1), (a_2, b_2), \ldots, (a_s, b_s)$. Let $R_i = [a_i]P + [b_i]Q$, for $i = 1, 2, \ldots, s$.

(ii) Divide $\mathcal{E}(\mathbb{F}_q)$ into s subsets $G_1, \ldots, G_s$ of roughly equal size.

(iii) Let $S_0 = [a_0]P + [b_0]Q$ for some random integers a_0, b_0.

(iv) Construct the sequence of points $S_1, S_2, \ldots$ as follows. For every $i \geq 0$, if $S_i \in G_{j_i}$ for some $1 \leq j_i \leq s$, set $S_{i+1} = S_i + R_{j_i}$.

(v) By the birthday paradox, after around $\sqrt{N}$ points have been constructed, there exist two points $S_i = [c_i]P + [d_i]Q$ and $S_j = [c_j]P + [d_j]Q$ such that $S_i = S_j$.

(vi) If $\gcd(d_j - d_i, N) = 1$, we have $m = -(c_j - c_i)/(d_j - d_i) \bmod N$. If $\gcd(d_j - d_i, N) \neq 1$, we have to choose a new point S_0 again to run the above algorithm.

For finite fields, there exists a subexponential-time algorithm, known as the index calculus algorithm, to solve the discrete logarithm problem. Essentially, this procedure exploits the facts that, in the case of a prime field $\mathbb{F}_p$, every element of the field can be regarded as an integer, which can then be factored into a product of prime numbers, while in the case of $\mathbb{F}_{p^e}$ where $e > 1$, each element can be expressed as a product of irreducible polynomials over $\mathbb{F}_p$. As such, the particular structure of the finite field plays a critical role in this algorithm.

An important question now confronts us: Do the structures of elliptic curves offer any "loophole" that can be exploited to solve the discrete logarithm problem? This question remains open today. In other words, for arbitrary elliptic curves, only the exponential-time general approaches are known to solve the discrete logarithm problem. Nonetheless, for certain classes of elliptic curves, there may exist algorithms that solve the discrete logarithm problem efficiently.

For example, in Subsection 3.5.5, we show that the discrete logarithm problem for supersingular elliptic curves can be reduced to a corresponding problem in some finite field, of a manageable size, for which the index calculus algorithm applies.

Anomalous elliptic curves over $\mathbb{F}_p$ are elliptic curves with exactly p points. It has been shown by Semaev [137], Satoh and Araki [135] and Smart [149] that the discrete logarithm problem on such curves can be transformed into a discrete logarithm problem in the additive group $\mathbb{Z}_p$, hence, making the problem trivial to solve. Furthermore, by using summation polynomials in the index calculus algorithm and the method of Weil descent, Claus Diem [48] showed that the discrete logarithm problem for elliptic curves over finite fields $\mathbb{F}_{q^n}$, where q and n are large, can be solved in subexponential time. The reader may refer to [48, 172] for details of these algorithms.

3.5 Pairings on an Elliptic Curve

Pairing on elliptic curves are an efficient and powerful tool for the study of elliptic curves. For applications to cryptography, pairings on elliptic curves play important roles in pairing-based cryptographic schemes and discrete logarithms.

Let G_1, G_2, G_3 be abelian groups. A **pairing** $w : G_1 \times G_2 \to G_3$ is a mapping that sends every pair of elements in $G_1 \times G_2$ to some element in G_3. Moreover, w is **bilinear** if it is linear in each of its two inputs. More specifically, let $g_1, g_1' \in G_1$ and $g_2, g_2' \in G_2$. Then $w(g_1 g_1', g_2) = w(g_1, g_2)w(g_1', g_2)$ and $w(g_1, g_2 g_2') = w(g_1, g_2)w(g_1, g_2')$. (Here, the group operations in G_1, G_2, G_3 are written multiplicatively, with multiplicative identity 1.)

In the case where $G_1 = G_2$, a pairing w is said to be **alternating** if, for all $g_1, g_2 \in G_1 = G_2$, $w(g_1, g_2) = w(g_2, g_1)^{-1}$. Furthermore, w is **nondegenerate** if $w(g_1, g_2) = 1$ for all $g_2 \in G_2$ implies that $g_1 = 1$.

For example, let V be the set of 2-tuples over any commutative ring R. Then the determinant function $\det(v, w)$, defined as the determinant of the 2×2 matrix with v and w as its rows, is a bilinear, alternating, and nondegenerate pairing from $V \times V$ to the additive group of the ring R.

In this section, we introduce two pairings from the product of two elliptic curves defined over $\mathbb{F}_q$ to $\overline{\mathbb{F}}_q^*$. Such pairings are useful as they often help to transfer operations or structures in elliptic curves to corresponding operations or structures in $\overline{\mathbb{F}}_q$.

From now on, we fix a positive integer m that is relatively prime to p, where p is the characteristic of $\mathbb{F}_q$.

3.5.1 The Weil Pairing

Let $\mathcal{E}$ be an elliptic curve over $\mathbb{F}_q$ and let Q be a point in $\mathcal{E}[m]$. Recall that a Miller function $f_{m,Q}$ satisfies $\mathrm{div}(f_{m,Q}) = mQ - mO$.

Lemma 3.5.1 *There exists a function $g_{m,Q}$ such that $g_{m,Q}^m = f_{m,Q} \circ [m]$.*

Proof. First of all, we note that, for any $T \in [m]^{-1}(Q)$ and $S \in [m]^{-1}(O)$, one has that $\nu_T(f_{m,Q} \circ [m]) = m$ and $\nu_S(f_{m,Q} \circ [m]) = -m$. Moreover, Q is the unique zero of $f_{m,Q}$ and O is its unique pole. Hence, $\mathrm{div}(f_{m,Q} \circ [m]) =$

$m \sum_{R \in \mathcal{E}[m]} (\tau_R(Q_0) - \tau_R(O))$, where Q_0 is any point such that $[m]Q_0 = Q$ and τ_R is the translation map in Example 3.2.3. By Proposition 3.1.22, the divisor $D = \sum_{R \in \mathcal{E}[m]} \tau_R(Q_0) - \tau_R(O)$ is principal since, as points on $\mathcal{E}$, $\sum_{R \in \mathcal{E}[m]} (\tau_R(Q_0) - \tau_R(O)) = \sum_{R \in \mathcal{E}[m]} Q_0 = [m^2]Q_0 = [m]Q = O$. Thus, we can find a function $g_{m,Q}$ with $\text{div}(g_{m,Q}) = D$. It follows that $mD = m \, \text{div}(g_{m,Q}) = \text{div}(g_{m,Q}^m) = \text{div}(f_{m,Q} \circ [m])$. By scaling, we obtain $g_{m,Q}^m = f_{m,Q} \circ [m]$.   □

We are now ready to define the Weil pairing $w_m(P, Q)$ from $\mathcal{E}[m] \times \mathcal{E}[m]$ to the group $\boldsymbol{\mu}_m$ of mth roots of unity in $\overline{\mathbb{F}}_q$ as follows.

Definition 3.5.2 The **Weil pairing** is defined to be a map $w_m : \mathcal{E}[m] \times \mathcal{E}[m] \to \boldsymbol{\mu}_m$ such that, for points P and Q in $\mathcal{E}[m]$, $w_m(P, Q) = (g_{m,Q} \circ \tau_P)/g_{m,Q}$. Equivalently, $w_m(P, Q) = g_{m,Q}(P + X)/g_{m,Q}(X)$ for any $X \in \mathcal{E}$ such that both X and $P + X$ are neither zeros nor poles of $g_{m,Q}$.

As τ_P^* fixes $f_{m,Q} \circ [m]$, for all $P \in \mathcal{E}[m]$, by Lemma 3.5.1, we have $g_{m,Q}^m \circ \tau_P = g_{m,Q}^m$ or $(g_{m,Q} \circ \tau_P)^m = g_{m,Q}^m$. It follows that $((g_{m,Q} \circ \tau_P)/g_{m,Q})^m = 1$. Therefore, $w_m(P, Q)$ is indeed an mth root of unity.

Remark 3.5.3 Note that the choice of $g_{m,Q}$ is not unique. In fact, the various $g_{m,Q}$'s differ by a constant multiple. It is easy to verify that $w_m(P, Q)$ is independent of the choice of $g_{m,Q}$.

Example 3.5.4 Consider the elliptic curve $\mathcal{E} : y^2 = x^3 - x$ over $\mathbb{F}_q$. Recall that $\mathbb{F}_q$ is of characteristic $\neq 2, 3$. The point $Q = (0, 0)$ is a 2-torsion point. Recall that $f_{2,Q} = x$, i.e., $\text{div}(x) = 2Q - 2O$. Now, for any point (x, y), $[2](x, y) = (\lambda^2 - 2x, \lambda(3x - \lambda^2) - y)$, where $\lambda = \frac{3x^2 - 1}{2y}$. Thus, $f = f_{2,Q} \circ [2]$ is given by $f = \lambda^2 - 2x = \frac{(3x^2 - 1)^2 - 8xy^2}{4y^2} = \frac{x^4 + 2x^2 + 1}{4y^2} = \frac{(x^2 + 1)^2}{4y^2}$. It follows that $g_{2,Q} = (x^2 + 1)/2y$.

Fix another point $P = (a, 0) \in \mathcal{E}[2] \setminus \{O\}$. We compute $S = (a, 0) + (b, c)$ using the addition formula. (Here, (b, c) is the point X in Definition 3.5.2.) By direct computation, the x-coordinate of S is $(ab - 1)(a + b)/(b - a)^2$ and its y-coordinate is $c(2a + b - 3a^2 b)/(b - a)^3$.

Table 3.2 gives the values of $w_2(P, Q)$ for $a = 0, 1, -1$.

TABLE 3.2: Values of $w_2(P, Q)$ for $a = 0, 1, -1$.

a	S	$g_{2,Q}(S)$	$w_2(P, Q)$
0	$(-1/b, c/b^2)$	$(b^2 + 1)/2c$	1
1	$((b+1)/(b-1), -2c/(b-1)^2)$	$-(b^2 + 1)/2c$	-1
-1	$(-(b-1)/(b+1), -2c/(b+1)^2)$	$-(b^2 + 1)/2c$	-1

As expected, each of the values is a square root of unity.

In the next theorem, we record some of the useful properties satisfied by w_m. The reader may consult books such as [144, Chapter III, Propositions 8.1 and 8.2] for the proofs.

Theorem 3.5.5 *The Weil pairing* $w_m : \mathcal{E}[m] \times \mathcal{E}[m] \to \overline{\mathbb{F}}_q$, *as defined in Definition 3.5.2, satisfies the following properties:*

(i) *The pairing* w_m *is bilinear, i.e., for all* $P_1, P_2, Q_1, Q_2 \in \mathcal{E}[m]$,

$$w_m(P_1 + P_2, Q_1) = w_m(P_1, Q_1)w_m(P_2, Q_1)$$
$$w_m(P_1, Q_1 + Q_2) = w_m(P_1, Q_1)w_m(P_1, Q_2).$$

(ii) *The pairing* w_m *satisfies* $w_m(T, T) = 1$ *for all* $T \in \mathcal{E}[m]$. *In particular, for* $P, Q \in \mathcal{E}[m]$, $w_m(P, Q) = w_m(Q, P)^{-1}$, *i.e.,* w_m *is alternating.*

(iii) *The pairing* w_m *is nondegenerate, i.e., if, for some* $P \in \mathcal{E}[m]$, *we have that* $w_m(P, Q) = 1$ *for all* $Q \in \mathcal{E}[m]$, *then* $P = O$.

(iv) *Let* m' *be another integer coprime to* p. *Let* $P \in \mathcal{E}[mm']$ *and* $Q \in \mathcal{E}[m] \subset \mathcal{E}[mm']$. *Then,* $w_{mm'}(P, Q) = w_m([m']P, Q)$.

(v) *Let* $\sigma : \mathcal{E} \to \mathcal{E}$ *be an isogeny. Then, for any* $P \in \mathcal{E}[m]$, $w_m(\sigma(P), \sigma(Q)) = w_m(P, Q)^{\deg(\sigma)}$.

(vi) *For any* τ *in the Galois group of* $\overline{\mathbb{F}}_q$ *over* $\mathbb{F}_q$, *we have*

$$w_m(\tau(P), \tau(Q)) = \tau(w_m(P, Q)).$$

Example 3.5.6 Refer to the pairing of Example 3.5.4.

(i) From the bilinearity, we have $w_2((1, 0), (-1, 0)) = w_2((0, 0) + (-1, 0), (-1, 0)) = w_2((0, 0), (-1, 0))w_2((-1, 0), (-1, 0)) = -1$.

(ii) Furthermore, let $P_0 \in \mathcal{E}[4]$ with $[2]P_0 = (0, 0)$. Then, it follows from Theorem 3.5.5(iv) that, for all $Q \in \mathcal{E}[2]$, $w_4(P_0, Q) = w_2([2]P_0, Q) = w_2((0, 0), Q)$.

Remark 3.5.7 Recall from Remark 3.3.6 that $\mathcal{E}[m]$ is a rank two $\mathbb{Z}_m$-module. Let P and Q generate $\mathcal{E}[m]$. We claim that $w_m(P, Q)$ is a primitive mth root of unity. Indeed, suppose this was not the case. Then there exists $d < m$ such that $1 = w_m(P, Q)^d = w_m([d]P, Q)$. Thus, for any integers a and b, $w_m([d]P, [a]P + [b]Q) = w_m([d]P, [a]P)w_m([d]P, [b]Q) = w_m(P, P)^{ad}w_m([d]P, Q)^b = 1$. This implies that $w_m([d]P, R) = 1$ for all $R \in \mathcal{E}[m]$. By the nondegeneracy of w_m (Theorem 3.5.5(iii)), we conclude that $[d]P = O$. This is a contradiction as P has order m. Hence, $w_m(P, Q)$ is a primitive mth root of unity.

Observe that the definition of the Weil pairing given in Definition 3.5.2 is not very useful for computations. This is because it is not immediately clear how we may compute the function $g_{m,Q}$. It will thus be useful if we can define the Weil pairing based on the Miller functions instead. This is possible as the next theorem shows. The proof uses the Weil reciprocity and may be found in books such as [172, Section 11.6.1].

Theorem 3.5.8 *Let m be a positive integer relatively prime to p, and let $P, Q \in \mathcal{E}[m]$. Fix some X that is not in the subgroup generated by P and Q. Let $D = (X \oplus Q) - X$ and $D' = (\ominus X \oplus P) - \ominus X$, where, from here till the end of this subsection, we use $\oplus$ and $\ominus$ for the group operations on $\mathcal{E}$ to avoid ambiguity. Then $w_m(P, Q) = f_{m,P}(D)/f_{m,Q}(D')$ where $f_{m,P}(D) = \frac{f_{m,P}(X \oplus Q)}{f_{m,P}(X)}$ and $f_{m,Q}(D') = \frac{f_{m,Q}(\ominus X \oplus P)}{f_{m,Q}(\ominus X)}$.*

Example 3.5.9 Consider the elliptic curve $\mathcal{E} : y^2 = x^3 + 30x + 34$ over $\mathbb{F}_{631}$. It can be shown that $\mathcal{E}$ has 650 $\mathbb{F}_{631}$-rational points. Furthermore, $\mathcal{E}[5] \subseteq \mathcal{E}(\mathbb{F}_{631})$ and is generated by $P = (121, 387)$ and $Q = (36, 60)$. Let $X = (0, 36)$ be a point of $\mathcal{E}$. One checks directly that X has order 130, so X is not in the subgroup generated by P and Q. In addition, $X \oplus Q = (176, 486)$ and $\ominus X \oplus P = (532, 300)$.

We compute the values $f_{5,P}(X \oplus Q), f_{5,P}(X), f_{5,Q}(\ominus X \oplus P)$, and $f_{5,Q}(\ominus X)$ via the Miller algorithm, by evaluating the functions at each step. Table 3.3 shows the values of $f_{5,P}(X \oplus Q), f_{5,P}(X), f_{5,Q}(\ominus X \oplus P)$, and $f_{5,Q}(\ominus X)$ at each stage of the Miller algorithm.

TABLE 3.3: Values of $f_{5,P}(X \oplus Q), f_{5,P}(X), f_{5,Q}(\ominus X \oplus P)$, and $f_{5,Q}(\ominus X)$.

i	$f_{5,P}(X \oplus Q)$	$f_{5,P}(X)$		i	$f_{5,Q}(\ominus X \oplus P)$	$f_{5,Q}(\ominus X)$
2	1	1		2	1	1
1	28	560		1	541	255
0	103	219		0	284	204

From Table 3.3, we obtain $f_{5,P}(X \oplus Q)/f_{5,P}(X) = 103/219 = 473$ and $f_{5,Q}(\ominus X \oplus P)/f_{5,Q}(\ominus X) = 284/204 = 88$. Thus, $w_5(P, Q) = 473/88 = 242$.

3.5.2   The Tate-Lichtenbaum Pairing

In this subsection, we present a special form of the **Tate-Lichtenbaum pairing** which is often used in cryptography. To this end, let m be a prime different from p, the characteristic of $\mathbb{F}_q$, and assume that $m|(q-1)$, so that $\mathbb{F}_q$ contains all the mth roots of unity. We further assume that $\mathcal{E}$ has a nonzero $\mathbb{F}_q$-rational point of order m. We also let $\mathcal{E}(\mathbb{F}_q)[m] = \mathcal{E}[m] \cap \mathcal{E}(\mathbb{F}_q)$ denote the group of all m-torsion points in $\mathcal{E}(\mathbb{F}_q)$.

Let $\boldsymbol{\mu}_m$ denote the subgroup of $\mathbb{F}_q^*$ of order m. Now, consider the two quotient groups: $G' = \mathcal{E}(\mathbb{F}_q)/m\mathcal{E}(\mathbb{F}_q)$ and $H = \mathbb{F}_q^*/(\mathbb{F}_q^*)^m \cong \boldsymbol{\mu}_m$, where $(\mathbb{F}_q^*)^m$

is obtained by raising every element of $\mathbb{F}_q^*$ to its mth power. Then the three groups $G = \mathcal{E}(\mathbb{F}_q)[m], G'$ and H all have exponent m.

We now define the **Tate-Lichtenbaum pairing** to be the map $t_m : G \times G' \to \boldsymbol{\mu}_m$ given by $t_m(P, Q) = f_{m,P}(Q + X)/f_{m,P}(X)$, where X is chosen so that both the numerator and denominator are well defined. Here, P is an m-torsion point in $\mathcal{E}(\mathbb{F}_q)[m]$, Q is a point in $\mathcal{E}(\mathbb{F}_q)$ which can be viewed as an equivalence class in $\mathcal{E}(\mathbb{F}_q)/m\mathcal{E}(\mathbb{F}_q)$, and t_m is well defined only up to multiplication by mth roots of unity. In other words, if we choose Q' so that $Q' - Q \in m\mathcal{E}(\mathbb{F}_q)$, then $t_m(P, Q)/t_m(P, Q')$ is an mth root of unity. Similarly, for any X, X', we have $f_{m,P}(Q + X)/f_{m,P}(X) = \zeta^m f_{m,P}(Q + X')/f_{m,P}(X')$ for some $\zeta \in \mathbb{F}_q^*$.

We state two important properties of the Tate-Lichtenbaum pairing and we again leave the proofs to the interested reader to check out from [172, Sections 3.4 and 11.3].

Theorem 3.5.10 (i) *The pairing t_m is bilinear, i.e., for all $P_1, P_2 \in G$ and $Q_1, Q_2 \in G'$,*

$$t_m(P_1 + P_2, Q_1) = t_m(P_1, Q_1)t_m(P_2, Q_1)$$
$$t_m(P_1, Q_1 + Q_2) = t_m(P_1, Q_1)t_m(P_1, Q_2).$$

(ii) *The pairing t_m is nondegenerate, i.e., if $t_m(P, Q) = 1$ for all $Q \in \mathcal{E}(\mathbb{F}_q)/m\mathcal{E}(\mathbb{F}_q)$, then $P = O$. Similarly, if $t_m(P, Q) = 1$ for all $P \in \mathcal{E}(\mathbb{F}_q)[m]$, then Q belongs to $m\mathcal{E}(\mathbb{F}_q)$, i.e., Q is O when viewed as a class in $\mathcal{E}(\mathbb{F}_q)/m\mathcal{E}(\mathbb{F}_q)$.*

To ensure that t_m is uniquely defined, we may define a modified **Tate pairing** to be $\hat{t}_m(P, Q) = t_m(P, Q)^{(q-1)/m}$. Recall that we have assumed that $m|(q - 1)$. Suppose further that $\mathcal{E}(\mathbb{F}_q)[m^2]$ and $\mathcal{E}(\mathbb{F}_q)[m]$ are isomorphic to $\mathbb{Z}_m$ (and hence, $m^2 \nmid |\mathcal{E}(\mathbb{F}_q)|$). In this case, $\mathcal{E}(\mathbb{F}_q)/m\mathcal{E}(\mathbb{F}_q)$ is isomorphic to $\mathcal{E}(\mathbb{F}_q)[m]$ (since all elements of $\mathcal{E}(\mathbb{F}_q)[m]$ can be viewed as distinct representatives of the equivalence classes of $\mathcal{E}(\mathbb{F}_q)/m\mathcal{E}(\mathbb{F}_q)$). Then it can be shown that $\hat{t}_m : \mathcal{E}(\mathbb{F}_q)[m] \times \mathcal{E}(\mathbb{F}_q)[m] \to \boldsymbol{\mu}_m$ is a perfect pairing, i.e., $\hat{t}_m(P, P) \neq 1$ for all $P \in \mathcal{E}(\mathbb{F}_q)[m] \setminus \{O\}$.

Remark 3.5.11 In cryptographic applications, the Tate pairing is often preferred to the Weil pairing for the following reasons:

- The computations involved are simpler, and hence more efficient. In general, the Tate pairing uses half the number of computations as compared to the Weil pairing.

- Since $\hat{t}_m(P, P) \neq 1$, it is more useful as nontrivial values of $\hat{t}_m([a]P, [b]P)$ can be computed.

However, we require $\mathbb{F}_q^*$ to contain the mth roots of unity and $\mathcal{E}(\mathbb{F}_q)$ to have a point of order m but not a point of order m^2.

3.5.3   Embedding Degrees

Observe that, for the Weil pairing, we work with points in the group $\mathcal{E}[m]$ which may not be a subgroup of $\mathcal{E}(\mathbb{F}_q)$. Furthermore, $w_m(P,Q) \in \overline{\mathbb{F}}_q$ is an mth root of unity. What is the smallest extension of $\mathbb{F}_q$ that we work with? Such considerations motivate the following definition.

Definition 3.5.12 For an integer m relatively prime to p, the **embedding degree** with respect to q and m is defined as the smallest integer d for which $q^d \equiv 1 \pmod{m}$. In other words, the embedding degree d is the order of q modulo m.

Indeed, the embedding degree with respect to q and m is the smallest degree of the field extension over $\mathbb{F}_q$ that contains all the mth roots of unity. More specifically, let $\boldsymbol{\mu}_m$ denote the group of mth roots of unity. Then d is the smallest integer such that $\boldsymbol{\mu}_m \subseteq \mathbb{F}_{q^d}$.

Let k be the smallest integer such that $\mathcal{E}[m] \subseteq \mathcal{E}(\mathbb{F}_{q^k})$, so that the computations of w_m are all carried out in $\mathbb{F}_{q^k}$. The next lemma shows that $d | k$.

Lemma 3.5.13 *Suppose that $\mathcal{E}[m] \subseteq \mathcal{E}(\mathbb{F}_{q^k})$. Then $q^k \equiv 1 \pmod{m}$.*

Proof. Let P and Q be points in $\mathcal{E}[m]$ that, together, generate $\mathcal{E}[m]$. According to Remark 3.5.7, $w_m(P,Q)$ is a primitive mth root of unity. Let $\zeta = w_m(P,Q)$. Let τ be any automorphism in the Galois group of $\overline{\mathbb{F}}_q$ over $\mathbb{F}_{q^k}$. Then Theorem 3.5.5(vi) yields $\tau(\zeta) = \tau(w_m(P,Q)) = w_m(\tau(P), \tau(Q)) = w_m(P,Q) = \zeta$. Hence, $\zeta \in \mathbb{F}_{q^k}$. Since ζ is a primitive mth root of unity, all the mth roots of unity lie in $\mathbb{F}_{q^k}$, and it follows that $q^k \equiv 1 \pmod{m}$.   □

Under certain additional assumptions, we can in fact show that $d = k$.

Theorem 3.5.14 *Let m be a positive integer relatively prime to p and suppose that there exists a point of order m in $\mathcal{E}(\mathbb{F}_q)$. Furthermore, suppose that $\gcd(m, q-1) = 1$. Let d be the embedding degree with respect to q and m. Then $\mathcal{E}[m] \subseteq \mathcal{E}(\mathbb{F}_{q^d})$.*

Proof. Let $P \in \mathcal{E}(\mathbb{F}_q)$ be a point of order m. Consider the Frobenius map ϕ_q on $\mathcal{E}$. Let T be a point in $\mathcal{E}[m]$ such that P and T generate $\mathcal{E}[m]$. We shall show that $\phi_{q^d}(T) = T$, thus showing that $\mathcal{E}[m] \subseteq \mathcal{E}(\mathbb{F}_{q^d})$. Note that $\phi_q(P) = P$ (since $P \in \mathcal{E}(\mathbb{F}_q)$) and $\phi_q(T) = [a]P + [b]T$ for some $a, b \in \mathbb{Z}_m$. By Theorem 3.5.5(v), $w_m(\phi_q(P), \phi_q(T)) = w_m(P,T)^{\deg(\phi_q)} = w_m(P,T)^q = w_m(P, [q]T)$. On the other hand, $w_m(\phi_q(P), \phi_q(T)) = w_m(P, [a]P + [b]T) = w_m(P, [b]T)$. Since $w_m(P,T)$ is a primitive mth root of unity (cf. Remark 3.5.7), comparing the two expressions yields $q \equiv b \pmod{m}$. Thus, $\phi_q(T) = [a]P + [q]T$.

Since ϕ_q is a group homomorphism (Theorem 3.2.9) and the multiplication maps commute with all isogenies, it follows that $\phi_q^2(T) = \phi_q([a]P + [q]T) = [a(1+q)]P + [q^2]T$. By applying ϕ_q recursively, we obtain $\phi_q^d(T) = [a(1 + q + \cdots + q^{d-1})]P + [q^d]T$. By the definition of d, $q^d \equiv 1 \pmod{m}$. Moreover, since

$\gcd(m, q-1) = 1$, it follows that $1 + q + \cdots + q^{d-1} \equiv 0 \pmod{m}$. Consequently, $\phi_{q^d}(T) = T$, as desired.   □

Remark 3.5.15 Theorem 3.5.14 shows that, if the embedding degree d is small, then computations such as pairing computations or finding multiples of points can be carried out in fields that are of reasonable size. This makes the computations more efficient. On the other hand, this fact is exploited to transform the elliptic curve discrete logarithm problem to the discrete logarithm problem in finite fields, which may be more feasible to solve if the embedding degree is small. (We will discuss this in more detail in Subsection 3.5.5.) In short, for cryptographic applications, we look for elliptic curves for which the embedding degrees are not so big as to make the computations cumbersome, but also not too small for the discrete logarithm problem to pose a threat. This consideration leads to the design of pairing friendly curves and the reader can find a survey of this topic in [60].

Example 3.5.16   (i) Suppose that $\mathcal{E}$ has $q+1$ $\mathbb{F}_q$-rational points and an $\mathbb{F}_q$-rational point of order m. In particular, $q \equiv -1 \pmod{m}$. Let $P \in \mathcal{E}[m]$. By Theorem 3.3.9(iii), $\phi_{q^2}(P) + [q]P = 0$, so $\phi_{q^2}(P) = -[q]P = P$. Since this holds for all $P \in \mathcal{E}[m]$, it follows that $\mathcal{E}[m] \subseteq \mathcal{E}(\mathbb{F}_{q^2})$. Hence, $d \leq 2$. Moreover, if m is odd and $q \not\equiv -1 \pmod{m^2}$, it follows from Theorem 3.5.14 that $d = 2$ since $\mathcal{E}[m] \not\subseteq \mathcal{E}(\mathbb{F}_q)$.

   (ii) Suppose that q is a square and that $\mathcal{E}$ has $N = q + 1 + \sqrt{q}$ points (which is possible by Theorem 3.3.10(iii)). Assume that $m|N$ and $m^2 \nmid N$. Then $m|(q^{3/2} - 1)$, so $m|(q^3 - 1)$. This shows that the embedding degree is 3.

   In fact, by looking at each of the cases in Theorem 3.3.10(ii)–(v), one can prove that the embedding degree of supersingular elliptic curves is at most 6.

3.5.4   Modified Weil Pairing on Supersingular Elliptic Curves

In many pairing-based cryptographic applications, it is essential that the pairing w used satisfies $w(P, P) \neq 1$ for some point P of order $m > 1$. As we have seen, the modified Tate pairing $\hat{t}_m$ fulfils this criterion while it does not hold true for the Weil pairing w_m. In this subsection, we introduce a modified Weil pairing when a certain distortion map exists.

Definition 3.5.17 Let m be a positive integer with $\gcd(m, q) = 1$ and let P be a point of order $m > 1$. An m-**distortion map** σ (with respect to P) from $\mathcal{E}$ to $\mathcal{E}$ is an isogeny such that $\{P, \sigma(P)\}$ generates $\mathcal{E}[m]$.

Let σ be an m-distortion map for $\mathcal{E}$ with respect to a point $P \in \mathcal{E}[m]$. Observe that, since σ is an isogeny on $\mathcal{E}$, it is defined for all points $Q \in \mathcal{E}[m]$ and we have $\sigma(Q) \in \mathcal{E}[m]$. We can define the modified Weil pairing $\hat{w}_m$ on

$\mathcal{E}[m] \times \mathcal{E}[m]$ as $\hat{w}_m(Q, Q') = w_m(Q, \sigma(Q'))$. Since P and $\sigma(P)$ generate $\mathcal{E}[m]$, it follows that $\hat{w}_m(P, P) \neq 1$.

We now give an example of a distortion map for supersingular elliptic curves.

Example 3.5.18 Let $q \equiv 3 \pmod 4$. By Example 3.3.18, the elliptic curve $\mathcal{E}$ with equation $y^2 = x^3 + x$ is supersingular. Let $\alpha \in \mathbb{F}_{q^2}$ such that $\alpha^2 = -1$. Note that $\alpha \notin \mathbb{F}_q$. Define the map σ on $\mathcal{E}$ by $\sigma((x, y)) = (-x, \alpha y)$ and $\sigma(O) = O$.

(i) First, we show that $\sigma((x, y))$ lies on $\mathcal{E}$. This is clear since $(\alpha y)^2 = \alpha^2 y^2 = -y^2 = -(x^3 + x) = (-x)^3 + (-x)$.

(ii) Clearly, σ is a morphism, and since $\sigma(O) = O$, it is an isogeny.

(iii) Let $m \geq 3$ with $\gcd(m, q) = 1$. Assume that there exists an $\mathbb{F}_q$-rational point P of order m. We claim that P and $\sigma(P)$ generate $\mathcal{E}[m]$. First of all, we note that σ is in fact an automorphism. Thus, $\sigma(P)$ also has order m. To prove our claim, it is sufficient to show that, for any $a, b \in \mathbb{Z}_m$ so that $[a]P = [b]\sigma(P)$, one must have $a = b = 0$. Since $P \in \mathcal{E}(\mathbb{F}_q)$, we can write $[a]P = (x_1, y_1)$ and $[b]P = (x_2, y_2) \in \mathcal{E}(\mathbb{F}_q)$, for some $x_1, y_1, x_2, y_2 \in \mathbb{F}_q$. Therefore, $(x_1, y_1) = \sigma((x_2, y_2)) = (-x_2, \alpha y_2)$. Since $\alpha \notin \mathbb{F}_q$, we must have $x_1 = x_2 = 0$ and $y_1 = y_2 = 0$. Hence $P = (0, 0)$, which has order 2, but this is not possible since $m \geq 3$.

Remark 3.5.19 Using a similar argument as in Example 3.5.18, we can define an m-distortion map for the curve $y^2 = x^3 + 1$ when $q \equiv 2 \pmod 3$. Specifically, suppose that $m > 3$ is relatively prime to p and suppose that $m | (q + 1)$. Define σ on $\mathcal{E}$ by $\sigma((x, y)) = (\alpha x, y)$ and $\sigma(O) = O$, where $\alpha^3 = 1$.

3.5.5 Pairings and the Discrete Logarithm Problem

As promised in Remark 3.5.15, in this subsection, we show that the discrete logarithm problem on an elliptic curve can be converted to a discrete logarithm problem in some finite field via the use of pairings. This idea was proposed by Menezes, Okamoto, and Vanstone (see [107]) and is commonly referred to as the MOV attack.

Let $P \in \mathcal{E}(\mathbb{F}_q)$ be of order m. Typically, m is a large prime. Suppose that $Q = [t]P$. Our aim is to find the value of $t \bmod N$, where $N = N(q)$ is the number of $\mathbb{F}_q$-rational points on $\mathcal{E}$.

Suppose that $\mathcal{E}[m] \subseteq \mathcal{E}(\mathbb{F}_{q^k})$. Lemma 3.5.13 says that $q^k \equiv 1 \pmod m$. Let N_k denote the number of $\mathbb{F}_{q^k}$-rational points on $\mathcal{E}$. The following procedure enables us to find t.

- Pick a random point $R \in \mathcal{E}(\mathbb{F}_{q^k})$.

- Let $R' = [N_k/m]R$. Then R' has order m.

- Let $\zeta = w_m(P, R')$. If $\zeta = 1$, choose a different R.

- Compute $\zeta' = w_m(Q, R')$.

- Find u such that $\zeta' = \zeta^u$.

- Then determine $t = u \bmod m$.

Observe that, in the above procedure, $\zeta' = w_m(Q, R') = w_m([t]P, R') = w_m(P, R')^t = \zeta^t$. This is why we have $t = u \bmod m$.

Remark 3.5.20 It follows from the MOV attack that, if the elliptic curve has a small embedding degree (such as in the case of supersingular elliptic curves), then the discrete logarithm problem on the curve may be carried over to one in a finite field of relatively small order. This is a potential drawback for using such curves since there exist subexponential-time algorithms to solve the discrete logarithm problem in finite fields via the index calculus method.

Remark 3.5.21 Let P be an $\mathbb{F}_q$-rational point of order m on an elliptic curve $\mathcal{E}$ over $\mathbb{F}_q$. Suppose further that $m | (q-1)$. Then the modified Tate pairing can be used to solve the discrete logarithm problem on $\mathcal{E}$ as well. This is due to the fact that $\hat{t}_m(P, P) \neq 1$ for all $P \in \mathcal{E}(\mathbb{F}_q)[m] \setminus \{O\}$. In particular, finding the value of t reduces to solving the discrete logarithm problem in $\mathbb{F}_q^*$ for $\hat{t}_m(P, Q)$ with respect to $\hat{t}_m(P, P)$.

3.6 Elliptic Curve Cryptography

As mentioned in the introduction, elliptic curves appear in many different aspects of public key cryptography. Having been equipped with the fundamental notions and results on elliptic curves as well as some computational considerations, we conclude this chapter with an overview of some of the key contributions of elliptic curves to public key cryptography.

3.6.1 The Elliptic Curve Factorization Method

We begin by showing how elliptic curves can be used to factor composite integers. This method was proposed by Hendrik W. Lenstra Jr. in 1987 [84] and is commonly termed the **elliptic curve factorization method** (or ECM for short).

Let $n = pq$ be a composite integer with p and q prime. For some integers $A, B \in \mathbb{Z}_n$, consider the curve $\mathcal{E}_n : y^2 = x^3 + Ax + B$ defined over $\mathbb{Z}_n$. Let $\mathcal{E}_n(\mathbb{Z}_n)$ denote the set of points $(x, y) \in \mathbb{Z}_n \times \mathbb{Z}_n$ such that $y^2 \equiv x^3 + Ax + B \pmod{n}$.

Strictly speaking, $\mathcal{E}_n$ is not an elliptic curve since elliptic curves are defined over fields. Nonetheless, there is no special reason to prevent us from carrying out the computations as we did for elliptic curves over finite fields.

Suppose that $P = (x, y)$ satisfies the equation $\mathcal{E}_n : y^2 = x^3 + Ax + B$ over $\mathbb{Z}_n$.

Recall that, in the addition and duplication formulas, expressions such as $(d - b)/(c - a)$ or $(3a^2 + A)/(2b)$ are involved. In other words, we need to compute the inverse of some elements modulo n. Since n is composite, not every integer less than n has an inverse. However, if by any chance we encounter a computation of $(c - a)^{-1} \bmod n$ such that $\gcd(c - a, n) > 1$, this will give us a factor of n, which, incidentally, is what we are after.

Here is a toy example.

Example 3.6.1 Let $n = 391$. Consider the curve given by $y^2 = x^3 + x - 1$ over $\mathbb{Z}_n$. It is easy to check that the point $P = (228, 14)$ lies on the curve. We compute $[2]P$ using the duplication formula, working modulo n. Using the notation in Theorem 3.1.19, we find that $M = -2$ and $[2]P = (330, 190)$. Letting $[3]P = [2]P + P = (330, 190) + (228, 14)$, we have $M = (190 - 14)/(330 - 228) \bmod 391$. However, $330 - 228 = 102$ and $\gcd(102, 391) = 17$. Hence, 17 is a factor of $n = 391$. In fact, $391 = 17 \cdot 23$.

A natural question arises. How often can we expect to encounter a nontrivial gcd when we add points? Equivalently, how do we pick the correct points to add to increase the chance of arriving at a nontrivial gcd?

Observe that the curve $\mathcal{E}_n$ can be viewed as an elliptic curve over $\mathbb{Z}_p$ and $\mathbb{Z}_q$ separately, and let $\mathcal{E}_p$ and $\mathcal{E}_q$ represent these curves, respectively. Let N_p (respectively, N_q) denote the number of $\mathbb{F}_p$-rational points (respectively, $\mathbb{F}_q$-rational points) on $\mathcal{E}_p$ (respectively, $\mathcal{E}_q$). For any point $P \in \mathcal{E}_n(\mathbb{Z}_n)$, let P_p be the point on $\mathcal{E}_p$ obtained by reducing the coordinates of P modulo p, and similarly for P_q. Then, $[N_p]P_p = O$ and $[N_q]P_q = O$. As such, if k is an integer such that $N_p | k$ or $N_q | k$, it will be very likely that, in the computation of the multiple $[k]P$, a nontrivial gcd will result. In view of this, we often take k to be a product of prime powers where the primes are less than some bound. In this way, if N_p (or N_q) is **L-smooth** for some L, i.e., all the prime factors of N_p (or N_q) are less than or equal to L, then we can expect it to be a factor of k.

Observe that, in Example 3.6.1, the curve $\mathcal{E}_{17} : y^2 = x^3 + x - 1$ over $\mathbb{F}_{17}$ has 18 points, while the curve $\mathcal{E}_{23} : y^2 = x^3 + x - 1$ over $\mathbb{F}_{23}$ has 20 points. By considering $k = 2^3 \cdot 3^3$ and $P = (1, 1)$, the factor 17 will be revealed in the process of computing $k[P]$.

We can now describe the elliptic curve factorization method as follows.

- Pick random x, y, A and let $B = y^2 - x^3 - Ax \bmod n$. If $\gcd(4A^3 + 27B^2, n) \neq 1$ or n, then we have factored n. If $\gcd(4A^3 + 27B^2, n) = n$, choose another set of x, y, A, and repeat this step.

- Fix a bound L.

- Let $P = (x, y)$.

- For $j = 2, 3, \ldots, L$, compute $[j]P$.

- If a nontrivial gcd is obtained, then we have factored n.

Remark 3.6.2 (i) Observe that this algorithm is very similar to the conventional Pollard $(p-1)$-method that works with the group $\mathbb{Z}_n^*$. Nonetheless, a big advantage of this approach with elliptic curves is that, each time a set of x, y, A is picked, a new elliptic curve is obtained with different N_p and N_q. By the Hasse-Weil bound, N_p and N_q are about the size of p and q, respectively. By varying N_p and N_q, we are more likely to hit on a smooth number so that $[k]P = O$.

(ii) With careful analysis, it can be shown that the elliptic curve factorization method has a subexponential-time complexity of

$$O\left(e^{(1+o(1))\sqrt{2\ln p \ln \ln p}}\right),$$

where p is the smallest prime factor of n. Hence, it can be quite an effective method to factor integers with relatively small factors (although it is slower than the number field sieve method for integers n whose factors are of the order of $\sqrt{n}$).

3.6.2 Discrete Logarithm-Based Elliptic Curve Schemes

Recall that, in general, the discrete logarithm problem on elliptic curves is considered much harder than its counterpart in finite fields. Hence, elliptic curves offer an attractive substitute for finite fields for the various discrete logarithm-based schemes. Indeed, elliptic curves over smaller fields can be used to offer a similar level of security, thereby making them very suitable for lightweight devices.

In this subsection, we describe a key exchange protocol based on elliptic curves as well as the elliptic curve variant of the ElGamal encryption scheme.

First, suppose that Alice and Bob want to derive a secret key from a point on an elliptic curve. They can launch the following protocol.

- Both Alice and Bob agree on a certain elliptic curve $\mathcal{E}$ over $\mathbb{F}_p$ and fix a particular point $P \in \mathcal{E}(\mathbb{F}_p)$ (of a large prime order).

- Alice picks an integer a randomly and computes $P_1 = [a]P$. She sends P_1 to Bob.

- Bob picks an integer b randomly and computes $P_2 = [b]P$. He sends P_2 to Alice.

- Alice computes $Q = [a]P_2 = [ab]P$.

- Bob computes $Q = [b]P_1 = [ab]P$.

- Their shared secret key is Q.

Suppose that Alice and Bob only require the x-coordinate to derive their final secret. Then it is not necessary for them to send the y-coordinates of P_1 and P_2 to each other. In this case, Alice and Bob will have to compute the y-value from the equation of the elliptic curve and they will obtain either the actual point or its negative. Eventually, they will obtain $\pm Q$ but, since the x-coordinates of Q and $-Q$ are the same, no confusion results.

Can a malicious intruder, Eve, who sees the exchange, find out Q? One possible means is for her to compute the discrete logarithm of P_1 with respect to P to obtain a. This will enable her to compute $Q = [a]P_2$. However, if $\mathcal{E}$ is not one of the elliptic curves for which discrete logarithm is easy, it will be extremely hard for her to do so.

Of course, what Eve seeks after is really the point Q. Can she determine Q from the knowledge of P, $[a]P$, and $[b]P$? So far, this seems to be as hard as the discrete logarithm problem itself. However, if she randomly picks a point R and wishes to test if $R = Q$, she may do so with the help of pairings using the following procedure.

- Choose a pairing (for example, the modified Weil pairing or the modified Tate pairing) w such that $w(P, P) \neq 1$.

- Compute $\zeta = w(P_1, P_2)$ and $\zeta' = w(R, P)$.

- If $\zeta = \zeta'$, then $Q = R$.

Let $R = [k]P$. Here, this procedure works since $w(P_1, P_2) = w([a]P, [b]P) = w(P, P)^{ab}$ and $w(R, P) = w([k]P, P) = w(P, P)^k$. Since $w(P, P) \neq 1$, we must have $k = ab$ so that $R = Q$.

Remark 3.6.3 The problem of determining Q from P, P_1, and P_2 is the computational **Diffie-Hellman problem**, while the problem of deciding if $R = Q$ is the **decisional Diffie-Hellman problem**. Thus, the decisional Diffie-Hellman problem for elliptic curves can be solved. Recall that both these problems are considered computationally hard for finite fields (see [42]).

Next, we describe an encryption scheme which is an analog of the ElGamal encryption scheme.

The first issue is to represent a message M as a point on an elliptic curve. We describe a possible method for an elliptic curve $\mathcal{E}$ over a prime field $\mathbb{F}_p$. Represent M as an integer with, say, $M < p/100$. Let $x_i = 100M + i$, where $i = 0, 1, \ldots, 99$. If there exists some y_i such that (x_i, y_i) is a point in $\mathcal{E}(\mathbb{F}_p)$, then let $P_M = (x_i, y_i)$. Note that the probability that this happens for each i is $1/2$. Conversely, given $P_M = (x, y)$, then $M = \lfloor x/100 \rfloor$.

Suppose that Alice wants to send her plaintext $P_M \in \mathcal{E}(\mathbb{F}_p)$ to Bob. The scheme can be described as follows.

- **Setup:** Bob chooses a large prime p and an elliptic curve $\mathcal{E}$ over $\mathbb{F}_p$ in which the discrete logarithm problem is hard to solve. He picks a point P in $\mathcal{E}(\mathbb{F}_p)$. Finally, he randomly selects an integer a and computes $Q = [a]P$.

  - Public key: $p, \mathcal{E}, P, Q$.
  - Private key: a.

- **Encryption:** For Alice to send P_M to Bob, she picks a random integer k and computes $M_1 = [k]P$. Then she computes $M_2 = P_M + [k]Q$. She sends (M_1, M_2) to Bob.

- **Decryption:** Upon receiving the pair (M_1, M_2) from Alice, Bob decrypts as follows. He computes $M_3 = M_2 - [a]M_1$. Then, $M_3 = P_M$.

It is straightforward to check that the scheme works since $M_2 - [a]M_1 = P_M + [k]Q - [a][k]P = P_M + [k][a]P - [a][k]P = P_M$.

In order for an eavesdropper Eve to decrypt the message, she will need either a or k, both of which require her to solve a discrete logarithm problem.

3.6.3 Pairing-Based Schemes

Excitement in elliptic curve cryptography reached a new climax when it was discovered that pairings could be used to design schemes for new scenarios that could not be considered previously. Two important examples are the tripartite key exchange and the identity-based encryption schemes.

We first look at the tripartite key exchange.

In this scenario, Alice, Bob, and Carl are trying to obtain a secret key S among themselves. Using the approach described in Subsection 3.6.2 requires two rounds of interaction between them. On the other hand, the bilinearity of pairings enables them to share a secret by just publishing a piece of public information each.

To this end, Alice, Bob, and Carl agree on an elliptic curve $\mathcal{E}$ over a fixed finite field $\mathbb{F}_q$. They also fix a point P on $\mathcal{E}(\mathbb{F}_q)$ of suitably large prime order. Suppose that there exists a pairing w on $\mathcal{E}$ such that $w(P, P) \neq 1$. Recall that this can be achieved via either the modified Tate pairing (under suitable conditions) or an appropriately defined modified Weil pairing.

The protocol is as follows.

- Alice, Bob, and Carl each pick a random integer, say a, b, and c, respectively. Alice, Bob, and Carl each compute $A = [a]P$, $B = [b]P$, and $C = [c]P$, respectively, and make it public.

- Alice computes $w(B, C)^a$, Bob computes $w(A, C)^b$, and Carl computes $w(A, B)^c$, which each uses as the desired common secret key.

Note that the values that Alice, Bob, and Carl compute independently are indeed the same, i.e., $S = w(P,P)^{abc} = w(A,B)^c = w(B,C)^a = w(A,C)^b$. Hence, S is their desired common secret key.

What can Eve, a malicious intruder, do? Evidently, if Eve can solve the elliptic curve discrete logarithm problem, she can compute S, just like what the others did since she will be able to find a, b, c then. We have seen that the elliptic curve discrete logarithm problem can be reduced to the discrete logarithm problem in finite fields. In fact, this is especially true here, since Eve can simply compute $\alpha = w(P,P)$ and $\beta = w(P,[a]P)$. Thus, she can obtain a if the discrete logarithm in the corresponding finite field can be solved. This suggests that, once again, elliptic curves with small embedding degrees (such as supersingular elliptic curves) should be avoided.

A novel use of elliptic curves in cryptography is in the invention of identity-based encryption. Here, identity-based encryption means that the public key of a user is linked to his identity. Examples of identity-based keys can include a person's e-mail address, personal name, etc. In such schemes, a trusted authority is required. A trusted authority is one who manages and publishes the public keys of all users.

Once again, we fix an elliptic curve $\mathcal{E}$ over a finite field $\mathbb{F}_p$, where p is a prime, and a point P in $\mathcal{E}(\mathbb{F}_p)$ of large prime order. We also fix a pairing w on $\mathcal{E}$ with $w(P,P) \neq 1$. Let Carl act as the trusted authority here. We now describe the main protocol.

- **Key Setup:**

    - Carl maintains a master secret $s \in \mathbb{Z}$. He computes $Q = [s]P$ and publishes Q to the public.

    - Suppose that e represents the identity of Bob. Let $E = [e]P$ be the public key of Bob.

    - Carl sends $E_0 = [s]E = [se]P$ to Bob as his secret key.

- **Encryption:** Suppose that Alice wants to send a secret message $M \in \mathbb{F}_p$ to Bob. She first picks a random integer $t \in \mathbb{Z}$, and sends $(U,V) = ([t]P, M + w(E,Q)^t)$ to Bob.

- **Decryption:** To obtain M, Bob computes $M = V - w(E_0, U)$.

Since $w(E_0, U) = w([s]E, [t]P) = w(E,P)^{st} = w(E,[s]P)^t = w(E,Q)^t$, the decryption indeed gives M.

We remark that we have only described the skeletal ideas of the identity-based encryption scheme. The actual scheme involves hash functions and other modifications to make the implementation more feasible and secure. Note how the bilinearity of the pairing comes in very handy here. As usual, this scheme is susceptible to attacks on both the elliptic curve discrete logarithm and elliptic curve Diffie-Hellman problems.

Apart from identity-based encryption schemes such as the one we just presented, identity-based signature schemes have been proposed as well. To find out more on this active area of research, the reader can check out the annual International Conference on Pairing-Based Cryptography (Pairing).

In a different direction, Boneh and Silverberg considered in [28] multi-linear maps, which are generalizations of bilinear maps such as the pairings discussed in this section. While such maps have cryptographic applications such as multipartite Diffie-Hellman key exchange and very efficient broadcast encryption, the explicit construction of such maps from algebraic geometry seems to have some serious obstacles, leading them to say that "such maps might have to either come from outside the realm of algebraic geometry, or occur as 'unnatural' computable maps arising from geometry." Very recently, Garg, Gentry, and Halevi [68] succeeded in constructing multilinear maps from ideal lattices and applied their construction to multipartitite Diffie-Hellman key exchange and non-interactive zero-knowledge proof systems. Their work has also led to the first construction of attribute-based encryption for general circuits [133, 67].

Chapter 4

Secret Sharing Schemes

A secret sharing scheme is a method of protecting a secret among a group of participants in such a way that only certain specified subsets of the participants (those belonging to an access structure) can reconstruct the secret. A secret sharing scheme is normally initialized by a trusted dealer who securely transfers a piece of information related to the secret, called a share, to each participant in the scheme. The first secret sharing schemes proposed independently by Shamir [142] and Blakley [21] were (t, n)-threshold schemes where the access structure consists of all subsets of at least t (out of a total number of n) participants. Secret sharing schemes for general access structures were introduced and constructed by Ito, Saito, and Nishizeki [79]. Secret sharing schemes, originally motivated by the problem in secure information storage, have become an indispensable basic cryptographic tool in any security environment where active entities are groups rather than individuals, e.g., general protocols for multiparty computation, Byzantine agreement, threshold cryptography, access control, and generalized oblivious transfer (cf. [9]).

4.1 The Shamir Threshold Scheme

Definition 4.1.1 Let t and n be positive integers such that $t \leq n$. Let $\mathcal{P} = \{P_1, \ldots, P_n\}$ be a group of n participants and let K denote the set of secrets. We assume P_i's share is selected from a set S_i. A (t, n)-**threshold scheme** (also called a t-**out-of-**n **secret sharing scheme**) is a pair of algorithms: the **share distribution algorithm** $\mathcal{D}$ and the **secret reconstruction algorithm** $\mathcal{C}$. For a secret from K and an element from the set of random inputs R, the share distribution algorithm applies the mapping

$$\mathcal{D} : K \times R \to S_1 \times \cdots \times S_n$$

to assign shares to the participants in $\mathcal{P}$. The secret reconstruction algorithm $\mathcal{C}$ takes the shares of a subset $A \subseteq \mathcal{P}$ of participants and returns the secret via the restriction of $\mathcal{C}$ to A, where

$$\mathcal{C} : \prod_{P_i \in \mathcal{P}} S_i \to K,$$

if and only if $|A| \geq t$.

A (t, n)-threshold scheme is **perfect** if, for all $(i_1, i_2, \ldots, i_{t-1})$ where $1 \leq i_j \leq n$, we have

$$\mathrm{prob}(\mathbf{K} = k \mid \mathbf{S}_{i_1} = s_{i_1}, \mathbf{S}_{i_2} = s_{i_2}, \ldots, \mathbf{S}_{i_{t-1}} = s_{i_{t-1}}) = \mathrm{prob}(\mathbf{K} = k),$$

where $\mathbf{K}, \mathbf{S}_{i_1}, \ldots, \mathbf{S}_{i_{t-1}}$ denote the random variables defined on the sets $K, S_{i_1}, \ldots, S_{i_{t-1}}$, respectively, and prob denotes the probability.

The best known (t, n)-threshold scheme using algebra is the **Shamir Threshold Scheme** [142], which we describe next.

Let $K = \mathbb{F}_q$, where $q \geq n+1$ is a prime power. Let $S_i = \mathbb{F}_q$, for $1 \leq i \leq n$. In the share distribution, the dealer chooses n distinct nonzero elements $x_i \in \mathbb{F}_q$, for $i = 1, \ldots, n$, and makes these elements public to all the participants. To share a secret $k \in K$, the dealer randomly chooses a secret polynomial of degree at most $t - 1$ over $\mathbb{F}_q$

$$f(x) = k + \sum_{j=1}^{t-1} a_j x^j,$$

where $a_j \in \mathbb{F}_q$ for all $1 \leq j \leq t - 1$, and then computes $y_i = f(x_i)$. For $1 \leq i \leq n$, the dealer gives the share y_i to P_i secretly.

Now suppose t participants $P_{i_1}, \ldots, P_{i_t}$ want to recover the secret. By pooling together their shares, they know t points $(x_{i_1}, y_{i_1}), \ldots, (x_{i_t}, y_{i_t})$ on the secret polynomial $f(x)$. Since $f(x)$ is a polynomial of degree at most $t-1$, knowing t points of $f(x)$ can uniquely determine the polynomial and recover $k = f(0)$. In other words, t points of $f(x)$ produce the following system of equations

$$k + a_1 x_{i_1} + a_2 x_{i_1}^2 + \ldots + a_{t-1} x_{i_1}^{t-1} = y_{i_1}$$

$$k + a_1 x_{i_2} + a_2 x_{i_2}^2 + \ldots + a_{t-1} x_{i_2}^{t-1} = y_{i_2}$$

$$\vdots \qquad\qquad \vdots$$

$$k + a_1 x_{i_t} + a_2 x_{i_t}^2 + \ldots + a_{t-1} x_{i_t}^{t-1} = y_{i_t}.$$

The above system has t linear equations and t unknowns $k, a_1, \ldots, a_{t-1}$. We can rewrite the system as follows:

$$A \begin{pmatrix} k \\ a_1 \\ \vdots \\ a_{t-1} \end{pmatrix} = \begin{pmatrix} y_{i_1} \\ y_{i_2} \\ \vdots \\ y_{i_t} \end{pmatrix},$$

where the coefficient matrix A is a Vandermonde matrix

$$
A = \begin{pmatrix}
1 & x_{i_1} & x_{i_1}^2 & \cdots & x_{i_1}^{t-1} \\
1 & x_{i_2} & x_{i_2}^2 & \cdots & x_{i_2}^{t-1} \\
\vdots & \vdots & \vdots & & \vdots \\
1 & x_{i_t} & x_{i_t}^2 & \cdots & x_{i_t}^{t-1}
\end{pmatrix}.
$$

Since the x_i's are all distinct, the determinant of A, $\det A = \prod_{1 \le k < j \le t} (x_{i_j} - x_{i_k})$, is nonzero. It follows that the system of equations has a unique solution for $k, a_1, \ldots, a_{t-1}$. We can then recover the secret k.

Next, we show that any $t-1$ participants can obtain no information about the secret. Assume the $t-1$ participants are $P_{i_1}, \ldots, P_{i_{t-1}}$, who want to recover the secret. Set $\mathbf{F} = \{f(x) \in \mathbb{F}_q[x] : f(x_{i_1}) = y_{i_1}, \ldots, f(x_{i_{t-1}}) = y_{i_{t-1}}, \deg(f) \le t-1\}$. We define $\theta : \mathbf{F} \to \mathbb{F}_q$ by $\theta(f(x)) = f(0)$. It is easy to verify that θ is one-to-one. Since $f(0) = k$ is the secret, it follows that they cannot rule out any of the possibilities for the secret in $\mathbb{F}_q$.

Therefore, the above scheme is a perfect (t, n)-threshold scheme.

Note that the above reconstruction algorithm outputs the full polynomial $f(x)$ when at least t shares are known. In other words, not only does it recover the secret k, it also recovers the random inputs $a_1, a_2, \ldots, a_{t-1}$. An alternative method to solving the system of linear equations is based on the **Lagrange Interpolation Formula**. The formula gives the explicit expression of the unique polynomial of degree at most $t-1$:

$$
f(x) = \sum_{j=1}^{t} y_{i_j} \prod_{1 \le \ell \le t, \ell \ne j} \frac{x - x_{i_\ell}}{x_{i_j} - x_{i_\ell}}.
$$

Since the t participants $P_{i_1}, \ldots, P_{i_t}$ only need to know the secret $k = f(0)$, they can simply compute $f(0)$ as follows:

$$
k = \sum_{j=1}^{t} y_{i_j} \prod_{1 \le \ell \le t, \ell \ne j} \frac{x_{i_\ell}}{x_{i_\ell} - x_{i_j}}.
$$

Letting

$$
b_j = \prod_{1 \le \ell \le t, \ell \ne j} \frac{x_{i_\ell}}{x_{i_\ell} - x_{i_j}} \quad \text{for all } 1 \le j \le t,
$$

we then have

$$
k = \sum_{j=1}^{t} b_j y_{i_j}.
$$

This means that the secret k is a linear combination of the t shares.

We give a toy example to illustrate the **Shamir Threshold Scheme**.

Example 4.1.2 Suppose $q = 7$, $t = 3$, $n = 6$, and the public values are

$x_i = i$, $1 \leq i \leq 6$. To share a secret, say 5, in $\mathbb{F}_7$, the dealer randomly selects a polynomial $f(x)$ of degree at most 2 with constant term 5, say, $f(x) = 5 + 3x + 2x^2$. The six shares are

$$s_1 = f(x_1) = 3, \quad s_2 = f(x_2) = 5, \quad s_3 = f(x_3) = 4,$$
$$s_4 = f(x_4) = 0, \quad s_5 = f(x_5) = 0, \quad s_6 = f(x_6) = 4.$$

Suppose P_1, P_3, and P_6 pool their shares together. They compute the secret by solving the following linear equations

$$k + a_1 + a_2 = 3$$
$$k + 3a_1 + 2a_2 = 4$$
$$k + 6a_1 + a_2 = 4.$$

Alternatively, according to the Lagrange Interpolation Formula, we have $b_1 = 6, b_2 = 6$, and $b_3 = 3$, and the secret is $k = b_1 s_1 + b_2 s_3 + b_3 s_6 = 5$.

4.2　Other Threshold Schemes

Besides the Shamir threshold scheme, there are also other threshold schemes known. In this section, we describe some of these well-known threshold schemes. We continue to assume that the set of participants is $\mathcal{P} = \{P_1, \ldots, P_n\}$.

4.2.1　The Karnin-Greene-Hellman (n, n)-Threshold Scheme

In the Shamir (t, n)-threshold scheme, it is required that the set of secrets can be "encoded" as a finite field $\mathbb{F}_q$ and $q \geq n + 1$. In [87], Karnin, Greene, and Hellman proposed a simple construction when $t = n$, for which these requirements are not necessary. Let $K = \mathbb{Z}_m$ and $S_i = \mathbb{Z}_m$ for $i = 1, 2, \ldots, n$ (where m is not necessarily a prime). The scheme works as follows.

To share a secret $k \in K$, the dealer independently selects $n - 1$ random elements $y_i \in \mathbb{Z}_m$, for $i = 1, 2, \ldots, n - 1$, and then computes $y_n = k - \sum_{i=1}^{n-1} y_i \bmod m$. The dealer then sends the share y_i to the participant P_i, for $i = 1, 2, \ldots, n$. Clearly, the n participants can recover the secret k by simply adding their shares over $\mathbb{Z}_m$,

$$k = \sum_{i=1}^{n} y_i \bmod m.$$

We show that any $n - 1$ participants have no information about the secret. Assume that $n - 1$ participants $P_1, \ldots, P_{i-1}, P_{i+1}, \ldots, P_n$ want to recover the secret. They can only compute $k - y_i = \sum_{j=1, j \neq i}^{n} y_j \bmod m$. Since y_i is a random value, knowing $k - y_i$ obtains no information about k. This means that the scheme is perfect.

The Karnin-Greene-Hellman (n, n)-threshold scheme can be easily generalized to any group. For instance, if $K = \{0, 1\}^\ell$, bit-wise XOR defines a group operation, we obtain a scheme where both the secret and the shares are strings of ℓ bits.

4.2.2 The Blakley Threshold Scheme

The **Blakley Threshold Scheme** [21] is based on an intuitive geometric idea: two non-parallel lines in a plane intersect at a unique point, while a single line cannot determine the point of intersection.

Let $\mathbf{P}^t(\mathbb{F}_q)$ denote the t-dimensional projective space over $\mathbb{F}_q$ defined in Chapter 1 (cf. Definition 1.1.3(ii)). Note that $\mathbf{P}^t(\mathbb{F}_q)$ is obtained by omitting the zero vector in $\mathbb{F}_q^{t+1}$ and identifying two vectors $\mathbf{x}$ and $\mathbf{x}'$ satisfying the relation $\mathbf{x} = \lambda \mathbf{x}'$, where λ is a nonzero element of $\mathbb{F}_q$. This defines an equivalence relation on $\mathbb{F}_q^{t+1} \setminus \{\mathbf{0}\}$. The set of equivalence classes, i.e., the lines through the origin of $\mathbb{F}_q^{t+1}$, are the points of $\mathbf{P}^t(\mathbb{F}_q)$; there are $(q^{t+1} - 1)/(q - 1)$ points in $\mathbf{P}^t(\mathbb{F}_q)$. Similarly, each i-dimensional subspace of $\mathbb{F}_q^{t+1}$ gives rise to an $(i - 1)$-dimensional subspace of $\mathbf{P}^t(\mathbb{F}_q)$, called an $(i - 1)$-**flat**. In particular, 0-**flats** (respectively, 1-**flats**, 2-**flats**, and $(t - 1)$-**flats**) are called **points** (respectively, **lines**, **planes**, and **hyperplanes**). Note that each point of $\mathbf{P}^t(\mathbb{F}_q)$ lies on $(q^t - 1)/(q - 1)$ hyperplanes.

The Blakley threshold scheme works as follows. To realize a (t, n)-threshold scheme, the secret is represented as a point P in $\mathbf{P}^t(\mathbb{F}_q)$. There are $(q^t - 1)/(q - 1)$ hyperplanes that contain P. The dealer randomly selects n distinct hyperplanes from these $(q^t - 1)/(q - 1)$ hyperplanes containing P, and distributes a hyperplane to each participant as his share. It is shown in [21] that if q is sufficiently large and n is not too large, then the probability that any t of the hyperplanes intersect in a point different from P is close to 0. Thus, any t out of n shares are sufficient to recover the secret in general. On the other hand, fewer than t hyperplanes will intersect only in some subspace containing P. Thus, fewer than t participants are able to recover the subspace, but still cannot figure out the secret P exactly. Note that the scheme is not perfect in general since the participants of an unauthorized subset (i.e., fewer than t participants) can have a better chance of guessing the secret than someone outside the group of participants. This means that an unauthorized subset of participants may obtain some partial information about the secret. Although the original scheme is not perfect, the geometric solution suggested by Blakley has gained much attention and has grown into an active area for the development of secret sharing schemes.

Simmons in [147] improved the Blakley scheme to make it perfect, where the Blakley scheme was reformulated in terms of affine spaces instead of projective spaces. Let $\mathbf{A}^t(\mathbb{F}_q)$ be the t-dimensional affine space over $\mathbb{F}_q$ defined in Definition 1.1.3(i). The cosets of $\{0\}$ in $\mathbb{F}_q^t$ are called **points**, those of 1-

dimensional (respectively, 2-dimensional and $(t - 1)$-dimensional) subspaces are called **lines** (respectively, **planes** and **hyperplanes**) and, in general, the cosets of i-dimensional subspaces are called i-**flats**. To realize a (t, n)-threshold scheme in $\mathbf{A}^t(\mathbb{F}_q)$, the secrets are encoded as the points of $\mathbf{A}^t(\mathbb{F}_q)$. To share a secret (a point) P, the dealer randomly chooses a line L_d that passes through P and makes L_d public (there are q points in each line). Then the dealer selects a hyperplane V such that V intersects L_d at only P. The shares of the secret are the points of V. A subset of participants can reconstruct the secret if and only if their shares can span V. Indeed, the shares of an unauthorized subset will only span a flat which intersects L_d in the empty set, so they gain no information about the secret.

For a detailed explanation of secret sharing schemes using projective and affine spaces, we refer the reader to [136, 147].

4.2.3   The Asmuth-Bloom Threshold Scheme

Asmuth and Bloom used the Chinese Remainder Theorem [4] to construct another threshold scheme. Let $p_0 < p_1 < \ldots < p_n$ be publicly known primes. The set of secrets is $\mathbb{Z}_{p_0}$ and the participant P_i is associated with the prime p_i, for $i = 1, 2, \ldots, n$. To share a secret $k \in \mathbb{Z}_{p_0}$, the dealer randomly selects an integer α such that $s = \alpha p_0 + k < \prod_{i=1}^{t} p_i$. For $i = 1, 2, \ldots, n$, the dealer then securely distributes to P_i the share

$$s_i = s \bmod p_i.$$

Assume that there are t participants $P_{i_1}, P_{i_2}, \ldots, P_{i_t}$ who want to recover the secret. They take their shares $s_{i_1}, s_{i_2}, \ldots, s_{i_t}$ to obtain the following system of congruences

$$s \equiv s_{i_1} \bmod p_{i_1}$$
$$\vdots$$
$$s \equiv s_{i_t} \bmod p_{i_t}.$$

By the Chinese Remainder Theorem, the above system of congruences has a unique solution $0 \leq s < \prod_{j=1}^{t} p_i \leq \prod_{j=1}^{t} p_{i_j}$, so the secret is computed as $k = s \bmod p_0$.

Obviously, any t participants can always reconstruct the secret from their shares. However, as pointed out in [4], the scheme is not perfect since, from the perspective of any $t - 1$ participants, the probabilities of their shares with respect to two different secrets are not the same, but asymptotically equal. A larger value of p_0 will eventually lead to a smaller difference between these two probabilities and this difference approaches zero when p_0 grows to infinity.

4.3   General Secret Sharing Schemes

Let $2^{\mathcal{P}}$ denote the family of all subsets of $\mathcal{P}$. A subset Γ of $2^{\mathcal{P}}$ with the property that, if $I \in \Gamma$ and $I \subseteq I'$, then $I' \in \Gamma$, is called **monotone increasing**. An **access structure** Γ is a monotone increasing subset of $2^{\mathcal{P}}$. The elements in Γ are called **authorized subsets** (of $\mathcal{P}$) and the elements not in Γ are called **unauthorized subsets**. The notion of access structures plays an important role in the theory of secret sharing. Informally, (perfect) secret sharing realizing a given access structure is a method of providing collective ownership of a secret by distributing the shares of the secret in such a way that any set of participants from the access structure is able to jointly recover the secret, whereas if a group of participants does not belong to the access structure, then the participants will get no information about the secret (in the information-theoretic sense). For example, the access structure of the Shamir (t, n)-threshold has the access structure defined by

$$\Gamma_{t,n} = \{I \subseteq \mathcal{P} : |I| \geq t\}.$$

Sometimes, we call $\Gamma_{t,n}$ the (t, n)-**threshold access structure**.

Example 4.3.1 Let $\mathcal{P} = \{P_1, P_2, P_3, P_4\}$ and

$$\Gamma = \left\{ \begin{array}{l} \{P_1, P_2, P_4\}, \{P_1, P_3, P_4\}, \{P_2, P_3\}, \\ \{P_1, P_2, P_3\}, \{P_2, P_3, P_4\}, \{P_1, P_2, P_3, P_4\} \end{array} \right\}.$$

Then Γ is an access structure of $\mathcal{P}$.

Similar to Definition 4.1.1 for threshold schemes, we have the following definition of secret sharing schemes for general access structures.

Definition 4.3.2 Let Γ be an access structure with n participants $\mathcal{P} = \{P_1, \ldots, P_n\}$ and let K denote the set of secrets. We assume P_i's share is selected from a set S_i. A **secret sharing scheme realizing** Γ is a pair of algorithms: the **share distribution algorithm** $\mathcal{D}$ and the **secret reconstruction algorithm** $\mathcal{C}$. For a secret from K and an element from the set of random inputs R, the share distribution algorithm applies the mapping

$$\mathcal{D} : K \times R \to S_1 \times \cdots \times S_n$$

to assign the shares to the participants in $\mathcal{P}$. The secret reconstruction algorithm $\mathcal{C}$ takes the shares of a subset $I \subseteq \mathcal{P}$ of participants and its restriction to I returns the secret if and only if $I \in \Gamma$, where

$$\mathcal{C} : \prod_{P_i \in \mathcal{P}} S_i \to K.$$

A secret sharing scheme over Γ is **perfect** if, for all $I = \{i_1, i_2, \ldots, i_j\} \notin \Gamma$, we have

$$\mathrm{prob}(\mathbf{K} = k \mid \mathbf{S}_{i_1} = s_{i_1}, \mathbf{S}_{i_2} = s_{i_2}, \ldots, \mathbf{S}_{i_j} = s_{i_j}) = \mathrm{prob}(\mathbf{K} = k).$$

4.3.1   Secret Sharing Schemes from Cumulative Arrays

Cumulative arrays for access structures were formally introduced in [80], but the idea behind them was implicitly used in the construction of [79] to show the existence of a perfect secret sharing scheme for *any* access structure.

Definition 4.3.3 Suppose $\mathcal{P} = \{P_1, \ldots, P_n\}$ is a set of n participants and Γ is an access structure on $\mathcal{P}$. Let $X = \{x_1, \ldots, x_d\}$ and let $\tau : \mathcal{P} \to 2^X$ be a mapping from $\mathcal{P}$ to the family of subsets of X. We call (τ, X) a **cumulative array** for Γ if the following condition is satisfied:

$$\bigcup_{P \in I} \tau(P) = X \quad \text{if and only if} \ \ I \in \Gamma.$$

Moreover, a cumulative array (τ, X) for Γ is called **minimal** if, for any cumulative array (τ', X') for Γ, $|X'| \geq |X|$.

Let (τ, X) be a cumulative array for an access structure Γ with $X = \{x_1, \ldots, x_d\}$. We may, without loss of generality, assume that the secret set K is an abelian group. To share a secret $k \in K$, the dealer constructs a Karnin-Greene-Hellman (d, d)-threshold scheme with d shares. In other words, the dealer randomly chooses $d - 1$ elements $k_1, \ldots, k_{d-1} \in K$ and computes $k_d = k - \sum_{i=1}^{d-1} k_i$, then distributes the shares to the participants according to (τ, X), i.e., the share of the participant P_i is

$$s_i = \{k_j \ : \ \text{if } x_j \in \tau(P_i)\}.$$

Then, obviously, the shares from the participants in $I \in 2^{\mathcal{P}}$ consist of all the d components $k_1, \ldots, k_d$ if and only if $I \in \Gamma$. We obtain the following theorem.

Theorem 4.3.4 *A cumulative array for an access structure Γ yields a perfect secret sharing scheme for Γ.*

Next, we show how to construct a cumulative array from any access structure. Let Γ be an access structure on $\mathcal{P}$. A subset $I \subseteq \mathcal{P}$ is called a **minimal authorized subset** if $I \in \Gamma$, but $J \notin \Gamma$ for any $J \subseteq I$ such that $J \neq I$. We denote by Γ_0 the set of all the minimal authorized subsets of Γ. Then Γ is uniquely determined by Γ_0. The set Γ_0 is called the **basis** of Γ. On the other hand, $I \subseteq \mathcal{P}$ is called a **maximal unauthorized subset** if $I \notin \Gamma$, but $I \cup \{P_j\} \in \Gamma$ for any $P_j \notin I$. We denote by $\Gamma^+ = \{U_1, \ldots, U_d\}$ the set of maximal unauthorized sets with respect to Γ, and define the mapping $\tau : \mathcal{P} \to 2^{\Gamma^+}$ by

$$\tau(P) = \{U_i \ : \ P \notin U_i, 1 \leq i \leq d\} \quad \text{for all } P \in \mathcal{P}.$$

Theorem 4.3.5 *For Γ, an access structure on $\mathcal{P}$, (τ, Γ^+) given above is a minimal cumulative array for Γ.*

Proof. First, we show that (τ, Γ^+) is a cumulative array. For each $I \notin \Gamma$, there exists an i such that $I \subseteq U_i$. It follows that, for each $P \in I$, $U_i \notin \tau(P)$, so

$$U_i \notin \bigcup_{P \in I} \tau(P),$$

which implies that $\cup_{P \in I} \tau(P) \neq \Gamma^+$.

On the other hand, for each $I \in \Gamma$, we know that $I \not\subseteq U_i$ for any U_i in Γ^+. It follows that, for each $U_i \in \Gamma^+$, there exists $P \in I$ such that $P \notin U_i$. We then have $\cup_{P \in I} \tau(P) = \Gamma^+$, proving that (τ, Γ^+) is indeed a cumulative array.

Next, we show that (τ, Γ^+) is minimal. Assuming that (θ, X) is a cumulative array for Γ, we need to show that $|\Gamma^+| \leq |X|$. For each $x \in X$, define

$$\alpha(x) = \{P : x \notin \theta(P)\}.$$

We claim that, for any $I \in \Gamma$ and any $x \in X$, $I \not\subseteq \alpha(x)$. For otherwise, if $I \subseteq \alpha(x)$, then $x \notin \theta(P)$ for each $P \in I$, which implies $x \notin \cup_{P \in I} \theta(P)$, a contradiction. This shows that, for each $x \in X$, $\alpha(x) \notin \Gamma$. On the other hand, for each $U \in \Gamma^+$, there exists $x \in X$ such that $x \notin \cup_{P \in U} \theta(P)$, we then have $U \subseteq \alpha(x)$. Combining the two observations above, we have shown that, for all $U \in \Gamma^+$, we have that $U = \alpha(x)$ for some $x \in X$. It follows that $|\Gamma^+| \leq |X|$, showing that (τ, Γ^+) is indeed minimal. $\square$

Since a superset of an authorized subset must again be an authorized set, we have the following corollary.

Corollary 4.3.6 *Let $\Gamma \subseteq 2^P$. There exists a perfect secret sharing scheme realizing Γ as access structure if and only if Γ is monotone increasing.*

We associate the cumulative array above with Table 4.1, where the binary

TABLE 4.1: Cumulative array of the access structure Γ.

Γ^+	U_1	U_2	$\cdots$	U_d
P_1	$c_{1,1}$	$c_{1,2}$	$\cdots$	$c_{1,d}$
P_2	$c_{2,1}$	$c_{2,2}$	$\cdots$	$c_{2,d}$
$\vdots$		$\vdots$		
P_n	$c_{n,1}$	$c_{n,2}$	$\cdots$	$c_{n,d}$

entries $c_{i,j}$ in the table are defined as follows:

$$c_{i,j} = \begin{cases} 1 & \text{if } P_i \notin U_j \\ 0 & \text{otherwise.} \end{cases}$$

In Example 4.3.1, we have $U_1 = \{P_1, P_2\}, U_2 = \{P_1, P_3\}, U_3 = \{P_1, P_4\}, U_4 = \{P_2, P_4\}$, and $U_5 = \{P_3, P_4\}$. The table for the corresponding cumulative array is shown in Table 4.2.

TABLE 4.2: Cumulative array from Example 4.3.1.

Γ^+	$\{P_1, P_2\}$	$\{P_1, P_3\}$	$\{P_1, P_4\}$	$\{P_2, P_4\}$	$\{P_3, P_4\}$
P_1	0	0	0	1	1
P_2	0	1	1	0	1
P_3	1	0	1	1	0
P_4	1	1	0	0	0

Assume the secret k is shared using a Karnin-Greene-Hellman $(5,5)$-threshold scheme such that $k = k_1 + k_2 + k_3 + k_4 + k_5$. We assign shares to the participants in accordance with the cumulative array as in Table 4.3.

TABLE 4.3: Secret sharing scheme corresponding to Example 4.3.1.

Γ^+	$\{P_1, P_2\}$ k_1	$\{P_1, P_3\}$ k_2	$\{P_1, P_4\}$ k_3	$\{P_2, P_4\}$ k_4	$\{P_3, P_4\}$ k_5
P_1	–	–	–	k_4	k_5
P_2	–	k_2	k_3	–	k_5
P_3	k_1	–	k_3	k_4	–
P_4	k_1	k_2	–	–	–

In other words, the share of P_1 is $s_1 = \{k_4, k_5\}$, the share of P_2 is $s_2 = \{k_2, k_3, k_5\}$, and so on.

We see that, in contrast to the Shamir threshold scheme, the size of each share of a secret sharing scheme constructed from a cumulative array is typically larger than the size of the secret. For example, in a (t, n)-threshold access structure Γ, the set of maximal unauthorized subsets is

$$\Gamma^+ = \{J \subseteq \mathcal{P} : |J| = t - 1\},$$

so $|\Gamma^+| = \binom{n}{t-1}$ and the share of each participant consists of $\binom{n-1}{t-1}$ components of k_i's. This means that the size of the share of each participant is $\binom{n-1}{t-1}$ times that of the secret. This is perhaps one of the reasons that cumulative arrays have not been the subject of much study in the context of secret sharing. It turns out, however, that cumulative arrays have applications in threshold cryptography, such as those studied in some recent works in sharing block cipher encryption [32, 101], threshold message authentication codes [100], and shared generation of pseudorandom functions [108, 169], just to mention a few.

We end this subsection with the notion of generalized cumulative arrays, introduced in [101] by Martin *et al.*

Definition 4.3.7 Let Γ be an access structure on $\mathcal{P}$. A **generalized cumulative array (GCA)** $(\tau_1, \ldots, \tau_\ell, X_1, \ldots, X_\ell)$ for Γ is a collection of disjoint finite sets $X_1, \ldots, X_\ell$, where each X_i is associated with a mapping $\tau_i \colon \mathcal{P} \to 2^{X_i}$, such that, for $I \subseteq \mathcal{P}$, we have

$$\bigcup_{P \in I} \tau_i(P) = X_i \text{ for some } i \ (1 \leq i \leq \ell) \text{ if and only if } I \in \Gamma.$$

Secret sharing schemes for Γ can be realized from GCAs in much the same way as from cumulative arrays. Again, we assume that the set of secrets, K, is an abelian group and $X_i = \{x_1^{(i)}, \ldots, x_{d_i}^{(i)}\}$. To share a secret $k \in K$, the dealer implements independently ℓ (d_i, d_i)-threshold schemes for the same secret k, where each (d_i, d_i)-threshold scheme consists of d_i shares $k_1^{(i)}, \ldots, k_{d_i}^{(i)} \in K$ satisfying $k = k_1^{(i)} + k_2^{(i)} + \cdots + k_{d_i}^{(i)}$, for each $1 \leq i \leq \ell$. The dealer then distributes the shares to the participants according to $(\tau_1, \ldots, \tau_\ell, X_1, \ldots, X_\ell)$. Precisely, the share of the participant P_t is

$$\bigcup_{i=1}^{\ell} \left\{ k_j^{(i)} : 1 \leq j \leq d_i \text{ and } x_j^{(i)} \in \tau_i(P_t) \right\}.$$

Then, it is easy to verify that there exists at least one (d_i, d_i)-threshold scheme for some i such that the participants from $I \in 2^{\mathcal{P}}$ have all the shares $k_1^{(i)}, \ldots, k_{d_i}^{(i)}$ if and only if $I \in \Gamma$.

The efficiency of a GCA can be measured by various parameters. In particular, the following values should be as small as possible:

(i) the number of shares for each $P \in \mathcal{P}$, measured as $\sum_{i=1}^{\ell} |\tau_i(P)|$;

(ii) the number of the total shares generated by the dealer, expressed by $\sum_{i=1}^{\ell} |X_i|$.

Various constructions for generalized cumulative arrays that improve the performance of secret sharing schemes are given in [101] and [99].

4.3.2 The Benaloh-Leichter Secret Sharing Scheme

In this subsection, we give another construction, due to Benaloh and Leichter [11], for general access structures. The construction is based on monotone Boolean circuits, so sometimes it is also called the **monotone circuit construction**.

The construction is based on a recursive approach: it begins with schemes for simple access structures, from which a scheme for a composition of those simple access structures is obtained. More explicitly, let Γ_1 and Γ_2 be two access structures on the same set of participants $\mathcal{P} = \{P_1, \ldots, P_n\}$, from which two new access structures $\Gamma_1 \vee \Gamma_2$ and $\Gamma_1 \wedge \Gamma_2$ are defined as follows:

$$\Gamma_1 \vee \Gamma_2 = \{A : A \in \Gamma_1 \text{ or } A \in \Gamma_2\}$$

and
$$\Gamma_1 \wedge \Gamma_2 = \{A \ : \ A \in \Gamma_1 \text{ and } A \in \Gamma_2\}.$$

Let $\Pi_i : K \times R_i \to S_1^{(i)} \times \cdots \times S_n^{(i)}$ be a scheme realizing Γ_i, for $i = 1, 2$, where the common set of secrets for both schemes is K.

First, a secret sharing scheme realizing $\Gamma_1 \vee \Gamma_2$ is constructed as follows. Let k be the secret to be shared. Let $\Pi_1(k, r_1) = (s_{11}, \ldots, s_{1n})$ and $\Pi_2(k, r_2) = (s_{21}, \ldots, s_{2n})$, where r_i is chosen uniformly and independently from R_i, for $i = 1, 2$. Now let

$$\Pi_\vee : K \times R_1 \times R_2 \to (S_1^{(1)} \times S_1^{(2)}) \times \cdots \times (S_n^{(1)} \times S_n^{(2)})$$
$$(k, r_1, r_2) \mapsto ((s_{11}, s_{21}), \ldots, (s_{1n}, s_{2n})),$$

where the set of shares for the participant P_i is $S_i^{(1)} \times S_i^{(2)}$ and the pair (s_{1i}, s_{2i}) is given to P_i as his share, for $1 \le i \le n$.

It is not difficult to prove that $\Pi_\vee$ is a perfect secret sharing scheme realizing $\Gamma_1 \vee \Gamma_2$ provided that Π_i is a perfect secret sharing scheme realizing Γ_i, for $i = 1, 2$.

Next, a secret sharing scheme realizing $\Gamma_1 \wedge \Gamma_2$ is constructed as follows. As before, let k be the secret to be shared. We may assume that K possesses an abelian group structure. Choose k_1 and k_2 randomly from K conditional on $k_1 + k_2 = k$ and then distribute shares for k_i using Π_i, for $i = 1, 2$. More explicitly, let $\Pi(k_1, r_1') = (s_{11}', \ldots, s_{1n}')$ and $\Pi(k_2, r_2') = (s_{21}', \ldots, s_{2n}')$, where r_i' is chosen uniformly and independently from R_i, for $i = 1, 2$. Let

$$\Pi_\wedge : K \times R_1 \times R_2 \to (S_1^{(1)} \times S_1^{(2)}) \times \cdots \times (S_n^{(1)} \times S_n^{(2)})$$
$$(k, r_1', r_2') \mapsto ((s_{11}', s_{21}'), \ldots, (s_{1n}', s_{2n}')),$$

where the set of shares for the participant P_i is $S_i^{(1)} \times S_i^{(2)}$ and the pair (s_{1i}', s_{2i}') is given to P_i as his share, for $1 \le i \le n$.

It can again be proved that $\Pi_\wedge$ is a perfect secret sharing scheme realizing $\Gamma_1 \wedge \Gamma_2$ provided that Π_i is a perfect secret sharing scheme realizing Γ_i, for $i = 1, 2$.

Now, for any access structure $\Gamma = \{A_1, \ldots, A_\ell\}$, it is theoretically possible to construct a secret sharing scheme realizing it by using the above method recursively, since $\Gamma = \{A_1\} \vee \cdots \vee \{A_\ell\}$ and there is a Karnin-Greene-Hellman $(|A_i|, |A_i|)$-threshold scheme for each $\{A_i\}$, $1 \le i \le l$.

To characterize the access structures that can be efficiently realized by the Benaloh-Leichter scheme, we need to view each access structure as a Boolean function in the following way. First, we identify each set $A \subseteq \mathcal{P}$ with its characteristic vector $\vartheta_A \in \{0, 1\}^n$, where the ith entry $\vartheta_A[i]$ equals 1 if and only if $P_i \in A$. On the other hand, to an access structure Γ, we associate the function $f_\Gamma : \{0, 1\}^n \to \{0, 1\}$, where $f_\Gamma(\vartheta_B) = 1$ if and only if $B \in \Gamma$. In this way, f_Γ and Γ are uniquely determined by each other, that is, f_Γ completely describes Γ and *vice versa*. The Boolean function f_Γ is monotone increasing since Γ is. Furthermore, $f_{\Gamma_1} \vee f_{\Gamma_2} = f_{\Gamma_1 \vee \Gamma_2}$ and $f_{\Gamma_1} \wedge f_{\Gamma_2} = f_{\Gamma_1 \wedge \Gamma_2}$.

Hence, if an access structure can be described by a small monotone formula, which is defined with **OR** and **AND** gates but *without* **NEGATION** gates and *without* negated variables, it can be efficiently computed by the Benaloh-Leichter scheme.

The following result shows that, for a given access structure Γ, the efficiency of the Benaloh-Leichter scheme is determined by the associated Boolean function f_Γ on which the construction is based.

Theorem 4.3.8 ([11]) *Let Γ be an access structure and assume that f_Γ can be computed by a monotone formula in which the variable x_i appears a_i times in the formula, for $1 \leq i \leq n$. Then there exists a secret sharing scheme from the Benaloh-Leichter construction realizing Γ in which the size of the share for P_i is the product of a_i and the size of the secret.*

It should be noted that an access structure can be realized with different schemes by different monotone Boolean functions, resulting in different sizes for the shares. We end this subsection with an illustration of this point.

Example 4.3.9 Let Γ be a $(3,5)$-threshold structure on $\mathcal{P} = \{P_1, P_2, P_3, P_4, P_5\}$ given by

$$
\Gamma = \left\{
\begin{array}{l}
\{P_1, P_2, P_3\}, \{P_1, P_2, P_4\}, \{P_1, P_2, P_5\}, \{P_1, P_3, P_4\}, \\
\{P_1, P_3, P_5\}, \{P_1, P_4, P_5\}, \{P_2, P_3, P_4\}, \{P_2, P_3, P_5\}, \\
\{P_2, P_4, P_5\}, \{P_3, P_4, P_5\}
\end{array}
\right\}.
$$

We give five examples of different Boolean expressions for Γ:

1. $\mathbf{C}_1 = x_1 x_2 x_3 + x_1 x_2 x_4 + x_1 x_2 x_5 + x_1 x_3 x_4 + x_1 x_3 x_5 + x_1 x_4 x_5 + x_2 x_3 x_4 + x_2 x_3 x_5 + x_2 x_4 x_5 + x_3 x_4 x_5$. This is the disjunctive normal form for Γ;

2. $\mathbf{C}_2 = (x_1 + x_2 + x_3)(x_1 + x_2 + x_4)(x_1 + x_2 + x_5)(x_1 + x_3 + x_4)(x_1 + x_3 + x_5)(x_1 + x_4 + x_5)(x_2 + x_3 + x_4)(x_2 + x_3 + x_5)(x_3 + x_4 + x_5)$. This is the conjunctive normal form for Γ;

3. $\mathbf{C}_3 = x_1 x_2 (x_3 + x_4 + x_5) + (x_1 + x_2 + x_3) x_4 x_5 + (x_1 + x_2) x_3 (x_4 + x_5)$;

4. $\mathbf{C}_4 = (x_1 + x_2)(x_3 + x_4) x_5 + (x_1 + x_2 + x_5)(x_1 + x_3)(x_1 + x_4)(x_2 + x_3)(x_2 + x_4)(x_3 + x_4 + x_5)$;

5. $\mathbf{C}_5 = x_1 (x_2 + x_3)(x_2 + x_4 + x_5)(x_3 + x_4 + x_5) + (x_1 + x_2 + x_3)(x_2 + x_4)(x_2 + x_5)(x_3 + x_4)(x_3 + x_5)(x_4 + x_5)$.

4.4 Information Rate

From Definition 4.3.2, a secret sharing scheme is perfect if the probability of any unauthorized subset of participants who, by pooling their shares together, can guess the secret correctly is not better than an outsider. A useful tool in studying perfectness is the notion of entropy, introduced by Shannon in 1948. The entropy can be thought of as a mathematical measure of information or uncertainty. We begin this section with a brief introduction to entropy.

4.4.1 Entropy

In this subsection, we review some basic concepts from information theory that are used in this book. For a complete treatment of the subject, the reader is referred to [44].

We denote by $\mathbf{X}$ a random variable defined on a finite set X with a given probability distribution $\{\text{prob}(x)\}_{x \in X}$. The **entropy** of $\mathbf{X}$ is defined by

$$H(\mathbf{X}) \stackrel{\text{def}}{=} - \sum_{x \in X} \text{prob}(\mathbf{X} = x) \log \text{prob}(\mathbf{X} = x)$$

(recall that log is $\log_2$).

The entropy $H(\mathbf{X})$ is a measure of the average information content of the elements in X or, equivalently, a measure of the average uncertainty one has about which element of the set X has been chosen when the choice of the elements from X is made according to the probability distribution $\{\text{prob}(x)\}_{x \in X}$.

The entropy has the following properties. The proofs are straightforward, so they are omitted here.

Proposition 4.4.1 *For any random variable $\mathbf{X}$ on a set X, the following properties hold:*

(i) $0 \le H(\mathbf{X}) \le \log |X|$;

(ii) $H(\mathbf{X}) = 0$ *if and only if there exists $x_0 \in X$ such that* $\text{prob}(\mathbf{X} = x_0) = 1$;

(iii) $H(\mathbf{X}) = \log |X|$ *if and only if* $\text{prob}(\mathbf{X} = x) = \frac{1}{|X|}$, *for all $x \in X$.*

Given two finite sets X, Y, and a joint probability distribution $\{\text{prob}(x,y)\}_{x \in X, y \in Y}$ on their Cartesian product, the entropy of the random variable $\mathbf{X} \times \mathbf{Y}$ on the set $X \times Y$ is defined to be:

$$H(\mathbf{X}, \mathbf{Y}) \stackrel{\text{def}}{=} - \sum_{(x,y) \in X \times Y} \text{prob}(\mathbf{X} = x, \mathbf{Y} = y) \log \text{prob}(\mathbf{X} = x, \mathbf{Y} = y).$$

The **conditional entropy** $H(\mathbf{X} \mid \mathbf{Y})$, also called the **equivocation** of $\mathbf{X}$ given $\mathbf{Y}$, is defined as

$$H(\mathbf{X} \mid \mathbf{Y})$$

$$\stackrel{\text{def}}{=} - \sum_{x \in X} \sum_{y \in Y} \text{prob}(\mathbf{Y} = y) \, \text{prob}(\mathbf{X} = x \mid \mathbf{Y} = y) \log \left(\text{prob}(\mathbf{X} = x \mid \mathbf{Y} = y) \right),$$

where $\text{prob}(\mathbf{X} = x \mid \mathbf{Y} = y)$ is the conditional probability of $\mathbf{X} = x$ given $\mathbf{Y} = y$. The conditional entropy can be written as

$$H(\mathbf{X} \mid \mathbf{Y}) = \sum_{y \in Y} \text{prob}(\mathbf{Y} = y) H(\mathbf{X} \mid \mathbf{Y} = y),$$

where

$$H(\mathbf{X} \mid \mathbf{Y} = y) = -\sum_{x \in X} \text{prob}(\mathbf{X} = x \mid \mathbf{Y} = y) \log \text{prob}(\mathbf{X} = x \mid \mathbf{Y} = y)$$

can be interpreted as the average uncertainty one has about which element of X has been chosen when the choice is made according to the probability distribution $\{\text{prob}(x \mid y)\}_{x \in X}$, i.e., when it is known that the value chosen from the set Y is y.

Some properties of the conditional entropy are listed in the following proposition. The proofs follow readily from the definitions, so we leave them as an exercise to the reader.

Proposition 4.4.2 *The following properties hold for any random variables* $\mathbf{X}$, $\mathbf{Y}$ *and* $\mathbf{Z}$:

(i) $0 \leq H(\mathbf{X} \mid \mathbf{Y}) \leq H(\mathbf{X})$;

(ii) $H(\mathbf{X} \mid \mathbf{Y}) = 0$ *if and only if, for every* $y \in Y$, *there exists* $x \in X$ *with* $\text{prob}(\mathbf{X} = x \mid \mathbf{Y} = y) = 1$;

(iii) $H(\mathbf{X} \mid \mathbf{Y}) = H(\mathbf{X})$ *if and only if* $\mathbf{X}$ *and* $\mathbf{Y}$ *are independent*;

(iv) $H(\mathbf{X}, \mathbf{Y}) = H(\mathbf{X}) + H(\mathbf{Y} \mid \mathbf{X}) = H(\mathbf{Y}) + H(\mathbf{X} \mid \mathbf{Y})$;

(v) $H(\mathbf{X} \mid \mathbf{Y}, \mathbf{Z}) \leq H(\mathbf{X} \mid \mathbf{Z})$.

If we have $n+1$ sets $X_1, \ldots, X_n, Y$ and a probability distribution on their Cartesian product, the conditional entropy $H(\mathbf{X}_1, \mathbf{X}_2, \ldots, \mathbf{X}_n \mid \mathbf{Y})$ of the joint space $\mathbf{X}_1 \times \cdots \times \mathbf{X}_n$ given $\mathbf{Y}$ is defined as

$$H(\mathbf{X}_1, \ldots, \mathbf{X}_n \mid \mathbf{Y})$$
$$\stackrel{\text{def}}{=} H(\mathbf{X}_1 \mid \mathbf{Y}) + H(\mathbf{X}_2 \mid \mathbf{X}_1, \mathbf{Y}) + \cdots + H(\mathbf{X}_n \mid \mathbf{X}_1, \ldots, \mathbf{X}_{n-1}, \mathbf{Y}). \tag{4.1}$$

In particular, we have

$$H(\mathbf{X}_1, \ldots, \mathbf{X}_n) = H(\mathbf{X}_1) + H(\mathbf{X}_2 \mid \mathbf{X}_1) + \cdots + H(\mathbf{X}_n \mid \mathbf{X}_1, \ldots, \mathbf{X}_{n-1}). \tag{4.2}$$

We state here another property of entropy that will be used in the next section.

Proposition 4.4.3 *For any random variables* $\mathbf{X}, \mathbf{Y}$ *and* $\mathbf{Z}$, *we have*

$$H(\mathbf{X} \mid \mathbf{Y}, \mathbf{Z}) \geq H(\mathbf{X} \mid \mathbf{Z}) - H(\mathbf{Z}).$$

Proof. From (4.2), we have that

$$H(\mathbf{X}, \mathbf{Y}, \mathbf{Z}) = H(\mathbf{Z}) + H(\mathbf{Y} \mid \mathbf{Z}) + H(\mathbf{X} \mid \mathbf{Y}, \mathbf{Z}),$$

so

$$
\begin{aligned}
& H(\mathbf{X} \mid \mathbf{Y}, \mathbf{Z}) + H(\mathbf{Z}) \\
&= H(\mathbf{X}, \mathbf{Y}, \mathbf{Z}) - H(\mathbf{Y} \mid \mathbf{Z}) \\
&\geq H(\mathbf{X}, \mathbf{Y}, \mathbf{Z}) - H(\mathbf{Y}) \\
&= H(\mathbf{X}) + H(\mathbf{Y} \mid \mathbf{X}) + H(\mathbf{Z} \mid \mathbf{X}, \mathbf{Y}) - H(\mathbf{Y}) \\
&\geq H(\mathbf{X}) + H(\mathbf{Y} \mid \mathbf{X}) - H(\mathbf{Y}) \\
&= H(\mathbf{X}, \mathbf{Y}) - H(\mathbf{Y}) \\
&= H(\mathbf{X} \mid \mathbf{Y}).
\end{aligned}
$$

$\square$

The **expectation** $E(g(x))$ of a function $g(x)$ on X is defined as

$$E(g(x)) \overset{\text{def}}{=} \sum_{x \in X} g(x) \operatorname{prob}(x),$$

where $\{\operatorname{prob}(x)\}_{x \in X}$ is the probability distribution on X.

Given random variables $\mathbf{X}$ and $\mathbf{Y}$, the **mutual information** $I(\mathbf{X}; \mathbf{Y})$ between $\mathbf{X}$ and $\mathbf{Y}$ is defined to be

$$I(\mathbf{X}; \mathbf{Y}) \overset{\text{def}}{=} H(\mathbf{X}) - H(\mathbf{X} \mid \mathbf{Y})$$

and it has the following properties:

 (i) $I(\mathbf{X}; \mathbf{Y}) = I(\mathbf{Y}; \mathbf{X})$;

 (ii) $I(\mathbf{X}; \mathbf{Y}) \geq 0$.

Given sets X, Y, Z and a joint probability distribution on their Cartesian product, the **conditional mutual information** $I(\mathbf{X}; \mathbf{Y} \mid \mathbf{Z})$ between $\mathbf{X}$ and $\mathbf{Y}$ given $\mathbf{Z}$ can be written as

$$I(\mathbf{X}; \mathbf{Y} \mid \mathbf{Z}) \overset{\text{def}}{=} E\left(\frac{\operatorname{prob}(\mathbf{X}=x, \mathbf{Y}=y \mid \mathbf{Z}=z)}{\operatorname{prob}(\mathbf{X}=x \mid \mathbf{Z}=z) \operatorname{prob}(\mathbf{Y}=y \mid \mathbf{Z}=z)} \right)$$

$$= H(\mathbf{X} \mid \mathbf{Z}) - H(\mathbf{X} \mid \mathbf{Z}, \mathbf{Y}).$$

If we have $n + 1$ sets $X_1, \ldots, X_n, Y$ and a probability distribution on their Cartesian product, the conditional mutual information $I(\mathbf{X}_1, \mathbf{X}_2, \ldots, \mathbf{X}_n; \mathbf{Y})$ of the joint space $\mathbf{X}_1 \times \mathbf{X}_2 \times \cdots \times \mathbf{X}_n$ given $\mathbf{Y}$ is defined as

$$I(\mathbf{X}_1, \mathbf{X}_2, \ldots, \mathbf{X}_n; \mathbf{Y}) \overset{\text{def}}{=} \sum_{i=1}^{n} I(\mathbf{X}_i; \mathbf{Y} \mid \mathbf{X}_1, \mathbf{X}_2, \ldots, \mathbf{X}_{i-1}).$$

4.4.2   Perfect Secret Sharing Schemes

For any secret sharing scheme, we have earlier assumed that there is a probability distribution on the set of secrets K, and we denote the entropy of this probability distribution by $H(\mathbf{K})$. Similarly, there is a probability distribution on the list of shares $S_I = \prod_{P_i \in I} S_i$ given to any specified subset of participants $I \subseteq \mathcal{P}$, and the entropy of this probability distribution is denoted by $H(\mathbf{S}_I)$. Using the notion of entropy, a perfect secret sharing scheme can be defined as follows.

Definition 4.4.4 Let Γ be an access structure on $\mathcal{P}$. A secret sharing scheme realizing the access structure Γ is **perfect** if the following two conditions hold:

(i) For any authorized subset of participants $I \in \Gamma$, we have

$$H(\mathbf{K} \mid \mathbf{S}_I) = 0,$$

i.e., the shares of an authorized subset of participants can uniquely determine the secret;

(ii) For any unauthorized subset of participants $I \notin \Gamma$, we have

$$H(\mathbf{K} \mid \mathbf{S}_I) = H(\mathbf{K}),$$

i.e., the shares of an unauthorized subset of participants give no information on the secret.

In this chapter, in particular, starting from Section 4.6 till the end of the chapter, we assume that a secret sharing scheme is perfect unless otherwise specified.

In the following, we show that, in a perfect secret sharing scheme, the uncertainty about the share of each participant is at least as great as the uncertainty about the secret. Similarly, the share for each participant is at least as long as the secret.

Theorem 4.4.5 *For any perfect secret sharing scheme realizing an access structure Γ and any participant $P \in \mathcal{P}$, we have*

$$H(\mathbf{S}_P) \geq H(\mathbf{K}),$$

where $H(\mathbf{S}_P)$ is the entropy of the share of the participant $P \in \mathcal{P}$.

Proof. Assume I is a maximal unauthorized subset of participants and $P \notin I$. Then, clearly, $I \cup \{P\} \in \Gamma$. We have the following identities

$$H(\mathbf{K} \mid \mathbf{S}_I) = H(\mathbf{K}) \quad \text{and} \quad H(\mathbf{K} \mid \mathbf{S}_I, \mathbf{S}_P) = 0.$$

It follows from Proposition 4.4.2(i) and (iv) that

$$\begin{aligned}
H(\mathbf{S}_P) &\geq H(\mathbf{S}_P \mid \mathbf{S}_I) \\
&= H(\mathbf{S}_P, \mathbf{S}_I) - H(\mathbf{S}_I) \\
&= H(\mathbf{K}, \mathbf{S}_P, \mathbf{S}_I) - H(\mathbf{K} \mid \mathbf{S}_I, \mathbf{S}_P) - H(\mathbf{S}_I) \\
&\quad \text{(since } H(\mathbf{K}, \mathbf{S}_P, \mathbf{S}_I) = H(\mathbf{S}_I, \mathbf{S}_P) + H(\mathbf{K} \mid \mathbf{S}_I, \mathbf{S}_P)) \\
&= H(\mathbf{K}, \mathbf{S}_P, \mathbf{S}_I) - H(\mathbf{S}_I) \\
&\quad \text{(since } H(\mathbf{K} \mid \mathbf{S}_I, \mathbf{S}_P) = 0) \\
&= H(\mathbf{K}, \mathbf{S}_P, \mathbf{S}_I) - H(\mathbf{K}, \mathbf{S}_I) + H(\mathbf{K} \mid \mathbf{S}_I) \\
&= H(\mathbf{K}, \mathbf{S}_P, \mathbf{S}_I) - H(\mathbf{K}, \mathbf{S}_I) + H(\mathbf{K}) \\
&\quad \text{(since } H(\mathbf{K} \mid \mathbf{S}_I) = H(\mathbf{K})) \\
&\geq H(\mathbf{K}) \\
&\quad \text{(since } H(\mathbf{K}, \mathbf{S}_P, \mathbf{S}_I) \geq H(\mathbf{K}, \mathbf{S}_I)).
\end{aligned}$$

This proves the desired result. □

We can also show that the size of the set of shares for each participant in a perfect secret sharing scheme is at least the size of the set of secrets. The following theorem is due to Kurosawa and Okada [92]

Theorem 4.4.6 *In any perfect secret sharing scheme realizing an access structure* Γ, *for any* $P \in \mathcal{P}$, *we have*

$$|S_P| \geq |K|,$$

where S_P *is the set of shares for* P.

Proof. Without loss of generality, for any $P \in \mathcal{P}$, we may assume that $\{P\} \notin \Gamma$. Let I be a minimal authorized subset in Γ such that $P \in I$. Let $J = I \setminus \{P\}$. Then J is an unauthorized subset. Therefore, for any $k \in K$ and any $s_J \in S_J$, we have

$$\text{prob}(\mathbf{K} = k \mid \mathbf{S}_J = s_J) = \text{prob}(\mathbf{K} = k).$$

Fix such an $s_J \in S_J$. Since I is an authorized subset, for any $k \in K$, there exists an s_P such that

$$\text{prob}(\mathbf{K} = k \mid \mathbf{S}_J = s_J, \mathbf{S}_P = s_P) = 1.$$

Thus, for the given s_J, the probability condition above actually induces a mapping $\theta : K \to S_P$ by $\theta(k) = s_P$. It is easy to verify that θ is one-to-one. Hence, $|S_P| \geq |K|$. □

Definition 4.4.7 Assume $\prod$ is a secret sharing scheme realizing the access structure Γ on a set of participants $\mathcal{P} = \{P_1, \ldots, P_n\}$.

(i) The **information rate** for $P_i \in \mathcal{P}$ is defined as

$$\rho_i = \frac{H(\mathbf{K})}{H(\mathbf{S}_i)}.$$

(ii) The **average rate** of the scheme $\prod$ is defined as

$$\tilde{\rho} = \frac{1}{n} \sum_{i=1}^{n} \rho_i.$$

(iii) The **information rate** of the scheme $\prod$ is defined as

$$\rho = \min_{1 \leq i \leq n} \rho_i.$$

From Theorems 4.4.5 and 4.4.6, we have the following relation for the information rates:

$$0 \leq \rho \leq \tilde{\rho} \leq 1.$$

Clearly, as shares in any secret sharing scheme need to be distributed to the participants through secure communication channels and to be stored by the participants, it is desirable to have the shares as short as possible. Indeed, the size of the shares is widely considered as the most important efficiency measure of a secret sharing scheme.

Definition 4.4.8 A perfect secret sharing scheme is called **ideal** if its information rate is $\rho = 1$. In other words, in an ideal secret sharing scheme, the size of the share for each participant is the same as the size of the secret.

Clearly, the Shamir (t, n)-threshold scheme is ideal, and so is the Karnin-Greene-Hellman (n, n)-threshold scheme.

Example 4.4.9 Consider the access structure Γ_{ustcon}, where the participants correspond to the edges of a complete undirected graph with m vertices $v_1, \ldots, v_m$, i.e., there are $n = \binom{m}{2}$ participants in the access structure, and a participant is an edge (v_i, v_j), where $i < j$. A set of participants (edges) is in the access structure if the set contains a path from v_1 to v_m. We construct a secret sharing scheme to realize this access structure as follows. Let $k \in \{0, 1\}$ be a secret. To share k, the dealer chooses $m - 2$ random bits $r_2, \ldots, r_{m-1}$ independently with uniform distribution. Furthermore, the dealer sets $r_1 = k$ and $r_m = 0$. The share of the participant (v_i, v_j) is $r_i \oplus r_j$, where $\oplus$ is the bit-XOR operation. It can be proved (see [9]) that the resulting secret sharing scheme is ideal.

A natural question to ask is: Given an access structure Γ, can we always find an ideal perfect secret sharing scheme to realize Γ? We address this problem in the rest of this subsection.

We first prove two lemmas.

Lemma 4.4.10 *Suppose Γ is an access structure on $\mathcal{P}$, $B \notin \Gamma$ and $A \cup B \in \Gamma$, where $A, B \subseteq \mathcal{P}$. Then, for any perfect secret sharing scheme realizing Γ, we have*

$$H(\mathbf{S}_A \mid \mathbf{S}_B) = H(\mathbf{K}) + H(\mathbf{S}_A \mid \mathbf{S}_B, \mathbf{K}).$$

Proof. From (4.1), we have

$$H(\mathbf{S}_A, \mathbf{K} \mid \mathbf{S}_B) = H(\mathbf{S}_A \mid \mathbf{S}_B, \mathbf{K}) + H(\mathbf{K} \mid \mathbf{S}_B)$$
$$H(\mathbf{S}_A, \mathbf{K} \mid \mathbf{S}_B) = H(\mathbf{K} \mid \mathbf{S}_A, \mathbf{S}_B) + H(\mathbf{S}_A \mid \mathbf{S}_B).$$

It follows that

$$H(\mathbf{S}_A \mid \mathbf{S}_B, \mathbf{K}) + H(\mathbf{K} \mid \mathbf{S}_B) = H(\mathbf{K} \mid \mathbf{S}_A, \mathbf{S}_B) + H(\mathbf{S}_A \mid \mathbf{S}_B).$$

Since $B \notin \Gamma$ and $A \cup B \in \Gamma$, we have

$$H(\mathbf{K} \mid \mathbf{S}_B) = H(\mathbf{K}) \quad \text{and} \quad H(\mathbf{K} \mid \mathbf{S}_A, \mathbf{S}_B) = 0,$$

so the result follows. □

Lemma 4.4.11 *Suppose Γ is an access structure and $A \cup B \notin \Gamma$, where $A, B \subseteq \mathcal{P}$. Then $H(\mathbf{S}_A \mid \mathbf{S}_B) = H(\mathbf{S}_A \mid \mathbf{S}_B, \mathbf{K})$.*

Proof. From the proof of Lemma 4.4.10, we know that

$$H(\mathbf{S}_A \mid \mathbf{S}_B, \mathbf{K}) + H(\mathbf{K} \mid \mathbf{S}_B) = H(\mathbf{K} \mid \mathbf{S}_A, \mathbf{S}_B) + H(\mathbf{S}_A \mid \mathbf{S}_B).$$

Since

$$H(\mathbf{K} \mid \mathbf{S}_B) = H(\mathbf{K}) \quad \text{and} \quad H(\mathbf{K} \mid \mathbf{S}_A, \mathbf{S}_B) = H(\mathbf{K}),$$

the result follows. □

Theorem 4.4.12 *Let*

$$\Gamma_0 = \{\{P_1, P_2\}, \{P_2, P_3\}, \{P_3, P_4\}\}$$

for $\mathcal{P} = \{P_1, P_2, P_3, P_4\}$. Then, in any perfect secret sharing scheme realizing Γ, the following inequality holds:

$$H(\mathbf{S}_2) + H(\mathbf{S}_3) \geq 3H(\mathbf{K}).$$

Proof. Let $A = \{P_3\}$ and $B = \{P_1, P_4\}$. Then we have $B \notin \Gamma$ and $A \cup B \in \Gamma$. Applying Lemma 4.4.10, we obtain

$$H(\mathbf{S}_3 \mid \mathbf{S}_1, \mathbf{S}_4) = H(\mathbf{K}) + H(\mathbf{S}_3 \mid \mathbf{S}_1, \mathbf{S}_4, \mathbf{K}).$$

We then have the following sequence of inequalities

$$
\begin{aligned}
H(\mathbf{K}) &= H(\mathbf{S}_3 \mid \mathbf{S}_1, \mathbf{S}_4) - H(\mathbf{S}_3 \mid \mathbf{S}_1, \mathbf{S}_4, \mathbf{K}) \\
&\leq H(\mathbf{S}_3 \mid \mathbf{S}_1, \mathbf{S}_4) \\
&\leq H(\mathbf{S}_3 \mid \mathbf{S}_1) \\
&= H(\mathbf{S}_3 \mid \mathbf{S}_1, \mathbf{K}) && \text{(by Lemma 4.4.11)} \\
&\leq H(\mathbf{S}_2, \mathbf{S}_3 \mid \mathbf{S}_1, \mathbf{K}) \\
&= H(\mathbf{S}_2 \mid \mathbf{S}_1, \mathbf{K}) + H(\mathbf{S}_3 \mid \mathbf{S}_1, \mathbf{S}_2, \mathbf{K}) \\
&\leq H(\mathbf{S}_2 \mid \mathbf{S}_1, \mathbf{K}) + H(\mathbf{S}_3 \mid \mathbf{S}_2, \mathbf{K}) \\
&= H(\mathbf{S}_2 \mid \mathbf{S}_1) - H(\mathbf{K}) + H(\mathbf{S}_3 \mid \mathbf{S}_2) - H(\mathbf{K}) && \text{(by Lemma 4.4.10)} \\
&\leq H(\mathbf{S}_2) + H(\mathbf{S}_3 \mid \mathbf{S}_2) - 2H(\mathbf{K}) \\
&\leq H(\mathbf{S}_2) + H(\mathbf{S}_3) - 2H(\mathbf{K}).
\end{aligned}
$$

Hence, the result follows. □

Corollary 4.4.13 *Let*

$$\Gamma_0 = \{\{P_1, P_2\}, \{P_2, P_3\}, \{P_3, P_4\}\}$$

for $\mathcal{P} = \{P_1, P_2, P_3, P_4\}$. *Then, for any perfect secret sharing scheme realizing* Γ, *the information rate satisfies* $\rho \leq 2/3$. *Consequently, there is no ideal secret sharing scheme realizing* Γ.

Proof. By the definition of ρ, we have

$$\frac{H(\mathbf{K})}{H(\mathbf{S}_2)} \geq \rho \text{ and } \frac{H(\mathbf{K})}{H(\mathbf{S}_3)} \geq \rho.$$

It follows that

$$\frac{H(\mathbf{K})}{H(\mathbf{S}_2) + H(\mathbf{S}_3)} \geq \frac{\rho}{2}.$$

By Theorem 4.4.12, we obtain

$$\frac{\rho}{2} \leq \frac{H(\mathbf{K})}{H(\mathbf{S}_2) + H(\mathbf{S}_3)} \leq \frac{1}{3}.$$

Hence, $\rho \leq \frac{2}{3}$. □

Let Γ^1, Γ^2, and Γ^3 be access structures on $\mathcal{P} = \{P_1, P_2, P_3, P_4\}$ with bases Γ_0^1, Γ_0^2, and Γ_0^3 as given below, respectively, by

$$\Gamma_0^1 = \{\{P_1, P_2\}, \{P_2, P_3\}, \{P_3, P_4\}, \{P_2, P_4\}\},$$

$$\Gamma_0^2 = \{\{P_1, P_2\}, \{P_2, P_3\}, \{P_1, P_3, P_4\}\},$$

$$\Gamma_0^3 = \{\{P_1, P_2\}, \{P_2, P_3\}, \{P_1, P_3, P_4\}, \{P_2, P_4\}\}.$$

In a similar way, we can show (see [156]) that the conditions in Theorem 4.4.12 are satisfied for Γ^1, Γ^2, and Γ^3. Therefore, the information rates of the perfect secret sharing schemes realizing Γ^1, Γ^2, and Γ^3 satisfy $\rho \leq \frac{2}{3}$.

The above discussion shows that there are access structures that cannot be realized by ideal perfect secret sharing schemes. In other words, there is at least one share whose size needs to be strictly larger than the size of the secret. Actually, one of the most important issues in the theory of secret sharing schemes is the size of shares. With the best known schemes, such as schemes based on cumulative arrays and monotone Boolean functions, most general access structures require the size of share to be exponential in the number of participants even if the size of the secret is only one bit. Beimel, in his thesis [8], conjectured:

There exists an $\epsilon > 0$ such that, for every positive integer n, there is an access structure Γ with n participants, for which every perfect secret sharing scheme realizing Γ requires shares of length exponential in the number of participants n, i.e., $2^{\epsilon n}$.

Proving or disproving this conjecture is one of the most important open questions in secret sharing schemes.

4.5   Quasi-Perfect Secret Sharing Schemes

We have seen that, in a perfect secret sharing scheme, the size of the set of shares for each participant is at least the size of the set of secrets, i.e., $|S_i| \geq |K|$, for all $1 \leq i \leq n$. Quasi-perfect secret sharing schemes are schemes in which the sizes of shares can be smaller than that of the secret. It was first introduced by Blakley and Meadows [22] under the name of ramp schemes.

Let $\mathcal{A} \subseteq 2^{\mathcal{P}}$. We say $\mathcal{A}$ is **monotone decreasing** if, for any $A \in \mathcal{A}$ and $B \subseteq A$, then $B \in \mathcal{A}$.

Definition 4.5.1 Let $\mathcal{A}, \Gamma \subseteq 2^{\mathcal{P}}$ be monotone decreasing and monotone increasing, respectively, such that $\mathcal{A} \cap \Gamma = \emptyset$. A secret sharing scheme is said to be **quasi-perfect** with adversary structure $\mathcal{A}$ and access structure Γ (sometimes, we simply say access structure $(\mathcal{A}, \Gamma)$, if no confusion arises) if the following conditions are satisfied:

(i) $H(\mathbf{K} \mid \mathbf{S}_A) = H(\mathbf{K})$ for any $A \in \mathcal{A}$;

(ii) $H(\mathbf{K} \mid \mathbf{S}_I) = 0$ for any $I \in \Gamma$;

(iii) $0 \leq H(\mathbf{K} \mid \mathbf{S}_B) \leq H(\mathbf{K})$ for any $B \in 2^{\mathcal{P}} \setminus (\mathcal{A} \cup \Gamma)$.

We say the scheme is **non-perfect** if there exists $B \in 2^{\mathcal{P}} \setminus (\mathcal{A} \cup \Gamma)$ such that

$$0 < H(\mathbf{K} \mid \mathbf{S}_B) < H(\mathbf{K}).$$

According to the definition, any perfect secret sharing scheme realizing an access structure Γ is quasi-perfect with the access structure pair $(\mathcal{A}, \Gamma)$, for any monotone decreasing $\mathcal{A} \subseteq 2^{\mathcal{P}} \setminus \Gamma$.

A typical example of a quasi-perfect scheme is a ramp scheme, introduced by Blakley and Meadows [22].

Definition 4.5.2 Let $0 \leq d < t \leq n$ be integers. A (d, t, n)-**ramp scheme** is a quasi-perfect secret sharing scheme realizing $(\mathcal{A}, \Gamma)$ such that

$$\mathcal{A} = \{A \subseteq \mathcal{P} : |A| \leq d\}$$
$$\Gamma = \{I \subseteq \mathcal{P} : |I| \geq t\}.$$

Sometimes, we write $\mathcal{A}$ and Γ as $\mathcal{A}_{d,n}$ and $\Gamma_{t,n}$, respectively.

Example 4.5.3 Let $0 \le d < t \le n$ be integers. Let the set of secrets be $K = \mathbb{F}_q^{t-d}$, where $q \ge n + (t - d)$, and let $x_1, \ldots, x_{n+(t-d)}$ be distinct in $\mathbb{F}_q$. To share a secret $(k_1, \ldots, k_{t-d}) \in K$, the dealer selects a random polynomial of degree at most $t - 1$,

$$f(x) = a_0 + a_1 x + \cdots + a_{t-1} x^{t-1}$$

such that $f(x_j) = k_{j-n}$ for $n+1 \le j \le n+(t-d)$. The share of the participant P_i is $f(x_i)$, for all $1 \le i \le n$. It is easy to see that t or more participants can recover the secret while d or fewer participants have no information about the secret. It is therefore a (d, t, n)-ramp scheme.

Theorem 4.5.4 *Let $\Gamma, \mathcal{A} \subseteq 2^{\mathcal{P}}$ be such that $\Gamma \cap \mathcal{A} = \emptyset$, and let $|K| \ge 2$. There exists a non-perfect secret sharing scheme with adversary structure $\mathcal{A}$ and access structure Γ on P for the set of secrets K if and only if $\mathcal{A}$ is monotone decreasing and Γ is monotone increasing.*

Proof. Assume that $\prod$ is a secret sharing scheme with adversary structure $\mathcal{A}$ and access structure Γ on $\mathcal{P}$ for the set of secrets K. For any $I \in \Gamma$ and $k \in K$, we have $H(\mathbf{K} \mid \mathbf{S}_I) = 0$ by the definition of $\prod$. Then, for any $I' \supseteq I$, we have

$$H(\mathbf{K} \mid \mathbf{S}_{I'}) \le H(\mathbf{K} \mid \mathbf{S}_I) = 0.$$

It follows that $I' \in \Gamma$. Therefore, Γ is monotone increasing. Similarly, $\mathcal{A}$ is monotone decreasing.

Conversely, suppose $\mathcal{A}$ and Γ are monotone decreasing and increasing, respectively. We shall construct a non-perfect secret sharing scheme with adversary structure $\mathcal{A}$ and access structure Γ on $\mathcal{P}$. We divide into two cases: $|K| > 2$ and $|K| = 2$.

If $|K| > 2$, without loss of generality, we may assume $K = \mathbb{Z}_{|K|} = \{0, 1, \ldots, |K| - 1\}$. For each $k \in K$, we write

$$k = 2k_1 + k_2,$$

where $k_2 \in \mathbb{Z}_2 = \{0, 1\}$, and $k_1 \in \mathbb{Z}_{\lfloor(|K|-1)/2\rfloor+1} = \{0, 1, \ldots, \lfloor(|K| - 1)/2\rfloor\}$.

Since Γ is monotone increasing, there exists a secret sharing scheme $\prod_1$ with access structure Γ to realize the secret $k_1 \in \mathbb{Z}_{\lfloor(|K|-1)/2\rfloor+1}$. Let $\Gamma_{\mathcal{A}}^c = 2^{\mathcal{P}} \setminus \mathcal{A}$. Then it is easy to see that $\Gamma_{\mathcal{A}}^c$ is monotone increasing, and there exists a secret sharing scheme $\prod_2$ with access structure $\Gamma_{\mathcal{A}}^c$ to realize the secret $k_2 \in \mathbb{Z}_2$. Assume that the shares of P_i, where $1 \le i \le n$, in $\prod_1$ and $\prod_2$ are s_{i1} and s_{i2}, respectively. We consider a secret sharing scheme $\prod$ for the set of secrets K as follows. To share a secret $k \in K$, we express $k = 2k_1 + k_2$ as above. Then we share k_1 by $\prod_1$ and k_2 by $\prod_2$ with n shares $(s_{11}, \ldots, s_{n1})$ and $(s_{12}, \ldots, s_{n2})$, respectively. Note that, in the scheme $\prod$, the share of P_i is $s_i = (s_{i1}, s_{i2})$, for $1 \le i \le n$. It follows that

(i) For any $I \in \Gamma$, we have $I \in \Gamma_{\mathcal{A}}^c$ since $\Gamma \subseteq \Gamma_{\mathcal{A}}^c$. It follows that

$$H(\mathbf{K}_1 \mid \mathbf{S}_I) = H(\mathbf{K}_2 \mid \mathbf{S}_I) = 0,$$

so $H(\mathbf{K} \mid \mathbf{S}_I) = 0$.

(ii) For any $A \in \mathcal{A}$, $H(\mathbf{K}_1 \mid \mathbf{S}_A) = H(\mathbf{K}_1)$ and $H(\mathbf{K}_2 \mid \mathbf{S}_A) = H(\mathbf{K}_2)$, so $H(\mathbf{K} \mid \mathbf{S}_A) = H(\mathbf{K})$.

(iii) For any $B \notin \Gamma \cup \mathcal{A}$, we have $H(\mathbf{K}_1 \mid \mathbf{S}_B) = H(\mathbf{K}_1)$ and $H(\mathbf{K}_2 \mid \mathbf{S}_B) = 0$. Therefore,
$$0 < H(\mathbf{K} \mid \mathbf{S}_B) < H(\mathbf{K}).$$

If $|K| = 2$, we consider the following distribution rule:

k	0	0	0	1	1	1
k_1	0	0	0	1	1	1
k_2	0	0	1	1	1	0

i.e., $k_1 = k$ and
$$\text{prob}(\mathbf{K}_2 = 0 \mid \mathbf{K} = 0) = \tfrac{2}{3},$$

$$\text{prob}(\mathbf{K}_2 = 1 \mid \mathbf{K} = 0) = \tfrac{1}{3},$$

$$\text{prob}(\mathbf{K}_2 = 1 \mid \mathbf{K} = 1) = \tfrac{2}{3},$$

$$\text{prob}(\mathbf{K}_2 = 0 \mid \mathbf{K} = 1) = \tfrac{1}{3}.$$

Using an identical argument as for the case $|K| > 2$, we can show that there exists a non-perfect secret sharing scheme with adversary structure $\mathcal{A}$ and access structure Γ on $\mathcal{P}$ with the set of secrets K. This completes the proof.   □

We see from Example 4.5.3 that, in a non-perfect secret sharing scheme, the size of the share of the participant can be smaller than the size of the secret. In the following, we derive a lower bound, due to Ogata and Kurosawa [118], on the sizes of shares in non-perfect secret sharing schemes.

Lemma 4.5.5 *In any secret sharing scheme with adversary structure $\mathcal{A}$ and access structure Γ, if $I \in \Gamma$, $A \in \mathcal{A}$, and $A \subset I$, then*
$$\sum_{P \in I \setminus A} H(\mathbf{S}_P) \geq H(\mathbf{K}).$$

Proof. Since
$$
\begin{aligned}
H(\mathbf{K} \mid \mathbf{S}_I) &= H(\mathbf{K} \mid \mathbf{S}_A, \mathbf{S}_{I \setminus A}) \\
&\geq H(\mathbf{K} \mid \mathbf{S}_A) - H(\mathbf{S}_{I \setminus A}) \qquad \text{(using Proposition 4.4.3)} \\
&\geq H(\mathbf{K} \mid \mathbf{S}_A) - \textstyle\sum_{P \in I \setminus A} H(\mathbf{S}_P),
\end{aligned}
$$
it follows that
$$\sum_{P \in I \setminus A} H(\mathbf{S}_P) \geq H(\mathbf{K} \mid \mathbf{S}_A) - H(\mathbf{K} \mid \mathbf{S}_I) = H(\mathbf{K}).$$

□

Theorem 4.5.6 *In any secret sharing scheme with adversary structure $\mathcal{A}$ and access structure Γ,*

$$\max_{P \in \mathcal{P}}\{H(\mathbf{S}_P)\} \geq \frac{H(\mathbf{K})}{\min\{|I \setminus A| \,:\, I \in \Gamma, A \in \mathcal{A}\}}.$$

Proof. First, assume $A \subset I$. Then, from Lemma 4.5.5,

$$|I \setminus A| \max_{P \in I \setminus A} H(\mathbf{S}_P) \geq \sum_{P \in I \setminus A} H(\mathbf{S}_P) \geq H(\mathbf{K}),$$

and so

$$|I \setminus A| \max_{P \in \mathcal{P}} H(\mathbf{S}_P) \geq H(\mathbf{K}).$$

Next, assume $A \not\subset I$. We define $I' = I \cup A$. Since Γ is monotone increasing, we know $I' \in \Gamma$ and $A \subseteq I'$, so

$$|I' \setminus A| \max_{P \in I' \setminus A} H(\mathbf{S}_P) \geq H(\mathbf{K}).$$

Clearly, $|I \setminus A| = |I' \setminus A|$. We obtain the desired result. $\square$

From Theorem 4.5.6, we know that, in a (d, t, n)-ramp scheme, $\max_{P \in \mathcal{P}}\{H(\mathbf{S}_P)\} \geq H(\mathbf{K})/(t - d)$ for any $P \in \mathcal{P}$. If $H(\mathbf{S}_P) = H(\mathbf{K})/(t - d)$ holds for all $P \in \mathcal{P}$, we say the ramp scheme is **optimal**.

4.6 Linear Secret Sharing Schemes

We first remind the reader that, starting from here till the end of this chapter, all secret sharing schemes are assumed to be perfect, unless otherwise specified.

Linear secret sharing schemes are schemes in which the secret is a linear combination of the shares of the participants. Most secret sharing schemes are linear, including the Shamir threshold scheme, the Blakley threshold scheme, the Asmuth-Bloom threshold scheme, the Benaloh-Leichter scheme, and the schemes from cumulative arrays and generalized cumulative arrays.

It is well known that linear secret sharing schemes are closely related to a linear algebraic model of computation called a **span program**, introduced by Karchmer and Wigderson [86]. The existence of an efficient construction of a linear secret sharing scheme for an access structure is essentially equivalent to the existence of a small monotone span program for the characteristic function of the access structure (or simply a small monotone span program for the access structure).

In this section, we establish this equivalence. For simplicity of exposition, we restrict the discussion to the case where the space of secrets K is $\mathbb{F}_q$. The case of $K = \mathbb{F}_q^d$ (where $d > 1$) may be treated similarly.

Definition 4.6.1 Let $K = \mathbb{F}_q$ be a finite field. We say a secret sharing scheme realizing an access structure Γ is **linear** over $\mathbb{F}_q$ if the following conditions are satisfied:

(i) The share of each participant is a vector over $\mathbb{F}_q$, i.e., for each i, there exists a constant d_i such that the share of P_i, denoted by $s_i = (s_{i,1}, s_{i,2}, \ldots, s_{i,d_i})$, is in $\mathbb{F}_q^{d_i}$;

(ii) The secret is a linear combination of shares from authorized subsets, i.e., for each $I \in \Gamma$, there exist constants $\{a_{i,j} : P_i \in I, 1 \le j \le d_i\}$ (which can be publicly computed) such that

$$k = \sum_{i \,:\, P_i \in I} \sum_{1 \le j \le d_i} a_{i,j} s_{i,j},$$

where the constants and the arithmetic are over $\mathbb{F}_q$.

The total **size** of the shares in the scheme is defined as $d = \sum_{i=1}^{n} d_i$.

It is easy to see that the Shamir (t, n)-threshold scheme is linear. Indeed, for the polynomial $f(x) = k + a_1 x + \cdots + a_{t-1} x^{t-1}$, we can compute k using the Lagrange Interpolation Formula

$$k = \sum_{j=1}^{t} s_{i_j} \prod_{1 \le \ell \le t, \ell \neq j} \frac{x_{i_\ell}}{x_{i_\ell} - x_{i_j}},$$

where $s_{i_j} = f(x_{i_j})$ is the share of the participant P_{i_j}. We have seen in Section 4.1 that, by letting

$$b_j = \prod_{1 \le \ell \le t, \ell \neq j} \frac{x_{i_\ell}}{x_{i_\ell} - x_{i_j}} \quad \text{for all } 1 \le j \le t,$$

we then have

$$k = \sum_{j=1}^{t} b_j s_{i_j}.$$

In other words, the secret k is a linear combination of any t shares.

Remark 4.6.2 Beimel [8] has shown that the linearity of the reconstruction function in linear secret sharing schemes is equivalent to the linearity of the share distribution function. More precisely, a secret sharing scheme is linear if

(i) The share of each participant is a vector over $\mathbb{F}_q$;

(ii) The share distribution function $\mathcal{D} : K \times R \to S_1 \times \cdots \times S_n$ is linear, i.e., each coordinate of the share of every participant is a linear combination of the secret k and the random input $r \in R$.

Let Γ be an access structure on $\mathcal{P} = \{P_1, \ldots, P_n\}$. Let M be a $d \times e$ matrix over $\mathbb{F}_q$, where $d \geq e$, let $\psi : \{1, \ldots, d\} \to \{1, \ldots, n\}$ be a surjective mapping and let ε be a nonzero vector in $\mathbb{F}_q^e$. Recall that we have assumed that $K = \mathbb{F}_q$.

Definition 4.6.3 The quadruple $\mathcal{M} = (K, M, \psi, \varepsilon)$ is called a **monotone span program** (**MSP** for short), with **labeling** ψ and **target vector** ε. The jth row of M is said to be labeled by P_i if $\psi(j) = i$. We say that the MSP $\mathcal{M}$ **computes the access structure** Γ if

$$I \in \Gamma \iff \varepsilon \in \text{span}(M_I),$$

where $\text{span}(M_I)$ is the subspace of K^e spanned by the rows of M, which are labeled by the members in I.

If $\varepsilon \in \text{span}(M_I)$ for some $I \subseteq \mathcal{P}$, then we say that $\mathcal{M}$ accepts I. The **size** of $\mathcal{M}$ is d, the number of rows of M.

Example 4.6.4 Let $\mathcal{P} = \{P_1, \ldots, P_6\}$ be a set of participants. Consider the access structure Γ on $\mathcal{P}$ defined by the basis

$$\Gamma_0 = \left\{ \begin{array}{l} \{P_1, P_2\}, \{P_3, P_4\}, \{P_5, P_6\}, \{P_1, P_5\}, \{P_1, P_6\}, \\ \{P_2, P_6\}, \{P_2, P_5\}, \{P_3, P_6\}, \{P_4, P_5\} \end{array} \right\}.$$

Define the matrix M over $\mathbb{F}_q$ with the labeling map ψ such that

$$M_{P_1} = \begin{pmatrix} 1 & 0 & 1 & 0 & 0 \\ 0 & 0 & 0 & 1 & 0 \\ 0 & 0 & 0 & 0 & 1 \end{pmatrix}$$

$$M_{P_2} = \begin{pmatrix} 0 & 0 & 1 & 0 & 0 \\ 0 & 0 & 0 & 1 & 0 \\ 0 & 0 & 0 & 0 & 1 \end{pmatrix}$$

$$M_{P_3} = \begin{pmatrix} 1 & 1 & 0 & 0 & 0 \\ 0 & 0 & 0 & 0 & 1 \end{pmatrix}$$

$$M_{P_4} = \begin{pmatrix} 0 & 1 & 0 & 0 & 0 \\ 0 & 0 & 0 & 1 & 0 \end{pmatrix}$$

$$M_{P_5} = \begin{pmatrix} 1 & 1 & 1 & 0 & 0 \\ 1 & 0 & 0 & 1 & 0 \end{pmatrix}$$

$$M_{P_6} = \begin{pmatrix} 0 & 1 & 1 & 0 & 0 \\ 1 & 0 & 0 & 0 & 1 \end{pmatrix}.$$

In other words, M_{P_i} consists of the rows of M labeled by P_i. It can be easily checked that $\mathcal{M} = (\mathbb{F}_q, M, \psi, \varepsilon)$ is an MSP computing Γ, where $\varepsilon = (1, 0, 0, 0, 0)$ and

$$
M = \begin{pmatrix} M_{P_1} \\ M_{P_2} \\ M_{P_3} \\ M_{P_4} \\ M_{P_5} \\ M_{P_6} \end{pmatrix}.
$$

Theorem 4.6.5 *Let K be a finite field. Assume there exists a monotone span program, of size d, over K computing an access structure Γ. Then there is a linear secret sharing scheme with the set of secrets K that realizes the access structure Γ, with d as the total size of the shares.*

Proof. Let $\mathcal{M} = (K, M, \psi, \varepsilon)$ be an MSP with e columns. We construct a linear sharing scheme as follows. The dealer randomly chooses a vector $\boldsymbol{\alpha} = (r_1, \ldots, r_e) \in K^e$ such that the inner product of $\boldsymbol{\alpha}$ and ε is the secret k, i.e., $\varepsilon \cdot \boldsymbol{\alpha} = k$. Consider the vector $M\boldsymbol{\alpha}^T$, and label each of its coordinates according to the labeling of the corresponding row in M. The share of P_i consists of all the coordinates of $M\boldsymbol{\alpha}^T$ that are labeled by P_i in M. We show that this gives rise to a linear secret sharing scheme realizing Γ.

Let I be an authorized subset in Γ. Since $\mathcal{M}$ computes Γ, we have $\varepsilon \in \mathrm{span}(M_I)$, i.e., ε is a linear combination of the rows labeled by the participants in I. We denote these rows by $M_{i_1}, \ldots, M_{i_u}$. It follows that there exist constants $b_{i_1}, \ldots, b_{i_u}$ in K such that

$$
\sum_{j=1}^{u} b_{i_j} M_{i_j} = \varepsilon.
$$

From the construction, it is easy to see that the participants in I hold the values

$$
M_{i_1} \cdot \boldsymbol{\alpha}, \ldots, M_{i_u} \cdot \boldsymbol{\alpha}.
$$

We have

$$
k = \varepsilon \cdot \boldsymbol{\alpha} = \left(\sum_{j=1}^{u} b_{i_j} M_{i_j} \right) \cdot \boldsymbol{\alpha} = \sum_{j=1}^{u} b_{i_j} (M_{i_j} \cdot \boldsymbol{\alpha}).
$$

This means that the participants in I can reconstruct the secret by applying a linear function to the coordinates of their shares.

Next, we show that, for any unauthorized subset $I \notin \Gamma$, the participants from I have no information about the secret. To this end, we show that, for any two secrets k and k', there is a one-to-one mapping between the shares of I with secret k and the shares of I with secret k'. From linear algebra, we know that $\varepsilon \notin \mathrm{span}(M_I)$ if and only if there exists $\boldsymbol{\beta} \in K^e$ such that $M_I \boldsymbol{\beta}^T = \mathbf{0}$ and $\varepsilon \cdot \boldsymbol{\beta} \neq 0$. Let $\boldsymbol{\sigma}_I$ be the possible vector of shares of I in accordance with secret k and let $\boldsymbol{\alpha}$ be the random vector for the shares generated by the dealer. We have $M_I \alpha^T = \sigma_I$ and $\varepsilon \cdot \alpha = k$. Define a mapping ϕ from the random vectors for the secret k to the random vectors for the secret k' as follows:

$$\phi(\alpha) = \alpha + \delta \beta,$$

where $\delta = (k' - k)/(\varepsilon \cdot \beta)$. Since α generates the share vector σ_I, the random vector $\phi(\alpha)$ generates the same share vector for I as well, because

$$M_I(\alpha + \delta \beta)^T = M_I \alpha^T + \delta M_I \beta^T = \sigma_I + \delta 0 = \sigma_I.$$

The secret corresponding to the random vector $\alpha + \delta \beta$ is

$$\varepsilon \cdot (\alpha + \delta \beta) = k + \delta \varepsilon \cdot \beta = k'.$$

It is easy to check that the mapping ϕ has an inverse, so it is one-to-one. The desired result follows. □

Now, we show the converse of Theorem 4.6.5, due to Beimel [8].

Theorem 4.6.6 *Let K be a finite field. Assume there exists a (perfect) linear secret sharing scheme over K realizing the access structure Γ in which the total size of the shares is d. Then there is a monotone span program over K of size d computing Γ.*

Proof. We construct the monotone span program $\mathcal{M} = (K, M, \psi, \varepsilon)$ as follows. Let

$$\mathcal{D} : K \times R \to S_1 \times \cdots \times S_n$$

be the share distribution algorithm of the linear secret sharing scheme, where R is the set of the random inputs and $S_i = K^{d_i}$. Let the share of P_i be $(s_{i,1}, \ldots, s_{i,d_i})$, and let $d = \sum_{i=1}^{n} d_i$.

M: Let M be a $d \times 2|R|$ matrix over K, where each row is labeled by a coordinate of shares, i.e., the i_jth row corresponds to the jth coordinate of the share vector of P_i. The columns are indexed by a pair $(k, r) \in \{0, 1\} \times R$. The entries are defined as

$$M_{i_j,(k,r)} = \mathcal{D}(k, r)_{i_j},$$

where $\mathcal{D}(k, r)_{i_j}$ denote the jth coordinate of the share vector of P_i, provided the secret and the random input for $\mathcal{D}$ are k and r, respectively.

ψ: The labeling mapping is defined by

$$\psi(i_j) = i, \quad \text{where } 1 \le i \le n \text{ and } 1 \le i_j \le d_i.$$

ε: We assume that the first $|R|$ columns of M are indexed by the secret 0 and the last $|R|$ columns of M are indexed by the secret 1. The target vector of $\mathcal{M}$ is defined as

$$\varepsilon = (0, \ldots, 0, 1, \ldots, 1),$$

where the first $|R|$ coordinates of ε are all 0 and the last $|R|$ coordinates are all 1.

We claim that this monotone span program computes Γ. In other words, we will prove that $I \in \Gamma$ if and only if $\varepsilon \in \text{span}(M_I)$.

Let $I \in \Gamma$. From the definition of a linear secret sharing scheme, we know that the secret is a linear combination of the coordinates of the shares of the participants in I. Since $\{0,1\}$ is a subset of K, it follows that there exist constants $\{a_{i,j} : P_i \in I, 1 \le j \le d_i\}$ such that, for every $(k,r) \in \{0,1\} \times R$, we have

$$k = \sum_{P_i \in I} \sum_{1 \le j \le d_i} a_{i,j} s_{i,j} = \sum_{P_i \in I} \sum_{1 \le j \le d_i} a_{i,j} M_{i_j,(k,r)}.$$

Let M_{i_j} be the i_jth row of M. Since the above equation holds for every column labeled by $(k,r) \in \{0,1\} \times R$, we obtain

$$\varepsilon = \sum_{P_i \in I} \sum_{1 \le j \le d_i} a_{i,j} M_{i_j},$$

so $\mathcal{M}$ computes Γ as required.

On the other hand, for any $I \subseteq \mathcal{P}$, if $\varepsilon \in \text{span}(M_I)$, then ε is a linear combination of the rows labeled by the participants from I. This implies that the participants from I can distinguish when the secret is 0 and when the secret is 1. The perfectness of the scheme implies that I must be in Γ. $\square$

We end this subsection with a short discussion on the notion of the dual access structure, which plays an important role in the next section.

Definition 4.6.7 Let Γ be an access structure on $\mathcal{P}$. The set $\Gamma^* = \{I : \bar{I} = \mathcal{P} \setminus I \notin \Gamma\}$ is called the **dual access structure of Γ**.

The following theorem, due to Gál [64] and Fehr [56], will be required in the proof of Theorem 4.7.3.

Theorem 4.6.8 *Assume K is a finite field. Let $\mathcal{M} = (K, M, \psi, \varepsilon)$ be an MSP of size d computing an access structure Γ. There exists an MSP $\mathcal{M}^* = (K, M^*, \psi, \varepsilon^*)$ of the same size d, computing its dual access structure Γ^*. Moreover, $\mathcal{M}^*$ can be efficiently constructed, and M and M^* satisfy $M^T M^* = \varepsilon^T \varepsilon^*$.*

Proof. Assume that M is a $d \times e$ matrix over K whose columns are linearly independent and that the target vector is $\varepsilon = (1, 0, \dots, 0) \in K^e$ (note that, if the columns of M are not linearly independent or ε is not in the given form, we can apply some suitable linear transformation to get a new MSP satisfying the required property to compute the same access structure as $\mathcal{M}$). Let $\mathbf{v}_0$ be a solution of the system of linear equations $M^T \mathbf{x} = \varepsilon^T$ and let $\mathbf{v}_1, \dots, \mathbf{v}_{d-e}$ be a basis of $\ker(M^T)$, where $\ker(M^T) = \{\mathbf{y} \in K^d : M^T \mathbf{y} = \mathbf{0}\}$. Set

$$M^* = (\mathbf{v}_0, \mathbf{v}_1, \dots, \mathbf{v}_{d-e}) \text{ and } \varepsilon^* = (1, 0, \dots, 0) \in K^{d-e+1}.$$

Note that M^* is a $d \times (d-e+1)$ matrix satisfying $M^T M^* = \varepsilon^T \varepsilon^* = E$, where E is an $e \times (d - e + 1)$ matrix with all entries equal to zero, except in its upper-left corner where the entry is 1. Furthermore, every solution of $M^T \mathbf{x} = \boldsymbol{\varepsilon}^T$ is a linear combination of the columns of M^* as follows

$$\mathbf{x} = \mathbf{v}_0 + k_1 \mathbf{v}_1 + \cdots + k_{d-e} \mathbf{v}_{d-e},$$

for some $k_1, \ldots, k_{d-e}$.

Now we show that the MSP $\mathcal{M}^* = (K, M^*, \psi, \boldsymbol{\varepsilon}^*)$ computes Γ^*. For any $I \in \Gamma$, there exists a vector $\boldsymbol{\lambda}$ such that $\boldsymbol{\lambda}_{\bar{I}} = \mathbf{0}$ and $M^T \boldsymbol{\lambda} = \boldsymbol{\varepsilon}^T$, where $\bar{I} = \mathcal{P} \setminus I$. Therefore, $\boldsymbol{\lambda}$ must be of the form $\boldsymbol{\lambda} = M^* \mathbf{k}$ with the first entry of $\mathbf{k}$ being 1. However, since $M_{\bar{I}}^* \mathbf{k} = \boldsymbol{\lambda}_{\bar{I}} = \mathbf{0}$ and $\mathbf{k} \cdot (\boldsymbol{\varepsilon}^*)^T = 1$, $\bar{I}$ cannot be accepted by $\mathcal{M}^*$.

Consider now a set I such that $\bar{I}$ is not accepted by $\mathcal{M}^*$. This means that $\boldsymbol{\varepsilon}^*$ is not in the subspace generated by the rows of $M_{\bar{I}}^*$ or, equivalently, there exists a vector $\mathbf{k}$ with $M_{\bar{I}}^* \mathbf{k} = \mathbf{0}$ and $\mathbf{k} \cdot (\boldsymbol{\varepsilon}^*)^T = 1$. If we set $\mathbf{a} = M^* \mathbf{k}$, then $\mathbf{a}_{\bar{I}} = \mathbf{0}$ and hence

$$M_I^T \mathbf{a}_I = M^T \mathbf{a} = M^T M^* \mathbf{k} = E\mathbf{k} = \boldsymbol{\varepsilon}^T.$$

Therefore $I \in \Gamma$.

Thus, we have shown that $I \in \Gamma$ if and only if $\bar{I}$ is not accepted by $\mathcal{M}^*$, proving the claim. □

4.7 Multiplicative Linear Secret Sharing Schemes

Multiplicative linear secret sharing and strongly multiplicative linear secret sharing were introduced by Cramer, Damgård, and Maurer in [45]. They play an important role in the design of secure multiparty computation protocols. Roughly speaking, multiplication is the property that the product of secrets can be recovered by a linear combination of the products of the shares of the individual participants. In this section, we study the multiplication properties of linear secret sharing schemes (LSSSs).

As we have seen, monotone span programs are equivalent to linear secret sharing schemes. Sometimes it is convenient to describe linear secret sharing schemes in terms of MSPs, which we do for this section.

Let $\mathcal{M} = (K, M, \psi, \boldsymbol{\varepsilon})$ be a linear secret sharing scheme realizing an access structure Γ. Given two vectors $\mathbf{x} = (x_1, \ldots, x_d)$, $\mathbf{y} = (y_1, \ldots, y_d) \in K^d$, we define $\mathbf{x} \diamond \mathbf{y}$ to be the vector containing all entries of the form $x_i \cdot y_j$ with $\psi(i) = \psi(j)$. More precisely, let

$$\mathbf{x} = (x_{11}, \ldots, x_{1d_1}, \ldots, x_{n1}, \ldots, x_{nd_n}),$$
$$\mathbf{y} = (y_{11}, \ldots, y_{1d_1}, \ldots, y_{n1}, \ldots, y_{nd_n}),$$

where $\sum_{i=1}^{n} d_i = d$, and $(x_{i1}, \ldots, x_{id_i})$, $(y_{i1}, \ldots, y_{id_i})$ are the entries labeled with P_i according to ψ. Then $\mathbf{x} \diamond \mathbf{y}$ is the vector composed of the $\sum_{i=1}^{n} d_i^2$ entries $x_{ij} y_{ik}$, where $1 \leq j, k \leq d_i, 1 \leq i \leq n$. We shall write the entries of $\mathbf{x} \diamond \mathbf{y}$ in some fixed order. We also define $\mathbf{x}^T \diamond \mathbf{y}^T = (\mathbf{x} \diamond \mathbf{y})^T$.

Definition 4.7.1 Let $\mathcal{M} = (K, M, \psi, \varepsilon)$ be a linear secret sharing scheme realizing an access structure Γ on $\mathcal{P}$. Then $\mathcal{M}$ is called **multiplicative** if there exists a **recombination vector** $\mathbf{z} \in K^{\sum_{i=1}^{n} d_i^2}$ such that, for all $k, k' \in K$ and $\boldsymbol{\rho}, \boldsymbol{\rho}' \in K^{e-1}$, we have

$$kk' = \mathbf{z} \left(M(k, \boldsymbol{\rho})^T \diamond M(k', \boldsymbol{\rho}')^T \right). \tag{4.3}$$

Moreover, $\mathcal{M}$ is **strongly multiplicative** if, for all $I \notin \Gamma$, $\mathcal{M}_{\bar{I}}$ is multiplicative, where $\mathcal{M}_{\bar{I}}$ denotes the MSP $\mathcal{M}$ restricted to the subset $\bar{I} = \mathcal{P} \setminus I$.

Note that, in the Shamir (t, n)-threshold scheme, if the secrets k and k' are shared using two polynomials $f(x)$ and $g(x)$ of degree at most $t - 1$, with the shares $(f(x_1), \ldots, f(x_n))$ and $(g(x_1), \ldots, g(x_n))$, respectively, then the shares of the secret kk' from the polynomial $h(x) = f(x)g(x)$ are $(f(x_1)g(x_1), \ldots, f(x_n)g(x_n))$. However, the degree of $h(x)$ is at most $2t - 2$. Thus, the reconstruction of $h(x)$ (and so kk') requires at least $2t - 1$ points of $(f(x_1)g(x_1), \ldots, f(x_n)g(x_n))$. It follows that $n \geq 2t - 1$, i.e., the Shamir (t, n)-threshold scheme is multiplicative provided $n \geq 2t - 1$, or equivalently, $t < n/2 + 1$. Similarly, we can show that the Shamir (t, n)-threshold scheme is strongly multiplicative provided $t < n/3 + 1$. In other words, the threshold access structure has the properties that *no two* (respectively, *no three*) unauthorized subsets can cover the full set of participants $\mathcal{P}$ in order to achieve the *multiplication* (respectively, *strong multiplication*) property. Such properties can be generalized to any access structure.

Definition 4.7.2 An access structure Γ on $\mathcal{P}$ is called Q^2 (respectively, Q^3) if no two (respectively, no three) sets from $\Gamma^c = \{I : I \subseteq \mathcal{P}, I \notin \Gamma\}$ can cover $\mathcal{P}$.

Theorem 4.7.3 *If there exists an MSP $\mathcal{M}$ of size d over K that computes a Q^2 access structure Γ, then there exists a multiplicative MSP $\mathcal{M}'$ of size at most $2d$ over K that computes Γ.*

The proof of Theorem 4.7.3 follows immediately from Lemmas 4.7.4 and 4.7.5 below.

Lemma 4.7.4 *If there exists an MSP $\mathcal{M}$ of size d over K that computes an access structure Γ, then there exists a multiplicative MSP $\mathcal{M}'$ of size at most $2d$ over K that computes $\Gamma \cup \Gamma^*$, where Γ^* is the dual of Γ.*

Proof. Assume $\mathcal{M} = (K, M, \psi, \varepsilon)$ is an MSP of size d computing an access structure Γ. Let $\mathcal{M}^* = (K, M^*, \psi, \varepsilon^*)$ be the MSP constructed in Theorem

4.6.8. Then $\mathcal{M}^*$ computes the dual access structure Γ^* with size d, and M and M^* satisfy $M^T M^* = \varepsilon^T \varepsilon^*$. We also denote the $d \times e$ matrix M as $M = (\mathbf{g}_0, \mathbf{g}_1, \ldots, \mathbf{g}_{e-1})$ and the $d \times (d-e+1)$ matrix M^* as $M^* = (\mathbf{v}_0, \mathbf{v}_1, \ldots, \mathbf{v}_{d-e})$. We construct $\mathcal{M}' = (K, M', \psi', \varepsilon')$ as follows.

Set

$$M' = \begin{pmatrix} \mathbf{g}_0 & \mathbf{g}_1 & \cdots & \mathbf{g}_{e-1} & \mathbf{0} & \cdots & \mathbf{0} \\ \mathbf{v}_0 & \mathbf{0} & \cdots & \mathbf{0} & \mathbf{v}_1 & \cdots & \mathbf{v}_{d-e} \end{pmatrix}.$$

Note that M' is a $2d \times d$ matrix.

The labeling mapping ψ' is constructed in accordance with ψ: $\psi'(i) = \psi(i)$ if $1 \le i \le d$ and $\psi'(i) = \psi(i-d)$ if $d+1 \le i \le 2d$. Let $\varepsilon' = (1, 0, \ldots, 0) \in K^d$. It is easy to see that $\mathcal{M}' = (K, M', \psi', \varepsilon')$ computes $\Gamma \cup \Gamma^*$.

We are left to show that $\mathcal{M}'$ is multiplicative. In other words, we will show that, for any $k, k' \in K$ and vectors $\boldsymbol{\rho}, \boldsymbol{\rho}' \in K^{d-1}$, kk' is a linear combination of the coordinates of $M'(k, \boldsymbol{\rho})^T \diamond M'(k', \boldsymbol{\rho}')^T$. We write $\boldsymbol{\rho} = (\boldsymbol{\rho}_1, \boldsymbol{\rho}_2)$ and $\boldsymbol{\rho}' = (\boldsymbol{\rho}'_1, \boldsymbol{\rho}'_2)$, where $\boldsymbol{\rho}_1, \boldsymbol{\rho}'_1 \in K^{e-1}$ and $\boldsymbol{\rho}_2, \boldsymbol{\rho}'_2 \in K^{d-e}$. We have

$$(k, \boldsymbol{\rho}_1) M^T M^* \begin{pmatrix} k' \\ \boldsymbol{\rho}'_2 \end{pmatrix} = (k, \boldsymbol{\rho}_1)(M^T M^*) \begin{pmatrix} k' \\ \boldsymbol{\rho}'_2 \end{pmatrix}$$

$$= (k, \boldsymbol{\rho}_1) E \begin{pmatrix} k' \\ \boldsymbol{\rho}'_2 \end{pmatrix}$$

$$= kk'.$$

Clearly, $(k, \boldsymbol{\rho}_1) M^T M^* (k', \boldsymbol{\rho}'_2)^T$ is a linear combination of the entries in the vector $M(k, \boldsymbol{\rho}_1)^T \diamond M^*(k', \boldsymbol{\rho}'_2)^T$, and so a linear combination of the entries in $M'(k, \boldsymbol{\rho})^T \diamond M'(k', \boldsymbol{\rho}')^T$, proving the result. $\square$

Lemma 4.7.5 *If Γ is a Q^2 access structure, then $\Gamma = \Gamma \cup \Gamma^*$.*

Proof. We show that, for any $I \in \Gamma^*$, we have $I \in \Gamma$. Indeed, otherwise, if $I \notin \Gamma$, then $\bar{I} \in \Gamma^*$ by the definition of Γ^*. Since $(\Gamma^*)^* = \Gamma$, it follows from $I \in \Gamma^*$ that $\bar{I} \notin \Gamma$. This shows that both I and $\bar{I}$ are not in Γ, which contradicts the assumption that Γ is Q^2. $\square$

Remark 4.7.6 The intuitive idea behind Theorem 4.7.3 can be described as follows. Let $\mathcal{M} = (K, M, \psi, \varepsilon)$ be a linear secret sharing scheme realizing an access structure Γ. From Theorem 4.6.8, we construct a linear secret sharing scheme $\mathcal{M}^* = (K, M^*, \psi, \varepsilon^*)$ for the dual access structure Γ^*. We then form a new secret sharing scheme as follows. We apply both $\mathcal{M}$ and $\mathcal{M}^*$ to share the same secret k independently. In other words, for each secret k, we have two share vectors $(a_1, \ldots, a_d)$ and $(a_1^*, \ldots, a_d^*)$ from $\mathcal{M}$ and $\mathcal{M}^*$, respectively. For another secret k', we obtain another two share vectors $(b_1, \ldots, b_d)$ and $(b_1^*, \ldots, b_d^*)$. We then show that kk' is a linear combination of the entries in $(a_1 b_1^*, a_2 b_2^*, \ldots, a_d b_d^*)$, by applying Theorem 4.6.8.

We define $\mathrm{msp}_K(\Gamma)$ to be the smallest size of an MSP that computes Γ, and define $\mu_K(\Gamma)$ (respectively, $\mu_K^*(\Gamma)$) to be the smallest size of a multiplicative (respectively, strongly multiplicative) MSP over K, which computes Γ (we adopt the convention that the value is ∞ if Γ cannot be computed). Clearly, we have

$$\mathrm{msp}_K(\Gamma) \le \mu_K(\Gamma) \le \mu_K^*(\Gamma).$$

Theorem 4.7.7 *For every finite field K and any access structure Γ, we have $\mu_K(\Gamma) < \infty$ if and only if Γ is Q^2, and $\mu_K^*(\Gamma) < \infty$ if and only if Γ is Q^3.*

Proof. We show that $\mu_K(\Gamma) < \infty$ if and only if Γ is Q^2.

Assume that Γ is Q^2. Since any access structure Γ can be realized with a linear secret sharing scheme (e.g., using the cumulative array construction), by Theorem 4.7.3, we know that Γ can be realized by a multiplicative secret sharing scheme, and so $\mu_K(\Gamma) < \infty$.

Conversely, assume that $\mathcal{M} = (K, M, \psi, \varepsilon)$ is a multiplicative MSP computing Γ. If Γ is not Q^2, there exists a set $I \subset \mathcal{P}$ such that $I \cup \bar{I} = \mathcal{P}$ and $I, \bar{I} \notin \Gamma$. It follows that neither the rows of M_I nor the rows of $M_{\bar{I}}$ can span ε. By the duality argument, there exist vectors $\boldsymbol{\kappa}$ and $\boldsymbol{\kappa}'$, both with first coordinate equal to 1, such that $M_I \boldsymbol{\kappa}^T = \mathbf{0}$ and $M_{\bar{I}} \boldsymbol{\kappa}'^T = \mathbf{0}$. By the multiplication property, we have $\mathbf{z}(M\boldsymbol{\kappa}^T \diamond M\boldsymbol{\kappa}'^T) = 1$, where $\mathbf{z}$ is the recombination vector. However, by the choice of $\boldsymbol{\kappa}$ and $\boldsymbol{\kappa}'$, we have $M\boldsymbol{\kappa}^T \diamond M\boldsymbol{\kappa}'^T = \mathbf{0}$ since $I \cup \bar{I} = \mathcal{P}$. We therefore have $\mathbf{z}(M\boldsymbol{\kappa}^T \diamond M\boldsymbol{\kappa}'^T) = 0$, a contradiction. Hence, Γ is Q^2.

Next, we prove the second statement on Q^3. Let Γ be a Q^3 access structure and $\mathcal{M} = (K, M, \psi, \varepsilon)$ a linear secret sharing scheme realizing it. For any $I \subseteq \mathcal{P}, I \notin \Gamma$, it is easy to see that $\mathcal{M}_{\bar{I}}$, the restriction of $\mathcal{M}$ to $\bar{I}$, realizes the access structure $\Gamma_{\bar{I}} = \{J \subseteq \bar{I} : J \in \Gamma\}$. The access structure $\Gamma_{\bar{I}}$ is Q^2 over $\bar{I}$ because Γ is Q^3 over $\mathcal{P} = \bar{I} \cup I$. Now we can transform $\mathcal{M}_{\bar{I}}$ into a multiplicative linear secret sharing scheme following the general construction of Theorem 4.7.3. The process applies to all $I \notin \Gamma$, so we obtain a strongly multiplicative linear secret sharing scheme realizing Γ (note that, in general, this construction results in a scheme of exponential size).

Conversely, assume that $\mathcal{M} = (K, M, \psi, \varepsilon)$ is a strongly multiplicative linear secret sharing scheme realizing Γ. If Γ is not Q^3, there exist $I_1, I_2, I_3 \notin \Gamma$ such that $I_1 \cup I_2 \cup I_3 = \mathcal{P}$. We may further assume $I_3 = \mathcal{P} \setminus (I_1 \cup I_2)$. Again, we know that $\mathcal{M}_{\bar{I}_3}$, the restriction of $\mathcal{M}$ to $\bar{I}_3$, realizes the access structure $\Gamma_{\bar{I}_3} = \{J \subseteq \bar{I}_3 : J \in \Gamma\}$. By the definition of a strongly multiplicative secret sharing scheme, we know that $\mathcal{M}_{\bar{I}_3}$ is multiplicative. This contradicts the assumption $I_1 \cup I_2 = \mathcal{P} \setminus I_3$, since both I_1 and I_2 are not in $\Gamma_{\bar{I}_3}$, proving the desired result. □

From Theorem 4.7.3, we know how to construct a multiplicative linear secret sharing scheme from a linear secret sharing scheme, with information rate decreasing at most by $1/2$. However, it remains open how to efficiently construct a strongly multiplicative linear secret sharing scheme from a linear secret sharing scheme or a multiplicative linear secret sharing scheme. In [181], Zhang *et al.* gave a new characterization of strong multiplication using so called 3-**multiplicative** linear secret sharing schemes.

It is easy to see that we have an induced labeling map ψ' : $\{1, \ldots, \sum_{i=1}^{n} d_i^2\} \to \{P_1, \ldots, P_n\}$ on the entries of $\mathbf{x} \diamond \mathbf{y}$, distributing the entry $x_{ij} y_{i\ell}$ to P_i, since both x_{ij} and $y_{i\ell}$ are labeled by P_i under ψ. For an MSP $\mathcal{M} = (K, M, \psi, \epsilon)$, write $M = (\mathbf{g}_1, \ldots, \mathbf{g}_e)$, where $\mathbf{g}_i \in K^d$, with $d = \sum_{i=1}^{n} d_i$, is the ith column vector of M, $1 \le i \le e$. We construct a new matrix $M_\diamond$ as follows:

$$M_\diamond = (\mathbf{g}_1 \diamond \mathbf{g}_1, \ldots, \mathbf{g}_1 \diamond \mathbf{g}_e, \mathbf{g}_2 \diamond \mathbf{g}_1, \ldots, \mathbf{g}_2 \diamond \mathbf{g}_e, \ldots, \mathbf{g}_e \diamond \mathbf{g}_1, \ldots, \mathbf{g}_e \diamond \mathbf{g}_e).$$

Obviously, $M_\diamond$ is a matrix over K with $\sum_{i=1}^{n} d_i^2$ rows and e^2 columns. For any two vectors $\mathbf{u}, \mathbf{v} \in K^e$, it is easy to verify that

$$(M\mathbf{u}^T) \diamond (M\mathbf{v}^T) = M_\diamond(\mathbf{u} \otimes \mathbf{v})^T,$$

where $\mathbf{u} \otimes \mathbf{v}$ denotes the tensor product with its entries written in a proper order. Define the induced labeling map ψ' on the rows of $M_\diamond$. We have the following lemma.

Lemma 4.7.8 *Let* $\mathcal{M} = (K, M, \psi, \varepsilon)$ *be a linear secret sharing scheme realizing an access structure* Γ, *and let* ψ' *be the associated labeling of* $M_\diamond$. *Then* $\mathcal{M}$ *is multiplicative if and only if* $\varepsilon \in \text{span}(M_\diamond)$, *where* $\varepsilon = (1, 0, \ldots, 0)$. *Moreover,* $\mathcal{M}$ *is strongly multiplicative if and only if* $\varepsilon \in \text{span}((M_\diamond)_{\overline{A}})$ *for all* $A \in \mathcal{A} = 2^\mathcal{P} \setminus \Gamma$.

Proof. By Definition 4.7.1, $\mathcal{M}$ is multiplicative if and only if $kk' = \mathbf{z}(M(k, \boldsymbol{\rho})^T \diamond M(k', \boldsymbol{\rho}')^T)$ for all $k, k' \in K$, $\boldsymbol{\rho}, \boldsymbol{\rho}' \in K^{e-1}$ and some recombination vector $\mathbf{z}$. Obviously,

$$M(k, \boldsymbol{\rho})^T \diamond M(k', \boldsymbol{\rho}')^T = M_\diamond((k, \boldsymbol{\rho}) \otimes (k', \boldsymbol{\rho}'))^T = M_\diamond(kk', \boldsymbol{\rho}'')^T, \quad (4.4)$$

where $(kk', \boldsymbol{\rho}'') = (k, \boldsymbol{\rho}) \otimes (k', \boldsymbol{\rho}')$. On the other hand, $kk' = \varepsilon(kk', \boldsymbol{\rho}'')^T$. Thus, $\mathcal{M}$ is multiplicative if and only if

$$(\varepsilon - \mathbf{z}M_\diamond)(kk', \boldsymbol{\rho}'')^T = 0. \quad (4.5)$$

As $k, k', \boldsymbol{\rho}$ and $\boldsymbol{\rho}'$ are arbitrary, equality (4.5) holds if and only if $\varepsilon - \mathbf{z}M_\diamond = \mathbf{0}$, i.e., $\varepsilon \in \text{span}(M_\diamond)$. The latter part of the lemma can be proved similarly. $\square$

Now we define 3-multiplicative linear secret sharing schemes. We extend the diamond product "$\diamond$" and define $\mathbf{x} \diamond \mathbf{y} \diamond \mathbf{z}$ to be the vector containing all entries of the form $x_i y_j z_\ell$ with $\psi(i) = \psi(j) = \psi(\ell)$, where the entries of $\mathbf{x} \diamond \mathbf{y} \diamond \mathbf{z}$ are written in some fixed order.

Definition 4.7.9 *Let* $\mathcal{M} = (K, M, \psi, \varepsilon)$ *be a linear secret sharing scheme realizing an access structure* Γ. *Then* $\mathcal{M}$ *is called* 3-**multiplicative** *if there* exists a recombination vector $\mathbf{z} \in K^{\sum_{i=1}^{n} d_i^3}$ such that, for all $k_1, k_2, k_3 \in K$ and $\boldsymbol{\rho}_1, \boldsymbol{\rho}_2, \boldsymbol{\rho}_3 \in K^{e-1}$, we have

$$k_1 k_2 k_3 = \mathbf{z} \left(M(k_1, \boldsymbol{\rho}_1)^T \diamond M(k_2, \boldsymbol{\rho}_2)^T \diamond M(k_3, \boldsymbol{\rho}_3)^T \right).$$

We can derive an equivalent definition for 3-multiplicative linear secret sharing schemes, similar to Lemma 4.7.8: $\mathcal{M}$ is 3-multiplicative if and only if $\boldsymbol{\varepsilon} \in \mathrm{span}(M \diamond M \diamond M)$, where $M \diamond M \diamond M$ is defined analogously to $M_\diamond$. The following theorem gives a necessary and sufficient condition for the existence of a 3-multiplicative linear secret sharing scheme.

Theorem 4.7.10 *For any access structure Γ, there exists a 3-multiplicative linear secret sharing scheme realizing Γ if and only if Γ is Q^3.*

Proof. Suppose $\mathcal{M} = (K, M, \psi, \boldsymbol{\varepsilon})$ is a 3-multiplicative linear secret sharing scheme realizing Γ. If Γ is not Q^3, then there exist $A_1, A_2, A_3 \in \mathcal{A} = 2^{\mathcal{P}} \setminus \Gamma$ such that $A_1 \cup A_2 \cup A_3 = \mathcal{P}$. It follows that there exists $\boldsymbol{\rho}_i \in K^{e-1}$ such that $M_{A_i}(1, \boldsymbol{\rho}_i)^T = \mathbf{0}$, for $1 \leq i \leq 3$. Since $A_1 \cup A_2 \cup A_3 = \mathcal{P}$, we have $M(1, \boldsymbol{\rho}_1)^T \diamond M(1, \boldsymbol{\rho}_2)^T \diamond M(1, \boldsymbol{\rho}_3)^T = \mathbf{0}$, which contradicts Definition 4.7.9.

On the other hand, from Theorem 4.7.13, we have a 3-multiplicative linear secret sharing scheme from a strongly multiplicative linear secret sharing scheme. By Theorem 4.7.7, the result follows. □

A trivial example of a 3-multiplicative linear secret sharing scheme is the Shamir threshold secret sharing scheme that realizes any Q^3 threshold access structure. Using an argument identical to that for the case of strongly multiplicative linear secret sharing schemes, we have a general construction for 3-multiplicative linear secret sharing schemes based on the Shamir threshold secret sharing schemes, with exponential complexity.

More generally, the notion of λ-**multiplicative** secret sharing schemes was introduced and studied in [181] and [6]. For any λ vectors $\mathbf{x}_i = (x_{i1}, \ldots, x_{id}) \in K^d$, $1 \leq i \leq \lambda$, we define $\diamond_{i=1}^\lambda \mathbf{x}_i$ to be the $\sum_{i=1}^n d_i^\lambda$-dimensional vector which contains entries of the form $\prod_{i=1}^\lambda x_{ij_i}$ with $\psi(j_1) = \cdots = \psi(j_\lambda)$.

Definition 4.7.11 Let $\mathcal{M} = (K, M, \psi, \boldsymbol{\varepsilon})$ be a linear secret sharing scheme realizing an access structure Γ, and let $\lambda > 1$ be an integer. Then $\mathcal{M}$ is λ-**multiplicative** if there exists a recombination vector $\mathbf{z}$ such that, for all $k_1, \ldots, k_\lambda \in K$ and $\boldsymbol{\rho}_1, \ldots, \boldsymbol{\rho}_\lambda \in K^{e-1}$, we have

$$\prod_{i=1}^{\lambda} k_i = \mathbf{z}(\diamond_{i=1}^\lambda M(k_i, \boldsymbol{\rho}_i)^T).$$

Moreover, $\mathcal{M}$ is **strongly λ-multiplicative** if, for all $A \notin \Gamma$, the restricted linear secret sharing scheme $\mathcal{M}_{\bar{A}}$ is λ-multiplicative.

Again, we can define a new matrix by taking the diamond product of λ

copies of M. This gives an equivalent definition for (strongly) λ-multiplicative linear secret sharing schemes. Moreover, since the Shamir threshold secret sharing scheme is trivially λ-multiplicative and strongly λ-multiplicative, a proper composition of the Shamir threshold secret sharing schemes results in a general construction for both λ-multiplicative linear secret sharing schemes and strongly λ-multiplicative linear secret sharing schemes. Let Q^λ be a straightforward extension of Q^2 and Q^3, i.e., an access structure Γ is Q^λ if the set of participants $\mathcal{P}$ cannot be covered by λ sets in $\mathcal{A} = 2^{\mathcal{P}} \setminus \Gamma$. The following corollary is easy to prove.

Corollary 4.7.12 *Let Γ be an access structure on $\mathcal{P}$. Then there exists a λ-multiplicative (respectively, strongly λ-multiplicative) linear secret sharing scheme realizing Γ if and only if Γ is Q^λ (respectively, $Q^{\lambda+1}$).*

Zhang *et al.* [181] showed the following results that establish the relationship between strongly multiplicative secret sharing schemes and 3-multiplicative secret sharing schemes. Their proofs are omitted here.

Theorem 4.7.13 *Let Γ be a Q^3 access structure and let $\mathcal{M} = (K, M, \psi, \varepsilon)$ be a strongly multiplicative linear secret sharing scheme realizing Γ. Suppose that $\mathcal{M}$ has size d and $|\psi^{-1}(P_i)| = d_i$, for $1 \leq i \leq n$. Then there exists a 3-multiplicative linear secret sharing scheme for Γ of size $O(d^2)$.*

Theorem 4.7.14 *Any 3-multiplicative linear secret sharing scheme is strongly multiplicative.*

4.8 Secret Sharing from Error-Correcting Codes

Secret sharing schemes are closely related to error-correcting codes. In this section, we show how to construct a secret sharing scheme from an error-correcting code.

Let C be a q-ary $[n+1, t, d]$-code, let $G = (\mathbf{g}_0, \mathbf{g}_1, \ldots, \mathbf{g}_n)$ be a generator matrix for C, and let $K = \mathbb{F}_q$ be the set of secrets. We give a construction for a secret sharing scheme for $\mathcal{P} = \{P_1, \ldots, P_n\}$, called EC-LSSS Construction, as follows:

(1) Let the generator matrix G be publicly known to everyone in the system.

(2) To share a secret $k \in K$, the dealer randomly selects a vector

$$\mathbf{r} = (r_0, r_1, \ldots, r_{t-1}) \in K^t$$

such that $k = \mathbf{r} \cdot \mathbf{g}_0$.

(3) Each participant P_i receives a share $s_i = \mathbf{r} \cdot \mathbf{g}_i$, for $i = 1, \ldots, n$.

It is easy to see that the shares of the participants $P_{i_1}, \ldots, P_{i_\ell}$ can reconstruct the secret k if $\mathbf{g}_0$ can be represented as a linear combination of $\mathbf{g}_{i_1}, \ldots, \mathbf{g}_{i_\ell}$. Indeed, assume $\mathbf{g}_0 = a_1 \mathbf{g}_{i_1} + \cdots + a_\ell \mathbf{g}_{i_\ell}$. Then we have

$$k = \mathbf{r} \cdot \mathbf{g}_0 = \mathbf{r} \cdot \left(\sum_{j=1}^{\ell} a_j \mathbf{g}_{i_j} \right) = \sum_{j=1}^{\ell} a_j \mathbf{r} \cdot \mathbf{g}_{i_j} = \sum_{j=1}^{\ell} a_j s_{i_j}.$$

Let

$$\Gamma_C = \{ I : I \subseteq \mathcal{P}, \ \mathbf{g}_0 \text{ is a linear combination of } \mathbf{g}_i, \ P_i \in I \}.$$

Then it is easy to check that Γ_C is a monotone increasing collection of subsets of $\mathcal{P}$.

Theorem 4.8.1 *For any q-ary $[n+1, t, d]$-code C, EC-LSSS Construction results in a linear secret sharing scheme realizing Γ_C.*

Proof. We show that EC-LSSS Construction gives rise to a linear MSP realizing Γ_C. We define $M = (\mathbf{g}_1, \ldots, \mathbf{g}_n)$ and set ψ to be the identity mapping on $\{1, \ldots, n\}$, i.e., $\psi(i) = i$, for $i = 1, \ldots, n$, and let $\varepsilon = \mathbf{g}_0$. Then it is easy to see that $\mathcal{M} = (K, M, \psi, \varepsilon)$ is an MSP realizing Γ_C. Using an argument identical to that for Theorem 4.6.5, we know that the secret sharing scheme from EC-LSSS Construction is a linear secret sharing scheme realizing Γ_C. □

An interesting question is the determination of the access structure Γ_C for any given code C, which turns out to be a hard question in general.

McEliece and Sarwate [104] observed that the Shamir secret sharing scheme is essentially the same as EC-LSSS Construction using a Reed-Solomon code. Recall that an $[n+1, t, d]$-Reed-Solomon code C is defined as

$$C = \{(f(0), f(x_1), \ldots, f(x_n)) : \deg(f) \leq t - 1, f(x) \in K[x]\},$$

where $x_1, \ldots, x_n$ are n distinct nonzero elements in K. The code C has a generator matrix

$$G = \begin{pmatrix} 1 & 1 & \cdots & 1 \\ 0 & x_1 & \cdots & x_n \\ 0 & x_1^2 & \cdots & x_n^2 \\ & & \vdots & \\ 0 & x_1^{t-1} & \cdots & x_n^{t-1} \end{pmatrix} = (\mathbf{g}_0, \mathbf{g}_1, \ldots, \mathbf{g}_n).$$

Since any t columns of G are linear independent and any $t + 1$ columns are linearly dependent in K^t, it is easy to see that Γ_C is the (t, n)-threshold access structure $\Gamma_{t,n}$. We have the following result immediately.

Theorem 4.8.2 *Let C be an $[n+1, t, d]$-Reed-Solomon code. Then* EC-LSSS Construction *results in a (t, n)-threshold scheme which is identical to the Shamir (t, n)-threshold scheme.*

More generally, threshold schemes can be constructed from MDS codes, which include the Reed-Solomon codes.

Theorem 4.8.3 *Let C be an $[n+1, t, d]$-MDS code over K. Then* EC-LSSS Construction *results in a (t, n)-threshold scheme.*

Proof. Assume that $G = (\mathbf{g}_0, \mathbf{g}_1, \ldots, \mathbf{g}_n)$ is a generator matrix of the $[n+1, t, d]$-MDS code C. It follows that any t columns of G are linearly independent, while any $t+1$ columns are linearly dependent over K. We then have $\mathbf{g}_0 \in \mathrm{span}(\mathbf{g}_{i_1}, \ldots, \mathbf{g}_{i_t})$ for any $\{i_1, \ldots, i_t\} \subseteq \{1, \ldots, n\}$, where $1 \le i_1 < i_2 < \cdots < i_t \le n$. Therefore $\Gamma_C = \Gamma_{t,n}$. By Theorem 4.8.1, the desired result follows. $\square$

Massey [102] observed that, for any linear code C, the access structure Γ_C can be determined by the so-called minimal codewords in $C^\perp$, the dual code of C.

Recall that, for any vector $\mathbf{x} = (x_0, x_1, \ldots, x_n)$, the **support** of $\mathbf{x}$ is defined as

$$\mathrm{supp}(\mathbf{x}) = \{i \,:\, x_i \neq 0\}.$$

Theorem 4.8.4 *For any $I = \{P_{i_1}, \ldots, P_{i_\ell}\} \subseteq \mathcal{P}$ and any linear code C, $I \in \Gamma_C$ if and only if $\{0, i_1, \ldots, i_\ell\} \supseteq \mathrm{supp}(\mathbf{c}^*)$ for some $\mathbf{c}^* \in C^\perp$.*

Proof. Let $G = (\mathbf{g}_0, \mathbf{g}_1, \ldots, \mathbf{g}_n)$ be a generator matrix of C. Observe that

$$I = \{P_{i_1}, \ldots, P_{i_\ell}\} \in \Gamma_C$$

$$\Longleftrightarrow \mathbf{g}_0 = \sum_{j=1}^\ell x_{i_j} \mathbf{g}_{i_j} \quad \text{for some } x_{i_1}, \ldots, x_{i_\ell}$$

$$\Longleftrightarrow (1, 0, \ldots, 0, -x_{i_1}, 0, \ldots, 0, -x_{i_\ell}, 0, \ldots, 0)(\mathbf{g}_0, \mathbf{g}_1, \ldots, \mathbf{g}_n)^T = \mathbf{0}$$

$$\Longleftrightarrow (1, 0, \ldots, 0, -x_{i_1}, 0, \ldots, 0, -x_{i_\ell}, 0, \ldots, 0)G^T = \mathbf{0}$$

$$\Longleftrightarrow (1, 0, \ldots, 0, -x_{i_1}, 0, \ldots, 0, -x_{i_\ell}, 0, \ldots, 0) \in C^\perp.$$

The desired result follows immediately. $\square$

Definition 4.8.5 For a given linear code C and codewords $\mathbf{c}_1, \mathbf{c}_2 \in C$, we say that $\mathbf{c}_2$ **covers** $\mathbf{c}_1$ if $\mathrm{supp}(\mathbf{c}_1) \subseteq \mathrm{supp}(\mathbf{c}_2)$. A nonzero codeword $\mathbf{c} \in C$ that covers only its scalar multiples, but no other nonzero codewords of C, is called a **minimal codeword**.

It is easy to see that there is a one-to-one correspondence between the set of minimal authorized subsets in Γ_C and the set of minimal codewords of the dual code $C^\perp$ whose first coordinate is 1.

Thus, Theorem 4.8.4 can be restated as follows.

Theorem 4.8.6 *For any $[n+1,t,d]$-code, the minimal authorized subsets of Γ_C are completely determined by the minimal codewords in $C^\perp$.*

Unfortunately, there is no known efficient algorithm to determine the minimal codewords in a linear code, so finding an *efficient* algorithm for computing Γ_C remains an interesting research problem.

We have, however, the following interesting result, due to Chen, Cramer, Goldwasser, de Haan, and Vaikuntanathan [39].

Theorem 4.8.7 *Let C be an $[n+1,t,d]$-code over a finite field K. Then* EC-LSSS Construction *results in a $(d^\perp - 2, n - d + 2, n)$-ramp scheme, where $d^\perp$ is the minimum distance of $C^\perp$.*

Proof. We first show that $\Gamma_C = \Gamma^*_{C^\perp}$, i.e., the access structure from C is the dual of the access structure from $C^\perp$. Indeed, $I \in \Gamma_C$ if and only if there is $\mathbf{c}^* = (c^*_0, c^*_1, \ldots, c^*_n) \in C^\perp$ with $c^*_0 = 1$ and $c^*_i = 0$ for those i's with $P_i \in \mathcal{P} \setminus I = \bar{I}$. Since $\mathbf{c}^*$ can be viewed as a share vector, with secret equal to 1 and shares equal to 0 for $\bar{I}$, in the secret sharing from EC-LSSS Construction for $C^\perp$, the existence of such a share vector is equivalent to $\bar{I} \notin \Gamma_{C^\perp}$.

Since there are at most $n + 1 - d^\perp$ zero c^*_i in $C^\perp$, we have

$$n - |I| = |\bar{I}| \le n + 1 - d^\perp,$$

which means $|I| \ge d^\perp - 1$. In other words, any subset of $\mathcal{P}$ of size smaller than or equal to $d^\perp - 2$ is an unauthorized subset. Similarly, any set of size $\le d - 2$ is not in $\Gamma_{C^\perp}$. We have proved that $\Gamma_C = \Gamma^*_{C^\perp}$. It follows that, for any I with $|I| \ge n - d + 2$, we have $I \in \Gamma_C$. This completes the proof. □

Using an argument identical to that for Theorem 4.8.7, we can prove the following more general result.

Corollary 4.8.8 *Let C be an $[n + 1, t, d]$-code over a finite field K. Then* EC-LSSS Construction *results in a (u, v, n)-ramp scheme, where $u = -2 + \min\{\text{wt}(\mathbf{c}^*) : \mathbf{c}^* \in C^\perp, c^*_0 = 1\}$ and $v = n + 2 - \min\{\text{wt}(\mathbf{c}) : \mathbf{c} \in C, c_0 = 1\}$.*

Recall that a self-dual code C is a linear code for which $C = C^\perp$.

Corollary 4.8.9 *Let C be a self-dual $[n + 1, t, d]$-code over a finite field K. Then* EC-LSSS Construction *results in a $(d-2, n+2-d, n)$-ramp scheme which is multiplicative.*

Proof. Since $d = d^{\perp}$ for self-dual codes, the secret sharing scheme is a $(d - 2, n + 2 - d, n)$-ramp scheme by Theorem 4.8.7. Since, for any $\mathbf{c}, \mathbf{c}' \in C$, we have $\mathbf{c} \cdot \mathbf{c}' = 0$, it follows that

$$c_0 c_0' = -c_1 c_1' - \cdots - c_n c_n',$$

from which the multiplication property follows. $\qquad\square$

4.9 Secret Sharing from Algebraic Geometry Codes

In the previous section, we have shown how a linear secret sharing scheme can be constructed from a linear code. However, the access structure of the secret sharing scheme from such a construction is hard to determine for most linear codes. In [38], Chen and Cramer applied techniques from algebraic geometry to construct linear secret sharing schemes with some nice properties: they are (d, t, n)-ramp schemes for which the parameters of d and t can be explicitly computed; they have the strong multiplication property; the size of the finite field $\mathbb{F}_q$ can be dramatically smaller than n, in contrast to the condition $q \geq n + 1$ for the Shamir scheme.

The construction of secret sharing schemes from algebraic curves proceeds as follows. The reader should refer to Chapter 1 for the notations and basic facts on algebraic curves.

(1) Let $\mathcal{X}$ be a smooth projective curve over $\mathbb{F}_q$ of genus g, and let $P, Q_0, Q_1, \ldots, Q_n$ be distinct rational points on $\mathcal{X}$. Let t be any fixed integer with $1 \leq t < n - 2g$. Define a divisor D as $D = (2g + t)P$. Then D has support $\{P\}$ and its degree is $2g + t$.

(2) To share a secret $k \in \mathbb{F}_q$, the dealer randomly selects an element $f \in \mathcal{L}(D)$ such that $f(Q_0) = k$.

(3) The share of the participant P_i is $f(Q_i) = s_i \in \mathbb{F}_q$, for $1 \leq i \leq n$.

Note that, in (2) of the construction above, the function $f \in \mathcal{L}(D)$ always exists since $\mathcal{L}(D)$ contains the constant functions. Moreover, since $\mathcal{L}(D)$ is a linear space over $\mathbb{F}_q$ of dimension $g + t + 1$ (see Theorem 1.4.7), we know that the choice of f with $f(Q_0) = k$ requires $g + t$ random elements from $\mathbb{F}_q$. For (3), the functions $f \in \mathcal{L}(D)$ only have a pole at P, so the values $f(Q_i)$ are well-defined and are from $\mathbb{F}_q$.

Lemma 4.9.1 *Let D be a divisor on $\mathcal{X}$ that is defined over $\mathbb{F}_q$ and let $\dim \mathcal{L}(D) > 0$. Then $f \in \mathcal{L}(D)$ is uniquely determined by the evaluations of f on any $\deg(D) + 1$ rational points on $\mathcal{X}$ outside the support of D.*

Proof. Let $d = \deg(D)$ and let $Q_1, \ldots, Q_{d+1}$ be $d+1$ rational points outside the support of D. We define

$$\phi : \mathcal{L}(D) \longrightarrow \mathbb{F}_q^{d+1}, \ f \mapsto (f(Q_1), \ldots, f(Q_{d+1})).$$

Then ϕ is an injective linear map of $\mathbb{F}_q$-vector spaces. Indeed, if $f, h \in \mathcal{L}(D)$ and $\phi(f) = \phi(h)$, then $f - h \in \mathcal{L}(D - (Q_1 + \cdots + Q_{d+1})) \subseteq \mathcal{L}(D)$ because the support of D is disjoint from the Q_i's. The degree of the divisor $D - (Q_1 + \cdots + Q_{d+1})$ is negative, so $f = h$.

Let f be an element in $\mathcal{L}(D)$. Given any $d+1$ rational points $Q_1, \ldots, Q_{d+1}$ on $\mathcal{X}$, for any rational point Q_0 different from $Q_1, \ldots, Q_{d+1}$ and outside the support of D, using elementary linear algebra, it can be shown that there exist coefficients $b_i \in \mathbb{F}_q$ such that

$$f(Q_0) = \sum_{i=1}^{d+1} b_i f(Q_i).$$

Hence, the desired result follows. $\square$

We denote by $\Gamma_{\mathcal{X}}$ the access structure of the secret sharing scheme from the above construction with a curve $\mathcal{X}$.

Lemma 4.9.2 *Let* $\mathcal{X}$ *be a smooth projective curve over* $\mathbb{F}_q$, *and let* $Q_0, Q_1, \ldots, Q_n$ *be* $n+1$ *distinct rational points on* $\mathcal{X}$. *Let* $I \subseteq \mathcal{P}$ *be a subset of participants. Then*

(i) $I \in \Gamma_{\mathcal{X}}$ *if and only if*

$$\dim \mathcal{L}\left(D - \left(Q_0 + \sum_{P_i \in I} Q_i\right)\right) = \dim \mathcal{L}\left(D - \sum_{P_i \in I} Q_i\right);$$

(ii) $I \notin \Gamma_{\mathcal{X}}$ *if and only if*

$$\dim \mathcal{L}\left(D - \left(Q_0 + \sum_{P_i \in I} Q_i\right)\right) < \dim \mathcal{L}\left(D - \sum_{P_i \in I} Q_i\right).$$

Proof. It is easy to see that (i) is equivalent to (ii). We show that (ii) is true. As mentioned before, the scheme can be viewed as a linear secret sharing scheme from the algebraic geometry code constructed from $\mathcal{L}(D)$ in Section 2.4. From Section 4.8, we know that $I \notin \Gamma_{\mathcal{X}}$ if and only if there exists $h \in \mathcal{L}(D)$ such that $h(Q_i) = 0$ for all $i \in I$ and $h(Q_0) = 1$.

Note that, in general, for any divisors E and E', if $E \leq E'$, then $\mathcal{L}(E) \subseteq \mathcal{L}(E')$. Since the support of D is disjoint from the Q_i's, we have

$$\mathcal{L}\left(D - \left(Q_0 + \sum_{P_i \in I} Q_i\right)\right) \subseteq \mathcal{L}\left(D - \sum_{P_i \in I} Q_i\right) \subseteq \mathcal{L}(D).$$

All functions h', if any, in the difference

$$\mathcal{L}\left(D - \sum_{P_i \in I} Q_i\right) \setminus \mathcal{L}\left(D - \left(Q_0 + \sum_{P_i \in I} Q_i\right)\right)$$

satisfy $h' \in \mathcal{L}(D), h'(Q_0) \neq 0$ and $h'(Q_i) = 0$ for all $P_i \in I$. By scaling h' such that $h'(Q_0) = 1$, we obtain the desired function h. Clearly, the difference of the above two spaces is nonempty if and only if their dimensions differ. $\qquad \square$

Theorem 4.9.3 *Let $\mathcal{X}$ be a smooth projective curve over $\mathbb{F}_q$ of genus g and let $1 \leq t < n - 2g$. The construction above yields a $(t, 2g + t + 1, n)$-ramp linear secret sharing scheme.*

Proof. Let $I \subseteq \mathcal{P}$ and $|I| = t$. We have

$$\deg\left(D - \left(Q_0 + \sum_{P_i \in I} Q_i\right)\right) = 2g - 1$$

and

$$\deg\left(D - \sum_{P_i \in I} Q_i\right) = 2g.$$

Therefore, from Theorem 1.4.7,

$$\dim \mathcal{L}\left(D - \left(Q_0 + \sum_{P_i \in I} Q_i\right)\right) = g < g + 1 = \dim \mathcal{L}\left(D - \sum_{P_i \in I} Q_i\right).$$

From Lemma 4.9.2, it follows that $I \notin \Gamma_{\mathcal{X}}$.

On the other hand, let $I \subseteq \mathcal{P}$ and $|I| = 2g + t + 1$. Then

$$\dim \mathcal{L}\left(\left(D - \sum_{P_i \in I} Q_i\right)\right) = 0,$$

since $D - \sum_{P_i \in I} Q_i$ is a divisor of negative degree. Therefore,

$$0 = \dim \mathcal{L}\left(D - \left(Q_0 + \sum_{P_i \in I} Q_i\right)\right) \leq \dim \mathcal{L}\left(D - \sum_{P_i \in I} Q_i\right) = 0.$$

From Lemma 4.9.2 again, we have $I \in \Gamma_{\mathcal{X}}$. $\qquad \square$

The above construction can be viewed as EC-LSSS Construction from the algebraic geometry code in Definition 2.4.2.

Thus, the resulting secret sharing scheme is indeed a perfect linear scheme, realizing some access structure on $\mathcal{P}$. While the determination of its exact access structure still remains open, Theorem 4.9.3 shows that we can pin down a pair consisting of an adversary structure $\mathcal{A}$ and an access structure Γ of the scheme. In other words, we have shown that the scheme obtained above from an algebraic curve is quasi-perfect with *explicit* access structure $(\mathcal{A}, \Gamma)$, where $\mathcal{A} = \mathcal{A}_{t,n}$ and $\Gamma = \Gamma_{2g+t+1,n}$. In certain applications, such as secure multiparty computation, it would be sufficient to compute an adversary structure $\mathcal{A}$ and an access structure Γ, instead of knowing the exact access structure, of the underlying secret sharing scheme (see [38]).

Next, we show that secret sharing schemes from algebraic curves have good multiplication properties. Note that Definition 4.7.1 for multiplicative linear secret sharing schemes can be generalized to quasi-perfect linear secret sharing schemes in a straightfoward manner. Here, we generalize the strong multiplication in Definition 4.7.1 to quasi-perfect linear secret sharing schemes.

Definition 4.9.4 Let $\mathcal{M} = (\mathbb{F}_q, M, \psi, \varepsilon)$ be a quasi-perfect linear secret sharing scheme with an adversary structure $\mathcal{A}$ and an access structure Γ. We say that $\mathcal{M}$ is **strongly multiplicative with respect to** $\mathcal{A}$ if, for any $I \subseteq \mathcal{P}$ with $I = \mathcal{P} \setminus A$ for some $A \in \mathcal{A}$, $\mathcal{M}_I$ is multiplicative. As before, $\mathcal{M}_I$ denotes the restriction of $\mathcal{M}$ to I.

Theorem 4.9.5 *Let X be a smooth projective curve over $\mathbb{F}_q$ of genus g and let $1 \le t < n - 2g$. The construction above yields a $(t, 2g + t + 1, n)$-ramp linear secret sharing scheme, which is multiplicative if $2t < n - 4g$, and strongly multiplicative with respect to $\mathcal{A}_{t,n}$ if $3t < n - 4g$.*

Proof. The proof is similar to that for the Shamir threshold scheme. We prove the strong multiplication case since the multiplication property can be treated in the same way. Note that, for any $f, h \in \mathcal{L}(D)$, we have $\mathrm{div}(fh) = \mathrm{div}(f) + \mathrm{div}(h)$. It follows that

$$0 \le (\mathrm{div}(f) + D) + (\mathrm{div}(h) + D) = \mathrm{div}(fh) + 2D.$$

We then have $fh \in \mathcal{L}(2D)$. Denote by $\mathcal{M}$ the secret sharing scheme based on X. Clearly, $\mathcal{M}$ is linear. By Lemma 4.9.1, it is easy to see that $\mathcal{M}$ is strongly multiplicative with respect to $\mathcal{A}_{t,n}$ if $n - t > \deg(2D) = 4g + 2t$. Indeed, let I be any set with $I \subseteq \{1, \ldots, n\}$ and $|I| = 4g + 2t + 1$. We define the following linear mappings

$$\phi_0 : \mathcal{L}(2D) \longrightarrow \mathbb{F}_q, \ \hat{f} \mapsto \hat{f}(Q_0),$$

$$\phi : \mathcal{L}(2D) \longrightarrow \mathbb{F}_q^{4g+2t+1}, \ \hat{f} \mapsto (\hat{f}(Q_i))_{i \in I},$$

and

$$\chi : \mathbb{F}_q^{4g+2t+1} \longrightarrow \mathcal{L}(2D),$$

such that $\chi\phi$ is the identity on $\mathcal{L}(2D)$.

Then, for all $f, h \in \mathcal{L}(D)$, if $s_i = f(Q_i)$ and $s_i' = h(Q_i)$ for all $i \in I$, we have

$$k \cdot k' = \phi_0\chi((s_i \cdot s_i')_{i \in I}),$$

where $k = f(Q_0), k' = h(Q_0)$. $\qquad\qquad\qquad\qquad\qquad\qquad\qquad\qquad\qquad\qquad$ □

Note that, in the Shamir threshold scheme, for a given finite field $\mathbb{F}_q$, the number of participants n, corresponding to the nonzero elements in $\mathbb{F}_q$, is bounded by $q - 1$. Actually, the Shamir threshold scheme is a special case of the construction from algebraic curves where the genus of the curve is 0. The number of participants n in the construction based on algebraic curves over $\mathbb{F}_q$ is determined by the number N_1 of $\mathbb{F}_q$-rational points of the curve, or, more precisely, $n \leq N_1 - 2$. For example, for elliptic curves (where $g = 1$), it is well known that $N_1 \leq q + \lfloor 2\sqrt{q} \rfloor + 1$ (cf. Corollary 1.5.4(iii)). It follows that the scheme can have up to $n = N_1 - 2 = q + \lfloor 2\sqrt{q} \rfloor - 1$ participants, an increase of up to $\lfloor 2\sqrt{q} \rfloor$ participants compared to the Shamir threshold scheme for the same field $\mathbb{F}_q$.

Example 4.9.6 Consider the Hermitian curve $y^r + y - x^{r+1} = 0$ over $\mathbb{F}_{r^2}$ in Example 1.5.5(ii). The genus of this curve is $r(r - 1)/2$, and there are $r^3 + 1$ $\mathbb{F}_{r^2}$-rational points. Let $r = 8$. Then the scheme can admit up to 511 participants, with up to $t = n - 2g - 1 = 511 - 56 - 1 = 454$ adversaries. Compare this scheme to the Shamir threshold scheme on the same field $\mathbb{F}_{64}$, which only allows at most 63 participants.

Since the Shamir $(t+1, n)$-threshold scheme can be viewed as a $(t, t+1, n)$-ramp scheme, it is multiplicative (respectively, strongly multiplicative) if and only if $t < n/2$ (respectively, $t < n/3$). We now show that, for the construction based on algebraic curves, the bounds on t can be asymptotically met even if the field $\mathbb{F}_q$ is small. Indeed, consider the family of curves $\{\mathcal{X}_i\}_{i \geq 1}$ over $\mathbb{F}_{q^2}$ in [65, 66] (see also Example 2.5.6) defined by

$$x_{i+1}^q + x_{i+1} = \frac{x_i^q}{x_i^{q-1} + 1}$$

for $i = 1, 2, \ldots$.

Let $g_i = g(\mathcal{X}_i)$ be the genus of $\mathcal{X}_i$, and let $n_i = N(\mathcal{X}_i)$ be the number of $\mathbb{F}_{q^2}$-rational points of $\mathcal{X}_i$. As shown in Example 2.5.6, we have $g_i/n_i \to 1/(q - 1)$. By Theorem 4.9.5, it follows that there is an infinite family of curves, yielding $(t_i, 2g_i + t_i + 1, n_i)$-ramp linear secret sharing schemes with strong multiplication for any $t_i < \left(\frac{1}{3} - \frac{4g_i}{3n_i} \right) n_i$, where $g_i/n_i \to 1/(q-1)$. Now, for any small $\epsilon > 0$, we can choose a prime power q such that $\frac{4}{3(q-1)} < \epsilon$. This implies there are infinite families of integers g, n, and t such that

$$\left(\frac{1}{3} - \epsilon \right) n \leq t < \left(\frac{1}{3} - \frac{4g}{3n} \right) n.$$

In other words, for any small $\epsilon > 0$, there exists a finite field $\mathbb{F}_{q^2}$ such that, for infinitely many n, there exists a scheme with strong multiplication and with $(\frac{1}{3} - \epsilon)n \leq t < \frac{1}{3}n$. Similarly, for each $\epsilon > 0$, there exists a finite field $\mathbb{F}_{q^2}$ such that, for infinitely many n, there exists a scheme with multiplication and with $(\frac{1}{2} - \epsilon)n \leq t < \frac{1}{2}n$.

Chapter 5

Authentication Codes

One of the main goals of a cryptographic system is to provide authentication, which simply means providing assurance about the content and origin of the communicated message. Traditionally, it was assumed that a secrecy system provides authentication by virtue of the secret key being only known to the intended communicants; this would prevent an enemy from constructing a fraudulent message. Simmons in [145] argued that the two goals of cryptography are independent. He showed that a system that provides perfect secrecy might not provide any protection against authentication threats. Similarly, a system can provide perfect authentication without concealing the message.

In an authentication system, a threat is an attempt by an enemy in the system to modify a communicated message or inject a fraudulent message into the communication channel. In a secrecy system, the attacker is passive, while, in an authentication system, the enemy is active and not only observes the communicated message and gathers information such as the plaintext and the ciphertext, but also actively interacts with the system to achieve its goal. This view of the system clearly explains Simmons's motivation for basing authentication systems on game theory [145].

The most important criteria that can be used to classify authentication systems are: (1) the relation between authenticity and secrecy, and (2) the framework for the security analysis. The first criterion divides authentication systems into those that provide **authentication with** and **without secrecy**. The second criterion divides authentication systems into systems with **unconditional security** and systems with **computational security**. Unconditional security implies that the enemy has unlimited resources, while in systems with computational security, the security relies on the required computation exceeding the enemy's computational power. In this book, we are mainly interested in unconditionally secure authentication codes without secrecy.

5.1 Authentication Codes

Authentication codes were invented by Gilbert, MacWilliams, and Sloane [69]. The general theory of unconditional authentication has been developed by Simmons ([146, 148]).

In the conventional model for unconditional authentication, there are three participants: a **transmitter**, a **receiver**, and an **opponent**. The transmitter wants to communicate some information to the receiver using a public channel which is subject to active attack, that is, the opponent can either impersonate the transmitter and insert a message in the channel, or replace a transmitted message with another. To protect against these threats, the transmitter and the receiver share a secret key. The key is then used in an authentication code (A-code for short).

The information that the transmitter wants to send is called a **source state**, denoted by s and taken from a finite source space $\mathcal{S}$. The source state is mapped into a (channel) message, denoted by m and taken from a message space $\mathcal{M}$.

Exactly how this mapping is performed is determined by the secret **key** (also called the **encoding rule**), which is denoted by e and taken from a key space $\mathcal{E}$. The key is secretly shared between the transmitter and the receiver.

Definition 5.1.1 An **authentication code (A-code)** is a quadruple $(\mathcal{S}, \mathcal{E}, \mathcal{M}, f)$, where f is a mapping from $\mathcal{S} \times \mathcal{E}$ to $\mathcal{M}$:

$$f : \mathcal{S} \times \mathcal{E} \longrightarrow \mathcal{M},$$

such that $f(s, e) = m = f(s', e)$ implies $s = s'$. Sometimes we use $(\mathcal{S}, \mathcal{E}, \mathcal{M})$ to denote an A-code, without specifying the mapping f.

In the definition above, an important property is that f satisfies the condition that $f(s, e) = m = f(s', e)$ implies $s = s'$. It follows that, for each $e \in \mathcal{E}$, $f(\cdot, e)$ induces an injective mapping from $\mathcal{S}$ to $\mathcal{M}$. In other words, two different source states cannot be mapped into the same message for a given key. In general, the mapping f can be a probabilistic mapping, i.e., $f(s, e)$ may take on one of several possible values determined by some probability distribution. This is called a **splitting**. Here, we are only interested in non-splitting A-code, i.e., where f is deterministic.

Given an A-code $(\mathcal{S}, \mathcal{E}, \mathcal{M}, f)$, in order to authenticate a source state, the transmitter and receiver follow the following protocol. First, they agree on a key $e \in \mathcal{E}$, which is selected according to the probability distribution on $\mathcal{E}$. At a later time, if the transmitter wants to communicate a source state $s \in \mathcal{S}$ to the receiver over an insecure channel, he computes $m = f(s, e)$ and sends m to the receiver. When the receiver receives the message m, he checks whether a source s such that $f(s, e) = m$ exists. If such an s exists, the message m is accepted as authentic (m is called *valid*). Otherwise, m is not authentic and is thus rejected.

We will study two different types of attacks that the opponent may carry out:

- **Impersonation attack:** the opponent introduces a message m into the channel, hoping to have it accepted as authentic by the receiver.

- **Substitution attack:** the opponent observes a message m in the channel, and then changes it to m', hoping for m' to be valid. We require that the messages m and m' correspond to different source states.

Associated with each of these attacks is a **deception probability**, which represents the probability that the opponent will successfully deceive the receiver. We assume that the opponent chooses the message that maximizes his chance of success. These probabilities are denoted by P_I for the impersonation attack and P_S for the substitution attack, respectively. They are formally defined as follows:

$$P_I = \max_{m \in \mathcal{M}} \text{prob}(m \text{ is valid})$$

and

$$P_S = \max_{m' \in \mathcal{M}, m' \neq m} \text{prob}(m' \text{ is valid} \mid m \text{ is valid}).$$

In order to compute P_I and P_S, we need to specify the probability distributions on $\mathcal{S}$ and $\mathcal{E}$. This induces a probability distribution on $\mathcal{M}$. We adopt

Kerckhoff's principle: *Everything in the system, except for the actual key, is public.*

In other words, we assume that the authentication code and the probability distributions on $\mathcal{S}$ and $\mathcal{E}$ are known to the opponent. The only information unknown to the opponent is the value of the key e.

We denote by $\mathcal{E}(m)$ the set of keys for which the message m is valid, i.e.,

$$\mathcal{E}(m) = \{e \in \mathcal{E} : \exists s \in \mathcal{S}, f(s, e) = m\}.$$

If we assume that the probability distributions on $\mathcal{S}$ and $\mathcal{E}$ are uniform, the deception probabilities can be expressed as

$$P_I = \max_{m \in \mathcal{M}} \frac{|\mathcal{E}(m)|}{|\mathcal{E}|}$$

and

$$P_S = \max_{m' \in \mathcal{M}, m' \neq m} \frac{|\mathcal{E}(m) \cap \mathcal{E}(m')|}{|\mathcal{E}(m)|}.$$

Since the opponent can choose between the two attacks, we define the **overall deception probability** of an A-code, denoted by P_D, as

$$P_D = \max\{P_I, P_S\}.$$

An A-code without secrecy is an A-code where the source state (i.e., plaintext) is concatenated with a tag to obtain the message which is sent through the channel.

Definition 5.1.2 An **A-code without secrecy** (or **systematic A-code**) is a quadruple $(\mathcal{S}, \mathcal{E}, \mathcal{T}, f)$, where f is a mapping from $\mathcal{S} \times \mathcal{E}$ to $\mathcal{T}$, where $\mathcal{S}$ is the set of source states, $\mathcal{E}$ is the set of keys, and $\mathcal{T}$ is the set of **authenticators** (or **tags**). Sometimes we use $(\mathcal{S}, \mathcal{E}, \mathcal{T})$ to denote an A-code without secrecy, without specifying the mapping f.

Given an A-code without secrecy $(\mathcal{S}, \mathcal{E}, \mathcal{T}, f)$, the transmitter and receiver follow the following protocol to achieve authentication. When the transmitter wants to send the information $s \in \mathcal{S}$ using a key $e \in \mathcal{E}$, which is secretly shared with the receiver, he transmits the message $m = (s, t)$, where $s \in \mathcal{S}$ and $t = f(s, e) \in \mathcal{T}$. (In terms of the notation in Definition 5.1.1, this means $\mathcal{M} = \mathcal{S} \times \mathcal{T}$.) When the receiver receives a message $m = (s, t)$, she checks the authenticity by verifying whether $t = f(s, e)$ or not, using the secret key $e \in \mathcal{E}$. If the equality holds, the message m is said to be *valid*.

Thus, for an A-code without secrecy $(\mathcal{S}, \mathcal{E}, \mathcal{T}, f)$ with uniformly distributed keys and source states, the deception probabilities can be expressed as

$$P_I = \max_{s,t} \frac{|\{e \in \mathcal{E} : t = f(s, e)\}|}{|\mathcal{E}|}$$

and

$$P_S = \max_{s,t} \max_{s' \neq s, t'} \frac{|\{e \in \mathcal{E} : t = f(s, e), t' = f(s', e)\}|}{|\{e \in \mathcal{E} : t = f(s, e)\}|}.$$

In authentication theory, we sometimes consider multiple transmissions, i.e., a key is used to authenticate multiple messages. An attack is said to be **spoofing of order** r if the opponent has seen r communicated messages and tries to construct a fraudulent message, under a single key. The opponent's chance of success in this case is denoted by P_r. In particular, $P_0 = P_I$ and $P_1 = P_S$.

Example 5.1.3 Let $\mathcal{S} = \{s = (s_1, \ldots, s_k) : s_i \in \mathbb{F}_q\}$. For each source state $s = (s_1, \ldots, s_k) \in \mathcal{S}$, we define a polynomial

$$s(x) = s_1 x + s_2 x^2 + \cdots + s_k x^k.$$

Let $\mathcal{E} = \{e = (a, b) : a, b \in \mathbb{F}_q\}$ and $\mathcal{T} = \mathbb{F}_q$. We define the mapping $f : \mathcal{S} \times \mathcal{E} \to \mathcal{T}$ by

$$f(s, e) = f((s_1, s_2, \ldots, s_k), (a, b)) = a + s(b),$$

where $s(b)$ is the value of $s(x)$ evaluated at $x = b$. Then $(\mathcal{S}, \mathcal{E}, \mathcal{T}, f)$ is an A-code without secrecy with

$$P_I = \frac{1}{q} \quad \text{and} \quad P_S = \frac{k}{q}.$$

Indeed, first we have $|\mathcal{E}| = q^2$ and

$$P_I = \max_{s,t} \frac{|\{e \in \mathcal{E} : t = f(s,e)\}|}{|\mathcal{E}|} = \max_{s,t} \frac{|\{(a,b) : s(b) + a = t\}|}{q^2}.$$

For a given value of b, a is uniquely determined by $a = t - s(b)$ for any value of (s,t) and thus $P_I = q/q^2 = 1/q$.

For the substitution attack, we have

$$P_S = \max_{s,t} \max_{s' \neq s, t'} \frac{|\{e \in \mathcal{E} : t = f(s,e), t' = f(s',e)\}|}{|\{e \in \mathcal{E} : t = f(s,e)\}|}$$

$$= \max_{s,t} \max_{s' \neq s, t'} \frac{|\{(a,b) : s(b) + a = t, s'(b) + a = t'\}|}{|\{(a,b) : s(b) + a = t\}|}$$

$$= \max_{s,t} \max_{s' \neq s, t'} \frac{|\{(a,b) : s(b) + a = t, (s - s')(b) + (t' - t) = 0\}|}{q}.$$

Now a is uniquely determined by $a = t - s(b)$ and since $(s - s')(x) + (t' - t)$ is a nonzero polynomial of degree at most k, it has at most k zeros. Thus, for any (s,t) and (s',t'), we have

$$|\{(a,b) : s(b) + a = t, (s - s')(b) + (t' - t) = 0\}| \leq k,$$

therefore $P_S = k/q$.

5.2 Bounds for A-Codes

In this section, we introduce some bounds for the security and efficiency of A-codes.

5.2.1 Information-Theoretic Bounds for A-Codes

We begin with a review of some fundamental bounds on A-codes which are obtained by using information theory. The security and efficiency of an A-code $(\mathcal{S}, \mathcal{E}, \mathcal{M}, f)$ (or $(\mathcal{S}, \mathcal{E}, \mathcal{T}, f)$ for A-code without secrecy) can be measured by a number of parameters: the deception probabilities P_I and P_S, the size of the key space $\mathcal{E}$, the size of the message space $\mathcal{M}$ (or the authentication tag space $\mathcal{T}$). The goal of authentication theory is to examine the relationships among these parameters and give constructions that, for the given source space and deception probabilities, have the shortest possible length for the key and the transmitted message.

We assume that the reader is familiar with the basic properties of entropy.

A brief review has been given in Subsection 4.4.1. For any set $\mathcal{Y}$, we let $\mathbf{Y}$ denote a random variable defined on $\mathcal{Y}$. We state without proof Simmons's information-theoretic bounds.

Theorem 5.2.1 (Simmons's bound) ([146, 30]) *For any A-code $(\mathcal{S}, \mathcal{E}, \mathcal{M}, f)$, we have*

(i) $P_I \geq 2^{-I(\mathbf{M};\mathbf{E})}$;

(ii) $P_S \geq 2^{-I(\mathbf{M}';\mathbf{E}|\mathbf{M})}$,

where $\mathbf{E}$ is the random variable defined on $\mathcal{E}$, and $\mathbf{M}'$ and $\mathbf{M}$ are random variables on $\mathcal{M}$ such that $m' \neq m$, with $m, m' \in \mathcal{M}$.

From Theorem 5.2.1, we obtain the following corollary.

Corollary 5.2.2 *In an A-code $(\mathcal{S}, \mathcal{E}, \mathcal{M}, f)$, we have*

$$P_S \geq 2^{-H(\mathbf{E}|\mathbf{M})}.$$

These bounds show how authentication codes provide protection. For the impersonation attack, we see that P_I is lower bounded by the mutual information between the transmitted message and the key. This means that, in order to have good protection against the impersonation attack, i.e., for P_I to be small, we must give away a lot of information about the key. On the other hand, from Corollary 5.2.2, we know that, in the substitution attack, P_S is lower bounded by the uncertainty about the key when a message has been observed. Thus, we cannot waste all the key entropy for protection against the impersonation attack, but some uncertainty about the key must remain for protection against the substitution attack.

A general form of Simmons's bounds for protection against spoofing of order r, proved independently by Rosenbaum [126] and Pei [121], is

$$P_r \geq 2^{-I(\mathbf{E};\mathbf{M}'|\mathbf{M}^r)},$$

where $I(\mathbf{E}; \mathbf{M}' \mid \mathbf{M}^r)$ is the conditional mutual information between the key and the message, given a string of r transmitted messages.

5.2.2 Combinatorial Bounds for A-Codes

In our model of A-codes, each source state s is mapped to a message m. We see that, among all the messages in $\mathcal{M}$, at least $|\mathcal{S}|$ messages must be authentic, since every source state is mapped to a different message in $\mathcal{M}$. Similarly, for the substitution attack, after observation of one valid message, at least $|\mathcal{S}| - 1$ of the remaining $|\mathcal{M}| - 1$ messages must be authentic. Thus, we have the following theorem.

Theorem 5.2.3 ([81]) *For any A-code* $(\mathcal{S}, \mathcal{E}, \mathcal{M})$, *we have*

$$P_I \geq \frac{|\mathcal{S}|}{|\mathcal{M}|} \quad and \quad P_S \geq \frac{|\mathcal{S}| - 1}{|\mathcal{M}| - 1}.$$

The above theorem shows that, in order to have good protection, i.e., for $P_D = \max\{P_I, P_S\}$ to be small, $|\mathcal{M}|$ must be chosen to be much larger than $|\mathcal{S}|$. For a fixed source space, an increase in the authentication protection implies an increase in the size of the message space.

By multiplying the bound in Theorem 5.2.1(i) and that in Corollary 5.2.2 together, we have

$$P_D^2 \geq P_I P_S \geq 2^{-I(\mathbf{M};\mathbf{E}) - H(\mathbf{E}|\mathbf{M})} = 2^{-H(\mathbf{E})}.$$

Since $H(\mathbf{E}) \leq \log|\mathcal{E}|$, we obtain the following famous **square root bound**.

Theorem 5.2.4 ([69]) *For any A-code* $(\mathcal{S}, \mathcal{E}, \mathcal{M})$, *we have*

$$P_D \geq \frac{1}{\sqrt{|\mathcal{E}|}}.$$

Moreover, the above bound can be tight only if $|\mathcal{S}| \leq \sqrt{|\mathcal{E}|} + 1$.

The square root bound gives a direct relationship between the size of the key space and the protection that we can expect to obtain. The following theorem follows directly.

Theorem 5.2.5 *In an A-code* $(\mathcal{S}, \mathcal{E}, \mathcal{M})$, *assuming that* $P_D = 1/q$, *then*

(i) $|\mathcal{E}| \geq q^2$;

(ii) $|\mathcal{M}| \geq q|\mathcal{S}|$.

We call an A-code **optimal** if the bounds (i) and (ii) of Theorem 5.2.5 are met with equality.

The following combinatorial bounds are the extension of Theorem 5.2.3 for A-codes to protect against spoofing of order r.

Theorem 5.2.6 ([162]) *We have*

(i) *in an A-code with secrecy* $(\mathcal{S}, \mathcal{E}, \mathcal{M})$,

$$P_r \geq \frac{|\mathcal{S}| - r}{|\mathcal{M}| - r}, \quad r = 0, 1, \ldots;$$

(ii) *in an A-code without secrecy* $(\mathcal{S}, \mathcal{E}, \mathcal{T})$,

$$P_r \geq \frac{1}{|\mathcal{T}|}, \quad r = 0, 1, \ldots.$$

An A-code that satisfies (i) or (ii) in Theorem 5.2.6 with equality, i.e., with $P_r = \frac{|\mathcal{S}|-r}{|\mathcal{M}|-r}$ for an A-code with secrecy and $P_r = \frac{1}{|\mathcal{T}|}$ for an A-code without secrecy, is said to provide **perfect protection for spoofing of order** r. The opponent's best strategy in spoofing of order r for such an A-code is to randomly select one of the remaining valid messages (codewords).

Next, we give some characterizations of A-codes from certain combinatorial objects. Given an A-code $(\mathcal{S}, \mathcal{E}, \mathcal{M}, f)$, we can associate an $|\mathcal{E}| \times |\mathcal{M}|$ matrix A, called the **incidence matrix**, where $A = (A_{e,m})_{e \in \mathcal{E}, m \in \mathcal{M}}$ is a binary matrix whose rows are labeled by the keys and columns by the messages, such that $A_{e,m} = 1$ if m is a valid message under e, and $A_{e,m} = 0$ otherwise.

An **authentication matrix** B of an A-code without secrecy $(\mathcal{S}, \mathcal{E}, \mathcal{T}, f)$ is a matrix of size $|\mathcal{E}| \times |\mathcal{S}|$ whose rows are labeled by the keys, columns by the source states, and $B_{e,s} = t$ if t is the authentication tag for the source state s under the key e, i.e., $B_{e,s} = t = f(s, e)$.

A-codes that provide perfect protection for all orders of spoofing up to r are said to be r-**fold secure**. These codes can be characterized by using combinatorial structures such as orthogonal arrays and t-designs.

Definition 5.2.7 An **orthogonal array** $\mathsf{OA}_\lambda(t, k, v)$ is an array with λv^t rows, each row of size k, with entries from a set X of v symbols, such that, in any t columns of the array, every t-tuple of elements of X occurs in exactly λ rows.

Example 5.2.8 The following table gives an $\mathsf{OA}_2(2, 5, 2)$ on the set $\{0, 1\}$:

$$
\begin{array}{ccccc}
0 & 0 & 0 & 0 & 0 \\
1 & 1 & 0 & 0 & 0 \\
0 & 0 & 0 & 1 & 1 \\
1 & 1 & 0 & 1 & 1 \\
1 & 0 & 1 & 0 & 1 \\
0 & 1 & 1 & 0 & 1 \\
1 & 0 & 1 & 1 & 0 \\
0 & 1 & 1 & 1 & 0.
\end{array}
$$

Definition 5.2.9 A **set system** is a pair $(X, \mathcal{B})$, where X is a set of elements called **points** and $\mathcal{B}$ is a collection of subsets of X, the members of which are called **blocks**. A set system can be described by an **incidence matrix**. Let $(X, \mathcal{B})$ be a set system, where $X = \{x_1, \ldots, x_N\}$ and $\mathcal{B} = \{B_1, \ldots, B_M\}$. The incidence matrix of $(X, \mathcal{B})$ is the $M \times N$ matrix $A = (a_{ij})$, where

$$
a_{ij} = \begin{cases} 1 & \text{if } x_j \in B_i \\ 0 & \text{if } x_j \notin B_i. \end{cases}
$$

Conversely, given an incidence matrix, we can define an associated set system in an obvious way.

Definition 5.2.10 A subset S of a set X is called a t-**subset** of X if S has cardinality t. A t-(v, k, λ) **design** is a set system $(X, \mathcal{B})$, where $|X| = v$ and $|B| = k$ for every $B \in \mathcal{B}$, and every t-subset of X occurs in exactly λ blocks in $\mathcal{B}$.

The incidence matrix of a t-(v, k, λ) design is the incidence matrix of the underlying set system $(X, \mathcal{B})$.

Example 5.2.11 The following table gives a 3-$(8, 4, 1)$ design on the set $\{0, 1, 2, 3, 4, 5, 6, 7\}$:

$$
\begin{array}{cccc}
7 & 0 & 1 & 3 \\
7 & 1 & 2 & 4 \\
7 & 2 & 3 & 5 \\
7 & 3 & 4 & 6 \\
7 & 4 & 5 & 0 \\
7 & 5 & 6 & 1 \\
7 & 6 & 0 & 2 \\
2 & 4 & 5 & 6 \\
3 & 5 & 6 & 0 \\
4 & 6 & 0 & 1 \\
5 & 0 & 1 & 2 \\
6 & 1 & 2 & 3 \\
0 & 2 & 3 & 4 \\
1 & 3 & 4 & 5 \\
\end{array}
$$

with incidence matrix

$$
\begin{pmatrix}
1 & 1 & 0 & 1 & 0 & 0 & 0 & 1 \\
0 & 1 & 1 & 0 & 1 & 0 & 0 & 1 \\
0 & 0 & 1 & 1 & 0 & 1 & 0 & 1 \\
0 & 0 & 0 & 1 & 1 & 0 & 1 & 1 \\
1 & 0 & 0 & 0 & 1 & 1 & 0 & 1 \\
0 & 1 & 0 & 0 & 0 & 1 & 1 & 1 \\
1 & 0 & 1 & 0 & 0 & 0 & 1 & 1 \\
0 & 0 & 1 & 0 & 1 & 1 & 1 & 0 \\
1 & 0 & 0 & 1 & 0 & 1 & 1 & 0 \\
1 & 1 & 0 & 0 & 1 & 0 & 1 & 0 \\
1 & 1 & 1 & 0 & 0 & 1 & 0 & 0 \\
0 & 1 & 1 & 1 & 0 & 0 & 1 & 0 \\
1 & 0 & 1 & 1 & 1 & 0 & 0 & 0 \\
0 & 1 & 0 & 1 & 1 & 1 & 0 & 0 \\
\end{pmatrix}.
$$

Theorem 5.2.12 ([152]) *Suppose we have an A-code without secrecy* $(\mathcal{S}, \mathcal{E}, \mathcal{T})$ *in which* $P_I = P_S = 1/|\mathcal{T}|$. *Then* $|\mathcal{E}| \geq |\mathcal{S}|(|\mathcal{T}| - 1) + 1$ *and equality occurs if and only if there exists an orthogonal array* $\mathrm{OA}_\lambda(2, |\mathcal{S}|, |\mathcal{T}|)$, *where* $\lambda = (|\mathcal{S}|(|\mathcal{T}| - 1) + 1)/|\mathcal{T}|^2$.

We say a source space is **r-fold uniform** if every string of r distinct source states has probability

$$\frac{1}{|\mathcal{S}|(|\mathcal{S}|-1)\cdots(|\mathcal{S}|-r+1)}.$$

The following result, due to Tombak and Safavi-Naini [162], relates A-codes with r-fold security to combinatorial structures.

Theorem 5.2.13 ([162]) *Let $(\mathcal{S}, \mathcal{E}, \mathcal{M})$ be an A-code with an r-fold uniform source. Then $(\mathcal{S}, \mathcal{E}, \mathcal{M})$ provides r-fold security against spoofing if and only if its incidence matrix is the incidence matrix of an $(r+1)$-$(|\mathcal{M}|, |\mathcal{S}|, \lambda)$ design.*

Two of the goals of authentication theory are to derive bounds for various parameters in A-codes and to construct A-codes with desired properties. For a review of different bounds and constructions for A-codes, the reader may refer to [81, 152, 85].

5.3　A-Codes and Error-Correcting Codes

A-codes are closely related to error-correcting codes (see [146] for more detailed explanation). In this section, we show their connections, based on the approach of Johansson, Kabatianskii and Smeets [83].

5.3.1　From A-Codes to Error-Correcting Codes

We begin with a construction of error-correcting codes from A-codes. To this end, we introduce the notion of an I-equitable A-code (see [81]).

Definition 5.3.1 An A-code without secrecy $(\mathcal{S}, \mathcal{E}, \mathcal{T}, f)$ is called I-**equitable** if it has the additional property:

$$\forall s \in \mathcal{S}, t \in \mathcal{T}, \quad P_I = \frac{|\{e : f(s,e) = t\}|}{|\mathcal{E}|}.$$

In other words, an I-equitable A-code without secrecy $(\mathcal{S}, \mathcal{E}, \mathcal{T}, f)$ has the property that, for any given $s \in \mathcal{S}$ and $t \in \mathcal{T}$, we have

$$|\{e \in \mathcal{E} : f(s,e) = t\}| = |\mathcal{E}|P_I. \tag{5.1}$$

It follows that, in an I-equitable A-code, $P_I = 1/|\mathcal{T}|$. These codes achieve the best protection against the impersonation attack in the sense that the bound $P_I \geq |\mathcal{S}|/|\mathcal{M}| = 1/|\mathcal{T}|$ in Theorem 5.2.3 is met with equality.

Theorem 5.3.2 *Let q be a prime power. If there exists an I-equitable A-code without secrecy $(\mathcal{S}, \mathcal{E}, \mathcal{T})$ with $P_I = 1/q$, then there exists a q-ary (n, M, d)-error-correcting code, where $n = |\mathcal{E}|$, $M = q(q-1)|\mathcal{S}| + q$, and $d = |\mathcal{E}|(1 - P_S)$.*

Proof. Let $n = |\mathcal{E}|$ and we set $\mathcal{E} = \{e_1, \ldots, e_n\}$. Let $\mathcal{T} = \{t_1, t_2, \ldots, t_q\}$. For each $s \in \mathcal{S}$, we define

$$\mathbf{c}[s] = (f(s, e_1), f(s, e_2), \ldots, f(s, e_n)) \in \mathcal{T}^n.$$

We then consider the set of codewords

$$C = \{\mathbf{c}[s] : s \in \mathcal{S}\} \subseteq \mathcal{T}^n.$$

For each codeword $\mathbf{c}[s]$, we define its composition as

$$\mathsf{comp}\,(\mathbf{c}[s]) = (\alpha_1, \alpha_2, \ldots, \alpha_q), \quad \text{where} \quad \alpha_i = \frac{1}{n}|\{j : \mathbf{c}[s]_j = t_i\}|. \tag{5.2}$$

Since $(\mathcal{S}, \mathcal{E}, \mathcal{T})$ is I-equitable, it follows from (5.1) that, for all $s \in \mathcal{S}$,

$$\mathsf{comp}\,(\mathbf{c}[s]) = (P_I, \ldots, P_I) = \left(\frac{1}{q}, \ldots, \frac{1}{q}\right).$$

On the other hand, we have

$$P_S = \max_{s,t} \max_{s' \neq s, t'} \frac{|\{e \in \mathcal{E} : t = f(s, e), t' = f(s', e)\}|}{|\{e \in \mathcal{E} : t = f(s, e)\}|}.$$

From (5.1), it follows that

$$|\{j : \mathbf{c}[s]_j = t, \mathbf{c}[s']_j = t'\}| \leq P_S P_I \cdot n. \tag{5.3}$$

By letting $t = t'$ and letting t run through $\mathcal{T}$, we obtain

$$d\,(\mathbf{c}[s], \mathbf{c}[s']) \geq n - q P_I P_S \cdot n = n(1 - P_S), \tag{5.4}$$

where $d(\mathbf{x}, \mathbf{y})$ is the Hamming distance between the vectors $\mathbf{x}$ and $\mathbf{y}$.

If we further assume that $\mathcal{T}$ is a finite field, and let $t' = \alpha t + \beta$ for arbitrary $\alpha \neq 0, \beta \in \mathcal{T}$, then (5.4) can be rewritten as

$$d\,(\mathbf{c}[s], \alpha \mathbf{c}[s] + \beta \mathbf{1}) \geq n - q P_I P_S \cdot n = n(1 - P_S). \tag{5.5}$$

This means that, for each codeword $\mathbf{c}[s]$ in the code C, we can form new codewords by using all the affine transformations $\phi : \mathbf{c} \mapsto \alpha \mathbf{c} + \beta \mathbf{1}$, where $0 \neq \alpha, \beta \in \mathcal{T}$. Since $P_S \neq 1$, no two codewords arising from these transformations can be the same and the distance property of (5.4) still holds. Since all codewords have constant composition, we can also add multiples of the codeword $\mathbf{1} = (1, \ldots, 1)$ to the code C, without changing the minimum distance. Thus, we have a new code C' given by

$$C' = \{\alpha \mathbf{c}[s] + \beta \mathbf{1} : \mathbf{c}[s] \in C, \alpha \neq 0, \beta \in \mathcal{T}\} \cup \{\gamma \mathbf{1} : \gamma \in \mathcal{T}\}$$

with the same distance property as C, i.e., the minimum distance d of the code C' satisfies $d \geq n(1 - P_S)$. The number of codewords in C' is $q(q-1)|\mathcal{S}| + q$, proving the desired result. $\qquad \square$

Theorem 5.3.3 *For an I-equitable A-code $(\mathcal{S}, \mathcal{E}, \mathcal{T})$ for which $P_I = P_S = 1/q$, the number of source states is upper bounded as*

$$|\mathcal{S}| \leq \frac{|\mathcal{E}| - 1}{q - 1}.$$

Proof. The code obtained from the A-code has parameters $(n, M, d) = (|\mathcal{E}|, q(q-1)|\mathcal{S}| + q, \theta|\mathcal{E}|)$, where $\theta = 1 - P_S = (q-1)/q$. Recall that $A_q(n, d)$ is the maximum number of codewords in an (n, M, d)-code. By shortening of codes and the Plotkin bound (Theorem 2.3.26), we have

$$A_q(n, \theta n) \leq qA_q(n-1, \theta n) \leq q\frac{\theta n}{\theta n - \theta(n-1)} = qn = q|\mathcal{E}|.$$

It follows immediately that $|\mathcal{S}| \leq \frac{|\mathcal{E}|-1}{q-1}$, proving the desired result. $\qquad \square$

5.3.2   From Linear Error-Correcting Codes to A-Codes

Next, we discuss how to obtain A-codes from error-correcting codes. Let C be a linear (n, M, d)-code over $\mathbb{F}_q$ with the following property

$$\forall \mathbf{c} \in C, \lambda \in \mathbb{F}_q, \quad \text{we have} \quad \mathbf{c} + \lambda\mathbf{1} \in C. \tag{5.6}$$

While obviously not all linear codes satisfy the condition in (5.6), the following lemma provides a way to tweak a given linear code to one that meets this condition, if the given code has a codeword of weight n.

Lemma 5.3.4 *If there exists a linear (n, M, d)-code C with a codeword $\mathbf{c}$ of Hamming weight n, then there exists a linear (n, M, d)-code C' such that $\mathbf{1} \in C'$.*

Proof. Let G be a generator matrix of the code C. We can do some elementary operations on G without changing the minimum distance, including multiplication of columns by nonzero scalars. Thus, if we multiply each column of G with the inverse of the coordinate of $\mathbf{c}$ in that column, we get a new code C' with the same parameters as C such that $\mathbf{1} \in C'$. $\qquad \square$

Clearly, a linear code containing $\mathbf{1}$ satisfies the condition in (5.6). From now on, we assume that C satisfies the condition in (5.6). Let $t, t' \in \mathbb{F}_q$ and $\mathbf{c} = (c_1, \ldots, c_n), \mathbf{c}' = (c'_1, \ldots, c'_n) \in C$.

If $d(\mathbf{c} - \mathbf{c}', (t - t')\mathbf{1}) \neq 0$, then

$$\left|\{j : c_j = t, c'_j = t'\}\right| \leq \left|\{j : c_j - c'_j = t - t'\}\right|$$

$$= n - d(\mathbf{c} - \mathbf{c}', (t - t')\mathbf{1})$$

$$\leq n - d.$$

Note that $d(\mathbf{c} - \mathbf{c}', (t - t')\mathbf{1}) = 0$ for some $t, t' \in \mathbb{F}_q$ if and only if $\mathbf{c} - \mathbf{c}' = \lambda\mathbf{1}$ for some $\lambda \in \mathbb{F}_q$.

We will see below that, if there are two codewords in C whose difference is a multiple of $\mathbf{1}$, it will result in an A-code with $P_S = 1$. We have to factor out these codewords from C. To this end, we define an equivalence relation R on C by

$$\mathbf{a}R\mathbf{b} \text{ if and only if } \mathbf{a} - \mathbf{b} = \lambda\mathbf{1} \text{ for some } \lambda \in \mathbb{F}_q.$$

Now we partition the code C into equivalence classes under the relation R. Then it is easy to see that each class contains exactly q elements.

We denote the set of equivalence classes induced by R on C by

$$\hat{C} = C/R = \{\hat{\mathbf{w}} : \mathbf{w} \in C\},$$

where $\hat{\mathbf{w}}$ denotes the equivalence class containing the codeword $\mathbf{w}$.

Let U be a code consisting of the representatives from the equivalence classes, that is, each class contributes one and only one element. Then it is not hard to see that U is an $(n, M/q, d)$-code.

Next, we construct a new code V as follows. Let $\mathbb{F}_q = \{0, \alpha_1, \ldots, \alpha_{q-1}\}$. Define

$$V = \{(\mathbf{u}, \mathbf{u} + \alpha_1\mathbf{1}, \mathbf{u} + \alpha_2\mathbf{1}, \ldots, \mathbf{u} + \alpha_{q-1}\mathbf{1}) : \mathbf{u} \in U\}. \tag{5.7}$$

Then V is an $(nq, M/q, qd)$-code.

Using an argument similar to that for Theorem 5.3.2 in Subsection 5.3.1, we construct an A-code $(\mathcal{S}, \mathcal{E}, \mathcal{T}, f)$ based on V as follows. Each codeword in V is associated with a source state, i.e.,

$$V = \{\mathbf{v}[s] : s \in \mathcal{S}\},$$

where $\mathbf{v}[s] = (f(s, e_1), f(s, e_2), \ldots, f(s, e_{nq}))$. From (5.7), we know that $\mathsf{comp}(\mathbf{v}) = (1/q, \ldots, 1/q)$ for any $\mathbf{v} \in V$. It follows that $P_I = 1/q$.

We compute the success probability of the substitution attack. Let s and s' be the two source states that maximize P_S. We consider their corresponding codewords $\mathbf{v}[s]$ and $\mathbf{v}[s']$ (and let $\mathbf{v}[s]_j$ and $\mathbf{v}[s']_j$ denote their jth coordinates, respectively). Assume that we have seen s with tag t from $\mathbf{v}[s]$, and we replace it with s' and tag t' from $\mathbf{v}[s']$. Then we have

$$P_S = \frac{\left|\{j : \mathbf{v}[s]_j = t, \mathbf{v}[s']_j = t'\}\right|}{\left|\{j : \mathbf{v}[s]_j = t\}\right|}$$

$$= \frac{\left|\{j : \mathbf{v}[s]_j = t, \mathbf{v}[s]_j - \mathbf{v}[s']_j = t - t'\}\right|}{\left|\{j : \mathbf{v}[s]_j = t\}\right|}$$

$$= \frac{\left|\{j : \mathbf{v}[s]_j - \mathbf{v}[s']_j = t - t'\}\right|}{n}.$$

Then P_S is the maximum value of the composition values in

$$\mathsf{comp}\left(\mathbf{v}[s] - \mathbf{v}[s']\right)$$

for $\mathbf{v}[s], \mathbf{v}[s'] \in V$.

Let α be the element of $\mathbb{F}_q$ with the maximum composition value. Consider $\mathsf{comp}\left(\mathbf{v}[s] - (\mathbf{v}[s'] + \alpha \mathbf{1})\right)$ in the code C. Then the maximum composition value is at the element 0 of $\mathbb{F}_q$ and the maximum composition value is $1 - d/n$. Thus, the maximum value of $\mathsf{comp}\left(\mathbf{v}[s] - \mathbf{v}[s']\right)$ over all pairs in the code V is $1 - d/n$ and therefore $P_S = 1 - d/n$.

We have obtained the following result.

Theorem 5.3.5 ([83]) *Let C be a linear (n, M, d)-code over $\mathbb{F}_q$ such that, if $\mathbf{c} \in C$, then $\mathbf{c} + \lambda \mathbf{1} \in C$ for all $\lambda \in \mathbb{F}_q$. Then there exists an A-code without secrecy $(\mathcal{S}, \mathcal{E}, \mathcal{T})$ with parameters*

$$|\mathcal{S}| = Mq^{-1}, \quad |\mathcal{E}| = nq, \quad |\mathcal{T}| = q, \quad P_I = 1/q, \quad and \quad P_S = 1 - d/n.$$

This construction is called the **q-twisted construction**.

Example 5.3.6 Let $\mathcal{P}_k$ be the set of all polynomials of degree $\leq k$ in $\mathbb{F}_q[x]$ and set $\mathbb{F}_q = \{0, \alpha_1, \ldots, \alpha_{q-1}\}$. We consider

$$C = \{(f(0), f(\alpha_1), \ldots, f(\alpha_{q-1})) : f \in \mathcal{P}_k\}.$$

Then C is a $(q, q^{k+1}, q - k)$-Reed-Solomon code. It is clear that $\mathbf{1} \in C$. Since C is linear, it follows that, if $\mathbf{c} \in C$, then $\mathbf{c} + \lambda \mathbf{1} \in C$ for all $\lambda \in \mathbb{F}_q$, so the condition (5.6) holds. By Theorem 5.3.5, we obtain an A-code $(\mathcal{S}, \mathcal{E}, \mathcal{T})$ with parameters

$$|\mathcal{S}| = q^k, \quad |\mathcal{E}| = q^2, \quad |\mathcal{T}| = q, \quad P_I = 1/q, \quad and \quad P_S = k/q,$$

which results in the same parameters as Example 5.1.3.

Recall that $A_q(n, d)$ denotes the maximum number of codewords in a q-ary code of length n with distance d. Let $A_q^*(n, d)$ denote the maximum number of codewords in a q-ary code C of length n and distance d, with the additional property that $\mathbf{c} \in C$ implies $\mathbf{c} + \lambda \mathbf{1} \in C$ for all $\lambda \in \mathbb{F}_q$. Let $S(|\mathcal{E}|, P_I, P_S)$ denote the maximum number of source states in an A-code with given $|\mathcal{E}|, P_I$, and P_S. From the relationship between error-correcting codes and A-codes we have just seen, it is interesting to further explore the relationship among these three values $A_q(n, d)$, $A_q^*(n, d)$, and $S(|\mathcal{E}|, P_I, P_S)$. From Theorem 5.3.5, we immediately have the following relationship between $A_q^*(n, d)$ and $S(|\mathcal{E}|, P_I, P_S)$:

$$S(n, 1/q, P_S) \geq A_q^* \left(n/q, (1 - P_S)n/q \right)/q.$$

Similar to the Gilbert-Varshamov bound (Theorem 2.3.10), we have the following result.

Lemma 5.3.7 *The quantity $A_q^*(n, d)$ satisfies*

$$A_q^*(n, d) \geq \frac{q^n}{V_q^n(d - 1)},$$

where $V_q^n(d - 1) = \sum_{i=0}^{d-1} \binom{n}{i}(q - 1)^i$ is the size of a usual Hamming sphere of radius $d - 1$ in $\mathbb{F}_q^n$ around a codeword.

Proof. Consider a code C with cardinality $A_q^*(n, d)$ satisfying the condition in (5.6). If $A_q^*(n, d)V_q^n(d-1) < q^n$, there exists a vector $\mathbf{x}$ of length n which does not lie in any of the spheres of radius $d - 1$ centered at a codeword of C. This implies that the vector $\mathbf{x} + \lambda\mathbf{1}$, for any $\lambda \in \mathbb{F}_q$, does not lie in any of these spheres either. Indeed, if $\mathbf{x} + \lambda\mathbf{1}$ is in the sphere around $\mathbf{c}$, then $d(\mathbf{x} + \lambda\mathbf{1}, \mathbf{c}) < d$ and so $d(\mathbf{x}, \mathbf{c} - \lambda\mathbf{1}) < d$. Therefore, $\mathbf{x}$ is in the sphere around $\mathbf{c} - \lambda\mathbf{1}$, which is a contradiction. We can hence add these q vectors $\mathbf{x} + \lambda\mathbf{1}$ ($\lambda \in \mathbb{F}_q$) as codewords and still have a code C' for which $\mathbf{c} \in C'$ implies $\mathbf{c} + \lambda\mathbf{1} \in C'$. This contradicts the maximality of $A_q^*(n, d)$. It follows that $A_q^*(n, d)V_q^n(d - 1) \geq q^n$, and the result follows. $\square$

5.4 Universal Hash Families and A-Codes

Universal hash families were introduced by Carter and Wegman [35], and were further studied by many authors. They have found numerous applications, such as in cryptography, complexity theory, search algorithms, and information retrieval, to mention a few. We refer to [154] for a good account of the development of this topic. In this section, we are interested in the application of universal hash families to authentication codes.

Consider a hash family $\mathcal{H}$, which is a set of N functions $h : A \rightarrow B$, where $|A| = k$ and $|B| = \ell$. Without loss of generality, we assume $k \geq \ell$ and we call $\mathcal{H}$ an $(N; k, \ell)$ **hash family**, while the elements h of $\mathcal{H}$ are called **hash functions** from A to B.

Definition 5.4.1 An $(N; k, \ell)$ hash family is called ϵ-**almost universal** (ϵ-**AU** for short) if, for any two distinct elements $a_1, a_2 \in A$, there are at most ϵN functions $h \in \mathcal{H}$ such that $h(a_1) = h(a_2)$.

Definition 5.4.2 An $(N; k, \ell)$ hash family is called ϵ-**almost strongly universal** (ϵ-**ASU** for short) if

(i) for any $a \in A$ and any $b \in B$, there exist exactly N/ℓ functions $h \in \mathcal{H}$ such that $h(a) = b$; and

(ii) for any two distinct elements $a_1, a_2 \in A$ and for any two (not necessarily distinct) elements $b_1, b_2 \in B$, there exist at most $\epsilon N/\ell$ functions $h \in \mathcal{H}$ such that $h(a_i) = b_i$, for $i = 1, 2$.

We can depict an $(N; k, \ell)$ hash family in the form of an $N \times k$ array H of ℓ symbols, where each row of the array corresponds to one of the functions in the family, the columns are indexed by the elements of A and the entries are the corresponding values of the hash functions at the elements of A. It is then easy to see that

(i) an $(N; k, \ell)$ hash family is ϵ-AU if and only if, for any two columns in H, there exist at most ϵN rows such that the entries in the two given columns are equal;

(ii) an $(N; k, \ell)$ hash family is ϵ-ASU if and only if each element occurs the same number of times in each column and, for any two columns in H, every possible ordered pair of elements occurs at most $\epsilon N/\ell$ times.

Remark 5.4.3 Note that ϵ-ASU implies ϵ-AU. Indeed, suppose $\mathcal{H}$ is an ϵ-ASU $(N; k, \ell)$ hash family, and a_1, a_2 are two elements in A. For each $b \in B$, there exist at most $\epsilon N/\ell$ functions $h \in \mathcal{H}$ such that $h(a_1) = h(a_2) = b$. Since there are ℓ choices for b, there are at most ϵN functions h such that $h(a_1) = h(a_2)$. Therefore $\mathcal{H}$ is an ϵ-AU $(N; k, \ell)$ hash family.

The following theorem, due to Bierbrauer, Johansson, Kabatianskii, and Smeets [14] and Stinson [154], establishes the equivalence between ϵ-AU hash families and error-correcting codes.

Theorem 5.4.4 *If there exists a q-ary (N, M, d)-code, then there exists an ϵ-AU $(N; M, q)$ hash family, where $\epsilon = 1 - d/N$. Conversely, if there exists an ϵ-AU $(N; M, q)$ hash family, then there exists a q-ary $(N, M, N(1 - \epsilon))$-code.*

Proof. Suppose $C = \{\mathbf{c}_1, \ldots, \mathbf{c}_M\}$ is a q-ary (N, M, d)-code. Construct an $N \times M$ array, H, in which the columns are the codewords in C. If we look at any two columns of H, we see that they contain different entries in at least d rows. Set $d = (1 - \epsilon)N$, we obtain an ϵ-AU $(N; M, q)$ hash family with $\epsilon = 1 - d/N$.

Conversely, taking the columns of the array associated with an ϵ-AU

$(N; M, q)$ hash family as codewords of a code, we obtain a q-ary (N, M, d)-code with $d \geq N(1 - \epsilon)$. □

Corollary 5.4.5 *Suppose q is a prime power and $1 \leq t \leq q$. Then there is a $\frac{t-1}{q}$-AU $(q; q^t, q)$ hash family.*

Proof. Recall that a Reed-Solomon code is a linear code with parameters $(q, q^t, q - t + 1)$ over $\mathbb{F}_q$. Applying Theorem 5.4.4, the result follows immediately. □

Example 5.4.6 ([154]) Let $q = 5$ and $k = 3$. We construct a $\frac{2}{5}$-AU $(5; 125, 5)$ hash family. Consider a $[5, 3, 3]$-Reed-Solomon code C over $\mathbb{F}_5$, with generator matrix

$$G = \begin{pmatrix} 1 & 1 & 1 & 1 & 1 \\ 1 & 2 & 4 & 3 & 0 \\ 1 & 4 & 1 & 4 & 0 \end{pmatrix}.$$

There are five functions in $\mathcal{H} = \{f_i : \mathbb{F}_5^3 \to \mathbb{F}_5, i \in \mathbb{F}_5\}$. Note that each codeword in C is obtained by computing $(a, b, c)G$, where $a, b, c \in \mathbb{F}_5$. The value of $f_i(a, b, c)$ is the ith coordinate from the codeword $(a, b, c)G$. We can write $G = (g_{ij})$, where $g_{ij} = 2^{ij} \bmod 5$, for $0 \leq i \leq 2$ and $0 \leq j \leq 3$, and $g_{04} = 1, g_{14} = g_{24} = 0$. Thus, we can easily obtain the following formula for the hash function f_i, for $i \in \mathbb{F}_5$:

$$f_i(a, b, c) = \begin{cases} a + b2^i + c4^i \bmod 5 & \text{if } 0 \leq i \leq 3 \\ a & \text{if } i = 4. \end{cases}$$

Theorem 5.4.7 ([14, 153]) *If there exists an A-code without secrecy $(\mathcal{S}, \mathcal{E}, \mathcal{T})$ with $P_I = 1/|\mathcal{T}|$ and $P_S = \epsilon$, then there exists an ϵ-ASU $(N; k, \ell)$ hash family with $N = |\mathcal{E}|, k = |\mathcal{S}|$ and $\ell = |\mathcal{T}|$. Conversely, if there exists an ϵ-ASU hash family with the above parameters, then there exists an A-code with the same parameters as above.*

Proof. Given an A-code without secrecy $(\mathcal{S}, \mathcal{E}, \mathcal{T}, f)$ with $P_I = 1/|\mathcal{T}|$ and $P_S = \epsilon$, to each key $e \in \mathcal{E}$ we can associate a unique function h_e from $\mathcal{S}$ to $\mathcal{T}$ defined by $h_e(s) = f(s, e)$. It is straightforward to verify that $\mathcal{H} = \{h_e : e \in \mathcal{E}\}$ is an ϵ-ASU hash family from $\mathcal{S}$ to $\mathcal{T}$.

Conversely, given an ϵ-ASU $(N; k, \ell)$ hash family $\mathcal{H}$ from A to B, we can associate an A-code without secrecy $(\mathcal{S}, \mathcal{E}, \mathcal{T})$, where $\mathcal{S} = A$, $\mathcal{T} = B$ and $|\mathcal{E}| = |\mathcal{H}|$, and each key $e \in \mathcal{E}$ corresponds to a unique hash function $h_e \in \mathcal{H}$ indexed by e. The (authentication) mapping $f : \mathcal{S} \times \mathcal{E} \to \mathcal{T}$ is defined by $f(s, e) = h_e(s)$. It is then easy to verify that the resulting A-code has $P_I = 1/\ell$ and $P_S = \epsilon$. □

Much research effort on A-codes has been devoted to constructions which ensure that the opponent's deception probabilities are bounded by $1/\ell$. In terms of ϵ-ASU hash families, this means $\epsilon = 1/\ell$. Such codes were shown to be equivalent to an orthogonal array, and from Theorem 5.2.12, we know that $|\mathcal{E}| \geq k(\ell - 1) + 1$. This means that, for a fixed security (i.e., $1/\ell$), the key size increases linearly as a function of the size of possible source states – a situation similar to the "one-time pad." Thus, for source states of large sizes, many bits of keys are required to store and "secretly" exchange.

The significance of ϵ-ASU hash families in the constructions of A-codes is that by not requiring the deception probability to be the theoretical minimum, i.e., $\epsilon > 1/\ell$, we can expect to reduce the key size significantly. As we will see later, by allowing $P_S > P_I = 1/\ell$ (i.e., $\epsilon > 1/\ell$), it is possible for the size of source states to grow exponentially in the key size. This observation is very important from the point of view of practice. We may deal with scenarios where we are satisfied with the deception probability slightly larger than $1/\ell$, but have limitation on the key storage.

5.4.1 ϵ-AU Hash Families

We derive bounds on ϵ-AU families from well-known bounds in coding theory. The following results are due to Sarwate [134] and Stinson [154].

Theorem 5.4.8 *If there exists an ϵ-AU $(N; k, \ell)$ hash family, then*

$$\epsilon \geq \frac{k - \ell}{\ell(k - 1)}.$$

Proof. When $\epsilon \geq 1/\ell$, it follows trivially that $\epsilon \geq \frac{k-\ell}{\ell(k-1)}$. We assume, henceforth, that $\epsilon < 1/\ell$. By the Plotkin bound in coding theory (cf. Theorem 2.3.26), for any q-ary (N, M, d)-code satisfying $(1 - \frac{1}{q})N < d$, we have

$$\frac{d}{N} \leq \frac{M(q - 1)}{(M - 1)q}.$$

Suppose $\mathcal{H}$ is an ϵ-AU $(N; k, \ell)$ hash family. Using Theorem 5.4.4, we construct an ℓ-ary $(N, k, N(1-\epsilon))$-code, which must satisfy the Plotkin bound. We then obtain

$$\frac{N(1 - \epsilon)}{N} \leq \frac{k(\ell - 1)}{(k - 1)\ell},$$

which implies the desired result. □

The following bound was proved by Stinson [153] using a nice variance technique similar to the proof of Fisher's inequality for balanced incomplete block designs.

Theorem 5.4.9 *If there exists an ϵ-AU $(N; k, \ell)$ hash family, then*

$$N \geq \frac{k(\ell - 1)}{k(\epsilon\ell - 1) + \ell^2(1 - \epsilon)}.$$

Proof. Suppose $\mathcal{H}$ is an ϵ-AU $(N; k, \ell)$ hash family from A to B. For every hash function $h \in \mathcal{H}$ and for every $y \in B$, define a block $Q_{hy} = h^{-1}(y)$ (it is possible that some blocks are empty). The set system $(A, \{Q_{hy} : h \in \mathcal{H}, y \in B\})$ satisfies the following properties:

1. there are k points and ℓN blocks;

2. the blocks can be partitioned into N parallel classes of ℓ blocks;

3. every pair of points occurs in at most ϵN blocks.

Let $C = Q_{hy}$ be any block, and set $\alpha = |C|$. For any block Q_{gx}, where $g \neq h$, define $\mu_{gx} = |C \cap Q_{gx}|$. Then we have the following relations, where each sum is taken over all $g \in \mathcal{H}, g \neq h$, and over all $x \in B$:

$$\sum 1 = \ell(N - 1)$$

$$\sum \mu_{gx} = \alpha(N - 1)$$

$$\sum \binom{\mu_{gx}}{2} \leq \binom{\alpha}{2}(\epsilon N - 1).$$

It follows that the mean of the values μ_{gx} is $\mu = \alpha/\ell$. If we compute the variance of the μ_{gx}, we obtain the following:

$$0 \leq \sum (\mu_{gx} - \mu)^2$$

$$= \sum \mu_{gx}^2 - 2\mu \sum \mu_{gx} + \mu^2 \ell(N - 1)$$

$$\leq \alpha(\alpha - 1)(\epsilon N - 1) + \alpha(N - 1) - \frac{\alpha^2(N - 1)}{\ell}.$$

This implies that

$$N \geq \frac{\alpha(\ell - 1)}{\ell\epsilon(\alpha - 1) + (\ell - \alpha)}.$$

Since each parallel class contains ℓ blocks that partition the k points, we can choose C such that $\alpha = |C| \geq k/\ell$. Since the right-hand side of the above inequality is an increasing function of α, the desired result follows. $\square$

The existence of an ϵ-AU hash family can be harnessed to obtain another new ϵ-AU hash family (possibly with a different ϵ). We discuss two such constructions below.

Theorem 5.4.10 *If there exists an ϵ-AU hash family $\mathcal{H}$ from A to B, then there exists an ϵ-AU hash family $\mathcal{H}_i$ from A^i to B^i with $|\mathcal{H}| = |\mathcal{H}_i|$, for any integer $i \geq 1$.*

Proof. For every $h \in \mathcal{H}$, we define a hash function $h_i : A^i \to B^i$ by

$$h_i(a_1, \ldots, a_i) = (h(a_1), \ldots, h(a_i)).$$

Then $\mathcal{H}_i = \{h_i : h \in \mathcal{H}\}$ is an ϵ-AU hash family as desired. $\qquad\square$

Theorem 5.4.11 *Suppose there exist an ϵ_1-AU $(N_1; k, \ell_1)$ hash family and an ϵ_2-AU $(N_2; \ell_1, \ell_2)$ hash family. Then there exists an $(\epsilon_1 + \epsilon_2)$-AU $(N_1 N_2; k, \ell_2)$ hash family.*

Proof. Suppose $\mathcal{H}_1 = \{h_1 : A_1 \to B_1\}$ and $\mathcal{H}_2 = \{h_2 : A_2 \to B_2\}$ are an ϵ_1-AU $(N_1; k, \ell_1)$ hash family and an ϵ_2-AU $(N_2; \ell_1, \ell_2)$ hash family, respectively, and $A_2 = B_1$. For each $(h_1, h_2) \in \mathcal{H}_1 \times \mathcal{H}_2$, we define a hash function $h : A_1 \to B_2$ by $h(a) = h_2(h_1(a))$. Let $\mathcal{H}$ be the set of all such hash functions. It is easy to see that the probability that $h(a) = h(a')$, for any two inputs $a, a' \in A_1$ and any $h \in \mathcal{H}$, is at most $\epsilon_1 + (1 - \epsilon_1)\epsilon_2 < \epsilon_1 + \epsilon_2$. Therefore, $\mathcal{H}$ is an $(\epsilon_1 + \epsilon_2)$-AU $(N_1 N_2; k, \ell_2)$ hash family. $\qquad\square$

Corollary 5.4.12 *Let q be a prime power and let $i \geq 1$ be an integer. Then there exists an $\frac{i}{q}$-AU $(q^i; q^{2^i}, q)$ hash family.*

Proof. By Corollary 5.4.5, there is a $\frac{1}{q}$-AU $(q; q^2, q)$ hash family. Applying Theorem 5.4.10, we have $\frac{1}{q}$-AU $(q; q^{2^j}, q^{2^{j-1}})$ hash families, for $1 \leq j \leq i$. Applying Theorem 5.4.11 to all these i hash families, we obtain the desired result. $\qquad\square$

5.4.2 ϵ-ASU Hash Families

Next, we move on to discuss bounds for ϵ-ASU hash families as well as constructions of ϵ-ASU hash families from known ϵ-AU hash families and ϵ-ASU hash families.

The following result is due to Stinson [154].

Theorem 5.4.13 *If there exists an ϵ-ASU $(N; k, \ell)$ hash family, then $\epsilon \geq 1/\ell$.*

Proof. Let $a_1, a_2 \in A$ be distinct elements. For any $b_1, b_2 \in B$, let

$$N_{b_1, b_2} = |\{h \in \mathcal{H} : h(a_1) = b_1, h(a_2) = b_2\}|.$$

Then $N_{b_1, b_2} \leq \epsilon N / \ell$ for all $b_1, b_2 \in B$, and

$$\sum_{b_1, b_2 \in B} N_{b_1, b_2} = N.$$

It follows that $\epsilon \geq 1/\ell$. $\qquad\square$

We call an ϵ-ASU $(N; k, \ell)$ **optimal** if equality holds in Theorem 5.4.13, i.e., $\epsilon = 1/\ell$. The following result shows that optimal ASU hash families are equivalent to orthogonal arrays.

Theorem 5.4.14 ([154]) *There exists a $\frac{1}{\ell}$-ASU $(N; k, \ell)$ hash family if and only if there exists an orthogonal array $\mathsf{OA}_\lambda(2, k, \ell)$ in which $\lambda = N/\ell^2$.*

Proof. We write the functions in a $\frac{1}{\ell}$-ASU $(N; k, \ell)$ hash family as an $N \times k$ array H of ℓ symbols, where each row of the array corresponds to one of the functions in the family. It is easy to see that H is an $\mathsf{OA}_{N/\ell^2}(2, k, \ell)$, and similarly for the converse. $\qquad \square$

The following bound was originally proved using combinatorial techniques in [153].

Theorem 5.4.15 *If there exists an ϵ-ASU $(N; k, \ell)$ hash family, then*

$$N \geq 1 + \frac{k(\ell - 1)^2}{\ell\epsilon(k-1) + \ell - k}.$$

Proof. Suppose $\mathcal{H}$ is an ϵ-ASU $(N; k, \ell)$ hash family from A to B. Let c be an element in A and let π be a permutation of B. For each $h \in \mathcal{H}$, we define a hash function $h_{c\pi}$ from A to B via

$$h_{c\pi}(a) = \begin{cases} h(a) & \text{if } a \neq c \\ \pi(h(a)) & \text{otherwise.} \end{cases}$$

It is easy to see that, for each $c \in A$ and each permutation π of B,

$$\mathcal{H}_{c\pi} = \{h_{c\pi} : h \in \mathcal{H}\}$$

is also an ϵ-ASU $(N; k, \ell)$ hash family from A to B.

We can apply the idea of construction for $\mathcal{H}_{c\pi}$ repeatedly. Then, for a given $b \in B$, we can obtain an ϵ-ASU $(N; k, \ell)$ hash family, also denoted by $\mathcal{H}$, in which there is one hash function h_0 such that $h_0(a) = b$ for all $a \in A$.

For every $h \in \mathcal{H}$, define $\mu_h = |\{a : b = h(a)\}|$. Then we have the following relations:

$$\sum_{h \in \mathcal{H}, h \neq h_0} 1 = N - 1$$

$$\sum_{h \in \mathcal{H}, h \neq h_0} \mu_h = k\left(\frac{N}{\ell} - 1\right)$$

$$\sum_{h \in \mathcal{H}, h \neq h_0} \binom{\mu_h}{2} \leq \left(\epsilon\frac{N}{\ell} - 1\right)\binom{k}{2}.$$

The mean of the values μ_h is

$$\mu = \frac{\sum_{h \in \mathcal{H}, h \neq h_0} \mu_h}{\sum_{h \in \mathcal{H}, h \neq h_0} 1} = \frac{k(N - \ell)}{\ell(N - 1)}.$$

We compute the variance of the μ_h, and obtain the following:

$$0 \le \sum_{h \in \mathcal{H}, h \neq h_0} (\mu_h - \mu)^2$$

$$= \sum_{h \in \mathcal{H}, h \neq h_0} \mu_h^2 - 2\mu \sum_{h \in \mathcal{H}, h \neq h_0} \mu_h + \mu^2(N-1)$$

$$\le k(k-1) \left(\epsilon \frac{N}{\ell} - 1\right) - \frac{k^2(\frac{N}{\ell} - 1)^2}{N-1} + k\left(\frac{N}{\ell} - 1\right).$$

By simplifying the above inequality, the desired result follows. □

Example 5.4.16 ([153]) Let q be a prime power, and let s, t be integers such that $s \ge t$. We show that there exists a $\frac{1}{q^t}$-ASU $(q^{s+t}; q^s, q^t)$ hash family.

Let $A = \mathbb{F}_{q^s}$ and $B = \mathbb{F}_{q^t}$. Then A and B are s-dimensional and t-dimensional vector spaces over $\mathbb{F}_q$, respectively. Let $\phi : A \to B$ be any surjective linear transformation. Then we have $|\phi^{-1}(b)| = q^{s-t}$ for every $b \in B$. Note that such a linear transformation always exists. For example, if the elements of A are represented as s-dimensional vectors over $\mathbb{F}_q$, then $\phi(a)$ could be defined as any t coordinates of a since $s \ge t$. Let

$$\mathcal{H} = \{h_{ab} : a \in A, b \in B\},$$

where $h_{ab}(x) = \phi(ax) + b$, for $x \in A$. Suppose $x_1, x_2 \in A$, $x_1 \neq x_2$, and $y_1, y_2 \in B$. We show that the number of hash functions h_{ab} such that $h_{ab}(x_1) = y_1$ and $h_{ab}(x_2) = y_2$ is exactly q^{s-t}. From the definition of $\mathcal{H}$, we obtain the following equation

$$\phi(ax_1) - \phi(ax_2) = y_1 - y_2$$

or equivalently,

$$\phi(a(x_1 - x_2)) = y_1 - y_2.$$

Since $x_1 - x_2$ is nonzero, it is invertible in $\mathbb{F}_{q^s}$. Thus, when a runs through $\mathbb{F}_{q^s}$, so does $a(x_1 - x_2)$. It follows that there are q^{s-t} choices of $a \in \mathbb{F}_{q^s}$ that satisfy the above equation. For each such a, a unique value of b is determined such that $h_{ab}(x_i) = y_i$, for $i = 1, 2$. Therefore, the functions constructed have the desired property.

A very useful method of constructing an ϵ-ASU hash family is to compose an AU hash family and an ASU hash family with appropriate parameters. The following theorem is due to Stinson [153], and independently to Bierbrauer, Johansson, Kabatianskii, and Smeets [14].

Theorem 5.4.17 *Suppose there exist an ϵ_1-AU $(N_1; k, \ell_1)$ hash family and an ϵ_2-ASU $(N_2; \ell_1, \ell_2)$ hash family. Then there exists an $(\epsilon_1 + \epsilon_2)$-ASU $(N_1 N_2; k, \ell_2)$ hash family.*

Proof. Without loss of generality, we assume $\mathcal{H}_1$ is an ϵ_1-AU $(N_1; k, \ell_1)$ hash family from A to B_1 and $\mathcal{H}_2$ is an ϵ_2-ASU $(N_2; \ell_1, \ell_2)$ hash family from B_1 to B_2. Define $\mathcal{H} = \{h_2 h_1 : h_1 \in \mathcal{H}_1, h_2 \in \mathcal{H}_2\}$. We show that $\mathcal{H}$ is an ϵ-ASU $(N_1 N_2; k, \ell_2)$ hash family from A to B_2 with $\epsilon \leq \epsilon_1 + \epsilon_2$.

Let $a_1, a_2 \in A$ $(a_1 \neq a_2)$ and let $b_1, b_2 \in B_2$. We compute an upper bound on the number of functions $h \in \mathcal{H}$ such that $h(a_1) = b_1$ and $h(a_2) = b_2$.

We first consider the case when b_1 and b_2 are the same, say $b_1 = b_2 = b$. Let $\mathcal{L} = \{h_1 \in \mathcal{H}_1 : h_1(a_1) = h_1(a_2)\}$ and let $\alpha = |\mathcal{L}|$. Since $\mathcal{H}_1$ is an ϵ_1-AU hash family, we have $\alpha \leq \epsilon_1 N_1$. For any $h_1 \in \mathcal{L}$, there are exactly N_2/ℓ_2 functions $h_2 \in \mathcal{H}_2$ such that $h_2(h_1(a_1)) = h_2(h_1(a_2)) = b$ (by property (i) of an ϵ-ASU hash family). On the other hand, for any $h_1 \in \mathcal{H}_1 \setminus \mathcal{L}$, there are at most $\epsilon_2 N_2/\ell_2$ functions $h_2 \in \mathcal{H}_2$ such that $h_2(h_1(a_1)) = h_2(h_1(a_2)) = b$ (by property (ii) of an ϵ-ASU hash family). Thus, the number of functions $h \in \mathcal{H}$ such that $h(a_1) = h(a_2) = b$ is at most

$$
\begin{aligned}
\alpha \times \frac{N_2}{\ell_2} + (N_1 - \alpha) \times \frac{\epsilon_2 N_2}{\ell_2} &= \frac{N_2(\alpha + (N_1 - \alpha)\epsilon_2)}{\ell_2} \\
&\leq \frac{N_2(\alpha + N_1 \epsilon_2)}{\ell_2} \\
&\leq \frac{N_2(N_1 \epsilon_1 + N_1 \epsilon_2)}{\ell_2} \\
&= \frac{N_1 N_2(\epsilon_1 + \epsilon_2)}{\ell_2}.
\end{aligned}
$$

If $b_1 \neq b_2$, since any $h_1 \in \mathcal{L}$ must give $h_2(h_1(a_1)) = h_2(h_1(a_2))$, the number of functions $h \in \mathcal{H}$ such that $h(a_1) = b_1$ and $h(a_2) = b_2$ is smaller. Thus we have proved property (ii) of an ϵ-ASU hash family. It is straightforward to see that property (i) of an ϵ-ASU hash family is also satisfied, hence we have proved that $\mathcal{H}$ is an ϵ-ASU hash family with $\epsilon \leq \epsilon_1 + \epsilon_2$. □

Corollary 5.4.18 ([14, 154]) *Suppose r and s are integers. Then there is a $\frac{1}{q^{r-1}}$-ASU $(q^{3r+2s}; q^{(r+s)(q^s(q-1)+1)}, q^r)$ hash family.*

Proof. Applying Corollary 5.4.5, replacing q and t by q^{r+s} and $q^s(q-1)+1$, respectively, we obtain a $\frac{q-1}{q^r}$-AU $(q^{r+s}; q^{(r+s)(q^s(q-1)+1)}, q^{r+s})$ hash family. By Example 5.4.16, we obtain a $\frac{1}{q^r}$-ASU $(q^{2r+s}; q^{r+s}, q^r)$ hash family. Applying Theorem 5.4.17, the desired result follows. □

5.5   A-Codes from Algebraic Curves

In this section, we describe a construction, given in [179], of ϵ-ASU hash families based on algebraic curves over finite fields which, in return, gives a construction of A-codes according to Theorem 5.4.7.

The construction proceeds as follows. Let $\mathcal{X}$ be a smooth projective curve over $\mathbb{F}_q$, $g = g(\mathcal{X})$ the genus of $\mathcal{X}$, $\mathbb{F}_q(\mathcal{X})$ the function field of $\mathcal{X}$, and $\mathcal{X}(\mathbb{F}_q)$ the set of all $\mathbb{F}_q$-rational points on $\mathcal{X}$.

Let M be a subset of $\mathcal{X}(\mathbb{F}_q)$, i.e., M is a set of $\mathbb{F}_q$-rational points of $\mathcal{X}$. Let D be a positive divisor with $M \cap \mathrm{Supp}(D) = \emptyset$. Choose an $\mathbb{F}_q$-rational point R in M and put $G = D - R$. Then $\deg(G) = \deg(D) - 1$, $\mathcal{L}(G) \subset \mathcal{L}(D)$ and $\mathbb{F}_q \cap \mathcal{L}(G) = \{0\}$, where $\mathcal{L}(G)$ and $\mathcal{L}(D)$ are the Riemann-Roch spaces of G and D, respectively. We have

$$\mathcal{L}(D) = \mathbb{F}_q \oplus \mathcal{L}(G) = \{\alpha + f : f \in \mathcal{L}(G), \alpha \in \mathbb{F}_q\}.$$

Each element $(P, \alpha) \in M \times \mathbb{F}_q$ can be associated with a map $h_{(P,\alpha)} : \mathcal{L}(G) \to \mathbb{F}_q$ defined by

$$h_{(P,\alpha)}(f) = f(P) + \alpha.$$

Lemma 5.5.1 *Let* $\mathcal{H} = \{h_{(P,\alpha)} : (P, \alpha) \in M \times \mathbb{F}_q\}$. *If* $\deg(D) \geq 2g + 1$, *then the cardinality of* $\mathcal{H}$ *is equal to* $q|M|$.

Proof. It is sufficient to prove that $\{h_{(P,\alpha)}\}_{(P,\alpha)\in M\times\mathbb{F}_q}$ are pairwise distinct. Assume that $h_{(P,\alpha)} = h_{(Q,\beta)}$ for (P, α) and (Q, β) in $M \times \mathbb{F}_q$, i.e.,

$$h_{(P,\alpha)}(f) = h_{(Q,\beta)}(f) \tag{5.8}$$

for all $f \in \mathcal{L}(G)$. In particular,

$$\alpha = h_{(P,\alpha)}(0) = h_{(Q,\beta)}(0) = \beta. \tag{5.9}$$

It follows from (5.8) and (5.9) that

$$f(P) = f(Q)$$

for all $f \in \mathcal{L}(G)$. This yields that

$$e(P) = e(Q) \tag{5.10}$$

for all $e \in \mathcal{L}(D)$ since $\mathcal{L}(D) = \mathbb{F}_q \oplus \mathcal{L}(G)$.

Suppose that P is different from Q. As $\deg(D - P) > \deg(D - P - Q) \geq 2g - 1$, we obtain, by the Riemann-Roch Theorem (cf. Theorem 1.4.7), that

$$\ell(D - P) = \deg(D) - g \quad \text{and} \quad \ell(D - P - Q) = \deg(D) - g - 1.$$

Moreover, we can choose a function u from the set $\mathcal{L}(D - P) \setminus \mathcal{L}(D - P - Q)$. Then it is clear that $u(P) = 0$ and $u(Q) \neq 0$, which contradicts (5.10). Hence $P = Q$. $\square$

Theorem 5.5.2 *Let $\mathcal{X}$ be a smooth projective curve over $\mathbb{F}_q$ and M a set of $\mathbb{F}_q$-rational points on $\mathcal{X}$. Suppose that D is a positive divisor with $\deg(D) \geq 2g+1$ and $M \cap \mathrm{Supp}(D) = \emptyset$. Then there exists an ϵ-ASU $(N; k, q)$ hash family with*

$$N = q|M|, \quad k = q^{\ell(D)-1} = q^{\deg(D)-g}, \quad \epsilon = \frac{\deg(D)}{|M|}.$$

Proof. Let $R \in M$ be an $\mathbb{F}_q$-rational point on $\mathcal{X}$ and put $G = D - R$. Define

$$A = \mathcal{L}(G), \quad B = \mathbb{F}_q,$$

and

$$\mathcal{H} = \{h_{(P,\alpha)} : (P, \alpha) \in M \times \mathbb{F}_q\}.$$

Let $N = |\mathcal{H}|$ (so $N = q|M|$ by Lemma 5.5.1). It is easy to verify that, for any $a \in A = \mathcal{L}(G)$ and $b \in B = \mathbb{F}_q$, there exist exactly $|M| = N/q$ pairs $(P, \alpha) \in M \times \mathbb{F}_q$ such that

$$h_{(P,\alpha)}(a) = a(P) + \alpha = b,$$

i.e., there exist exactly N/q functions $h_{(P,\alpha)} \in \mathcal{H}$ such that $h_{(P,\alpha)}(a) = b$.

Now let a_1, a_2 be two distinct elements of A and b_1, b_2 two elements of B. We consider

$$m = \max_{\substack{a_1 \neq a_2 \in A \\ b_1, b_2 \in B}} \left| \{h_{(P,\alpha)} \in \mathcal{H} : h_{(P,\alpha)}(a_1) = b_1, h_{(P,\alpha)}(a_2) = b_2\} \right|$$

$$= \max_{\substack{a_1 \neq a_2 \in A \\ b_1, b_2 \in B}} \left| \{(P, \alpha) \in M \times \mathbb{F}_q : a_1(P) + \alpha = b_1, a_2(P) + \alpha = b_2\} \right|$$

$$= \max_{\substack{a_1 \neq a_2 \in A \\ b_1, b_2 \in B}} \left| \{(P, \alpha) \in M \times \mathbb{F}_q : (a_1 - a_2 - b_1 + b_2)(P) = 0, a_2(P) + \alpha = b_2\} \right|.$$

As $a_1 - a_2 \in \mathcal{L}(G) \setminus \{0\}$ and $b_1 - b_2 \in \mathbb{F}_q$, we know that $0 \neq a_1 - a_2 - b_1 + b_2 \in \mathcal{L}(D)$. Thus, there are at most $\deg(D)$ distinct zeros of $a_1 - a_2 - b_1 + b_2$ in M (see proof of Theorem 2.4.3). Since α is uniquely determined by P from the equality $a_2(P) + \alpha = b_2$, we have that at most $\deg(D)$ pairs $(P, \alpha) \in M \times \mathbb{F}_q$ satisfy

$$(a_1 - a_2 - b_1 + b_2)(P) = 0 \quad \text{and} \quad a_2(P) + \alpha = b_2,$$

i.e.,

$$m \leq \deg(D) = \frac{\deg(D)}{|M|} \frac{N}{q}.$$

Hence, we can take $\epsilon = \deg(D)/|M|$. This completes the proof. □

Theorem 5.5.2 gives a construction of ϵ-ASU hash families based on general algebraic curves over finite fields. In the examples below, we apply Theorem 5.5.2 to some special curves to obtain ϵ-ASU families with nice parameters.

Example 5.5.3 Consider the projective line $\mathcal{L}$ over $\mathbb{F}_q$. Then $g = g(\mathcal{L}) = 0$.

(a) Let d be an integer such that $1 \le d \le q$, and P an $\mathbb{F}_q$-rational point of $\mathcal{L}$. Put

$$D = dP \quad \text{and} \quad M = \mathcal{L}(\mathbb{F}_q) \setminus \{P\}.$$

Then $\deg(D) = d \ge 2g+1$, $|M| = q$, and $M \cap \mathrm{Supp}(D) = \emptyset$. By Theorem 5.5.2, we obtain an ϵ-ASU $(N; k, q)$ hash family with

$$N = q^2, \quad k = q^d, \quad \epsilon = \frac{d}{q},$$

which has the same parameters as Example 5.3.6.

(b) Let d be an integer satisfying $2 \le d \le q$. Put $M = \mathcal{L}(\mathbb{F}_q)$. As there always exists an irreducible polynomial of degree d over $\mathbb{F}_q$, we can find a positive divisor D such that $\deg(D) = d$ and $M \cap \mathrm{Supp}(D) = \emptyset$. Then $\deg(D) = d \ge 2g+1$ and $|M| = q+1$. By Theorem 5.5.2, we obtain an ϵ-ASU $(N; k, q)$ hash family with

$$N = q(q+1), \quad k = q^d, \quad \epsilon = \frac{d}{q+1}.$$

Example 5.5.4 Let $q = p^u$ for some prime p. Put

$$N_q = \begin{cases} q + \lfloor 2\sqrt{q} \rfloor & \text{if } p | \lfloor 2\sqrt{q} \rfloor \text{ and } u \ge 3 \text{ odd} \\ q + \lfloor 2\sqrt{q} \rfloor + 1 & \text{otherwise.} \end{cases}$$

It is proved in [141] and [173] that there exists an elliptic curve $\mathcal{E}$ over $\mathbb{F}_q$ with N_q $\mathbb{F}_q$-rational points (cf. Corollary 3.3.12). Let d be an integer such that $3 \le d \le N_q - 1$, and let P be an $\mathbb{F}_q$-rational point of $\mathcal{E}$. Put

$$D = dP \quad \text{and} \quad M = \mathcal{E}(\mathbb{F}_q) \setminus \{P\}.$$

Then $\deg(D) = d \ge 2g+1$, $|M| = N_q - 1$, and $M \cap \mathrm{Supp}(D) = \emptyset$. By Theorem 5.5.2, we obtain an ϵ-ASU $(N; k, q)$ hash family with

$$N = q(N_q - 1), \quad k = q^{d-1}, \quad \epsilon = \frac{d}{N_q - 1}.$$

Example 5.5.5 Let q be a square prime power and let $r = \sqrt{q}$. Consider the Hermitian curve $\mathcal{H}$ over $\mathbb{F}_q$ defined by

$$y^r + y = x^{r+1}.$$

Then the number of $\mathbb{F}_q$-rational points of $\mathcal{H}$ is equal to $r^3 + 1 = q\sqrt{q} + 1$ and the genus of $\mathcal{H}$ is $g = \sqrt{q}(\sqrt{q} - 1)/2$. Choose an $\mathbb{F}_q$-rational point P and put $D = dP$ for an integer d such that $2g + 1 = q - \sqrt{q} + 1 \le d \le r^3 = q\sqrt{q}$. Define M to be the set $\mathcal{H}(\mathbb{F}_q) \setminus \{P\}$. Then $\deg(D) = d \ge 2g+1$, $|M| = q\sqrt{q}$, and $M \cap \mathrm{Supp}(D) = \emptyset$. By Theorem 5.5.2, we obtain an ϵ-ASU $(N; k, q)$ hash family with

$$N = q^2\sqrt{q}, \quad k = q^{d - \sqrt{q}(\sqrt{q}-1)/2}, \quad \epsilon = \frac{d}{q\sqrt{q}}.$$

Example 5.5.6 Consider the curves $\{\mathcal{X}_m\}_{m\geq 1}$ over $\mathbb{F}_{q^2}$ given in Example 2.5.6 by:

$$x_{m+1}^q + x_{m+1} = \frac{x_m^q}{x_m^{q-1} + 1}$$

for $m = 1, 2, \ldots$. Then the number of $\mathbb{F}_{q^2}$-rational points of $\mathcal{X}_m$ is more than $(q^2 - q)q^{m-1}$, and the genus g_m of $\mathcal{X}_m$ is less than q^m, for all $m \geq 1$. Choose an integer c between 2 and $q-1$ (c is independent of m), an $\mathbb{F}_{q^2}$-rational point P_m of $\mathcal{X}_m$ and put $D_m = cq^m P_m$. Let M_m be a subset of $\mathcal{X}_m(\mathbb{F}_{q^2}) \setminus \{P_m\}$ with

$$|M_m| = (q^2 - q)q^{m-1} = q^m(q - 1).$$

By Theorem 5.5.2, we obtain a sequence of ϵ-ASU $(N_m; k_m, q^2)$ hash families with

$$N_m = q^m(q^3 - q^2), \quad k_m = q^{2cq^m - g_m} \geq q^{2(c-1)q^m}, \quad \epsilon = \frac{\deg(D_m)}{|M_m|} = \frac{c}{q-1}.$$

From Theorem 5.4.7, we can translate ϵ-ASU hash families, in particular those from Theorem 5.5.2, to A-codes directly. For example, rephrasing the construction of Example 5.5.6 in terms of A-codes, we have the following result.

Corollary 5.5.7 *Let q be a prime power, and let c and m be integers satisfying $2 \leq c \leq q-1$ and $m \geq 1$. There exists an A-code without secrecy $(\mathcal{S}, \mathcal{E}, \mathcal{T})$ with the following parameters*

$$|\mathcal{S}| = q^{2(c-1)q^m}, \quad |\mathcal{E}| = q^m(q^3 - q^2), \quad |\mathcal{T}| = q^2,$$

and with deception probabilities

$$P_I = \frac{1}{q^2}, \quad P_S = \frac{c}{q - 1}.$$

For example, if we take $q = 2^{20}$, $c = 2$ and $m = 1$, we can obtain an A-code with 40×2^{20} bits of source state, 80 bits of key, and 40 bits of tag, while the probabilities of the impersonation and substitution attacks are bounded by approximately 2^{-40} and 2^{-19}, respectively.

5.6 Linear Authentication Codes

A source state $s \in \mathcal{S}$ in an A-code without secrecy $(\mathcal{S}, \mathcal{E}, \mathcal{T}, f)$ can be uniquely associated with a mapping $\phi_s : \mathcal{E} \to \mathcal{T}$ defined by $\phi_s(e) = f(s, e)$, for all $e \in \mathcal{E}$. Then the A-code $(\mathcal{S}, \mathcal{E}, \mathcal{T}, f)$ can be characterized by the family of mappings $\Phi = \{\phi_s : s \in \mathcal{S}\}$. In a conventional authentication code, the key space $\mathcal{E}$ and the authenticator space $\mathcal{T}$ do not have any algebraic structures. We now consider A-codes in which $\mathcal{E}$ and $\mathcal{T}$ have some additional algebraic structures. In particular, $\mathcal{E}$ and $\mathcal{T}$ are vector spaces over a finite field $\mathbb{F}_q$, and Φ is a family of $\mathbb{F}_q$-linear mappings from $\mathcal{E}$ to $\mathcal{T}$. These codes are called **linear A-codes**. As shown in [171], linear A-codes are useful in constructing distributed authentication schemes.

Definition 5.6.1 An A-code without secrecy $(\mathcal{S}, \mathcal{E}, \mathcal{T}, f)$ is **linear over** $\mathbb{F}_q$ if

(i) $\mathcal{E}$ and $\mathcal{T}$ are finite dimensional vector spaces over $\mathbb{F}_q$; and

(ii) for every $s \in \mathcal{S}$, ϕ_s, defined by $\phi_s(e) = f(s, e)$, is an $\mathbb{F}_q$-linear transformation from $\mathcal{E}$ to $\mathcal{T}$.

We identify $\mathcal{S}$ with $\Phi = \{\phi_s : s \in \mathcal{S}\}$, and write the A-code as $(\Phi, \mathcal{E}, \mathcal{T}, f)$ to emphasize that the source states are represented as linear transformations. We may assume that $\mathcal{E} = \mathbb{F}_q^n$ and $\mathcal{T} = \mathbb{F}_q^m$. Given a basis $\mathbf{e}_1, \mathbf{e}_2, \ldots, \mathbf{e}_n$ of $\mathcal{E}$ and a basis $\mathbf{a}_1, \mathbf{a}_2, \ldots, \mathbf{a}_m$ of $\mathcal{T}$, a linear transformation $\phi \in \Phi$ can be represented by a *unique* $n \times m$ matrix A over $\mathbb{F}_q$ such that $\phi(\mathbf{e}) = \mathbf{e}A$, for all $\mathbf{e} \in \mathcal{E}$. We let $\ker(\phi) = \{\mathbf{e} \in E : \phi(\mathbf{e}) = \mathbf{0}\}$. Then $\ker(\phi)$ is a subspace of $\mathcal{E}$ and its dimension is denoted by $\dim(\ker(\phi))$.

We compute the success probabilities of the impersonation and substitution attacks for a linear A-code. For the impersonation attack, we have

$$
\begin{aligned}
P_I &= \max_{\phi \in \Phi} \max_{\mathbf{a} \in \mathcal{T}} \frac{|\{\mathbf{e} : \phi(\mathbf{e}) = \mathbf{a}\}|}{|\mathcal{E}|} \\
&= \max_{\phi \in \Phi} \frac{|\{\mathbf{e} : \phi(\mathbf{e}) = \mathbf{0}\}|}{|\mathcal{E}|} \\
&= \max_{\phi \in \Phi} q^{\dim(\ker(\phi)) - n} \\
&= q^{\gamma - n},
\end{aligned}
$$

where $\gamma = \max_{\phi \in \Phi}\{\dim(\ker(\phi)) : \phi \in \Phi\}$. Clearly, $\gamma \leq n - m$ and, if equality holds, then P_I achieves the maximal value. In this case, each ϕ is onto, i.e., $\phi(\mathcal{E}) = \mathcal{T}$, for all $\phi \in \Phi$.

For the substitution attack, we have

$$
\begin{aligned}
P_S &= \max_{\substack{\phi, \phi' \in \Phi \\ \phi \neq \phi'}} \max_{\mathbf{a}, \mathbf{a}' \in \mathcal{T}} \frac{|\{\mathbf{e} : \phi(\mathbf{e}) = \mathbf{a}, \phi'(\mathbf{e}) = \mathbf{a}'\}|}{|\{\mathbf{e} : \phi(\mathbf{e}) = \mathbf{a}\}|} \\
&= \max_{\substack{\phi, \phi' \in \Phi \\ \phi \neq \phi'}} \max_{\mathbf{a}, \mathbf{a}' \in \mathcal{T}} \frac{|\{\mathbf{e} : \phi(\mathbf{e}) = \mathbf{a}\} \cap \{\mathbf{e} : \phi'(\mathbf{e}) = \mathbf{a}'\}|}{|\{\mathbf{e} : \phi(\mathbf{e}) = \mathbf{0}\}|}.
\end{aligned}
$$

In order to compute P_S, we need the following lemma.

Lemma 5.6.2 *For any $\phi, \phi' \in \Phi$ and any $\mathbf{a}, \mathbf{a}' \in \mathcal{T}$, we have either*

(i) $|\{\mathbf{e} : \phi(\mathbf{e}) = \mathbf{a}\} \cap \{\mathbf{e} : \phi'(\mathbf{e}) = \mathbf{a}'\}| = 0$; *or*

(ii) $|\{\mathbf{e} : \phi(\mathbf{e}) = \mathbf{a}\} \cap \{\mathbf{e} : \phi'(\mathbf{e}) = \mathbf{a}'\}| = |\{\mathbf{e} : \phi(\mathbf{e}) = \mathbf{0}\} \cap \{\mathbf{e} : \phi'(\mathbf{e}) = \mathbf{0}\}|$.

Proof. Assume that $|\{\mathbf{e} : \phi(\mathbf{e}) = \mathbf{a}\} \cap \{\mathbf{e} : \phi'(\mathbf{e}) = \mathbf{a}'\}| \neq 0$, then there exists $\mathbf{e}_0 \in \{\mathbf{e} : \phi(\mathbf{e}) = \mathbf{a}\} \cap \{\mathbf{e} : \phi'(\mathbf{e}) = \mathbf{a}'\}$. We define a function τ from $\{\mathbf{e} : \phi(\mathbf{e}) = \mathbf{a}\} \cap \{\mathbf{e} : \phi'(\mathbf{e}) = \mathbf{a}'\}$ to $\{\mathbf{e} : \phi(\mathbf{e}) = \mathbf{0}\} \cap \{\mathbf{e} : \phi'(\mathbf{e}) = \mathbf{0}\}$ by $\tau(\mathbf{e}) = \mathbf{e} - \mathbf{e}_0$. It is easy to see that τ is one-to-one, which implies $|\{\mathbf{e} : \phi(\mathbf{e}) = \mathbf{a}\} \cap \{\mathbf{e} : \phi'(\mathbf{e}) = \mathbf{a}'\}| \leq |\{\mathbf{e} : \phi(\mathbf{e}) = \mathbf{0}\} \cap \{\mathbf{e} : \phi'(\mathbf{e}) = \mathbf{0}\}|$. On the other hand, we can define a function π from $\{\mathbf{e} : \phi(\mathbf{e}) = \mathbf{0}\} \cap \{\mathbf{e} : \phi'(\mathbf{e}) = \mathbf{0}\}$ to $\{\mathbf{e} : \phi(\mathbf{e}) = \mathbf{a}\} \cap \{\mathbf{e} : \phi'(\mathbf{e}) = \mathbf{a}'\}$ by $\pi(\mathbf{e}) = \mathbf{e} + \mathbf{e}_0$. Again, π is one-to-one, which implies $|\{\mathbf{e} : \phi(\mathbf{e}) = \mathbf{0}\} \cap \{\mathbf{e} : \phi'(\mathbf{e}) = \mathbf{0}\}| \leq |\{\mathbf{e} : \phi(\mathbf{e}) = \mathbf{a}\} \cap \{\mathbf{e} : \phi'(\mathbf{e}) = \mathbf{a}'\}|$, completing the proof of the lemma. $\qquad\square$

From Lemma 5.6.2, P_S can be rewritten as

$$P_S = \max_{\substack{\phi, \phi' \in \Phi \\ \phi \neq \phi'}} \frac{|\{\mathbf{e} : \phi(\mathbf{e}) = \mathbf{0}\} \cap \{\mathbf{e} : \phi'(\mathbf{e}) = \mathbf{0}\}|}{|\{\mathbf{e} : \phi(\mathbf{e}) = \mathbf{0}\}|}.$$

It follows that both P_I and P_S must be the reciprocals of some powers of q, i.e., $P_I = q^{-t}$ and $P_S = q^{-d}$ for some integers t and d with $t \geq d$, and so the performance of a linear A-code over $\mathbb{F}_q$ can be determined by the parameters $|\Phi|, n, m, t$, and d. For given t and d (which correspond to the security level of the A-code), and n and m (which correspond to the key size and the length of the tag), we would like to have $|\Phi|$ as large as possible. Equivalently, for given t, d, and $|\mathcal{S}|$ (the number of sources), we would like to construct a linear A-code with $|\Phi| = |\mathcal{S}|$ such that n and m are as small as possible.

5.6.1   Interpreting a Linear A-Code as a Family of Subspaces

Given an A-code without secrecy $\mathbf{A} = (\mathcal{S}, \mathcal{E}, \mathcal{T}, f)$, we may, without loss of generality, assume $(\mathcal{T}, +)$ is an abelian group. Let $\mathcal{E}^* = \mathcal{E} \times \mathcal{T}$. We define a new A-code $\mathbf{A}^* = (\mathcal{S}, \mathcal{E}^*, \mathcal{T}, f^*)$ with

$$f^* : \mathcal{S} \times \mathcal{E}^* \to \mathcal{T}$$

defined by $f^*(s, (e, a)) = f(s, e) + a$.

Lemma 5.6.3 *Let* $\mathbf{A} = (\mathcal{S}, \mathcal{E}, \mathcal{T}, f)$ *be an A-code without secrecy. Then* $\mathbf{A}^* = (\mathcal{S}, \mathcal{E}^*, \mathcal{T}, f^*)$ *defined above is an I-equitable A-code without secrecy and* $P_S^* \leq P_S$, *where* P_S^* *and* P_S *are the probabilities of substitution attacks in* $\mathbf{A}^*$ *and* $\mathbf{A}$, *respectively.*

Proof. For any $s \in \mathcal{S}$ and $a \in \mathcal{T}$, we have

$$
\begin{aligned}
P_I^* &= \frac{|\{(e,b) : f^*(s,(e,b)) = a\}|}{|\mathcal{E} \times \mathcal{T}|} \\
&= \frac{|\bigcup_{b \in \mathcal{T}} \{e \in \mathcal{E} : f(s,e) = a - b\}|}{|\mathcal{E}||\mathcal{T}|} \\
&= \frac{|\mathcal{E}|}{|\mathcal{E}||\mathcal{T}|} \\
&= \frac{1}{|\mathcal{T}|}.
\end{aligned}
$$

Hence, $(\mathcal{S}, \mathcal{E}^*, \mathcal{T}, f^*)$ is I-equitable.

Furthermore,

$$
\begin{aligned}
P_S^* &= \max_{\substack{s, s' \in \mathcal{S} \\ s \neq s'}} \max_{a, a' \in \mathcal{T}} \frac{|\{(e,b) : f(s,e) = a - b\} \cap \{(e,b) : f(s',e) = a' - b\}|}{|\{(e,b) : f(s,e) = a - b\}|} \\
&\leq \max_{\substack{s, s' \in \mathcal{S} \\ s \neq s'}} \max_{c, c' \in \mathcal{T}} \frac{|\{e : f(s,e) = c\} \cap \{e : f(s',e) = c'\}|}{|\{e : f(s,e) = c\}|} \\
&= P_S.
\end{aligned}
$$

$\square$

The I-equitable property means that, for any choice of s and a, (s,a) has the least chance of success for the impersonation attack, and minimizes P_I. In view of Lemma 5.6.3, we only consider I-equitable linear A-codes.

We further assume that $P_S < 1$. Then any source state $\phi \in \Phi$ of a linear A-code $(\Phi, \mathcal{E}, \mathcal{T})$ can be interpreted as a surjective linear mapping from $\mathcal{E}$ to $\mathcal{T}$. Indeed, for given $\phi_0 \in \Phi$, let $L_0 = \mathrm{Im}(\phi_0) \subseteq \mathcal{T}$. If there exists $\phi \in \Phi$ and $\phi \neq \phi_0$ such that $\mathrm{Im}(\phi) \neq L_0$, since the A-code is I-equitable, we know that $\dim(\mathrm{Im}(\phi)) = \dim(L_0)$. It follows that there exists an isomorphism θ from $\mathrm{Im}(\phi)$ to L_0 and $\theta\phi$ is an $\mathbb{F}_q$-linear mapping from $\mathcal{E}$ to $\mathcal{T}$ with $\ker(\theta\phi) = \ker(\phi_0)$. Note that $\theta\phi \notin \Phi$. Otherwise, if ϕ is authenticated, the authenticated message $(\phi, \phi(e))$ can be substituted with $(\theta\phi, \theta(\phi(e))$ that the receiver will always accept as authentic. This contradicts the assumption $P_S < 1$. Thus, we can simply replace ϕ by $\theta\phi$ without changing the parameters of the A-code, and the procedure can be repeatedly carried out until each element in Φ is an onto linear mapping from $\mathcal{E}$ to L_0. We then take $\mathcal{T}$ to be L_0.

Let $V(n,q)$ denote an n-dimensional linear space over $\mathbb{F}_q$.

Definition 5.6.4 A linear A-code without secrecy $(\mathcal{S}, \mathcal{E}, \mathcal{T})$ is called an $[n, M, t, d]$-**linear A-code** if $|\mathcal{S}| = M, |\mathcal{E}| = q^n, P_I = 1/q^t$, and $P_S = 1/q^d$.

Theorem 5.6.5 *There exists an $[n, M, t, d]$-linear A-code if and only if there exists a family $\mathcal{L}$ of subspaces of $V(n,q)$ such that*

(i) $|\mathcal{L}| = M$;

(ii) $\dim(L) = n - t$, *for all* $L \in \mathcal{L}$;

(iii) $\dim(L \cap L') \leq n - (t + d)$, *for all* $L, L' \in \mathcal{L}, L \neq L'$.

Proof. Consider an $[n, M, t, d]$-linear A-code $\mathbf{A} = (\Phi, \mathcal{E}, \mathcal{T})$ and let

$$\mathcal{L} = \{\ker(\phi) : \phi \in \Phi\}.$$

Since $\mathbf{A}$ is I-equitable, $P_I = 1/q^{n-\dim(\ker(\phi))} = q^{-t}$, and so $\dim(\ker(\phi)) = n - t$, for all $\phi \in \Phi$. From Lemma 5.6.2, we know that

$$
\begin{aligned}
P_S &= \max_{\substack{\phi,\, \phi' \,\in\, \Phi \\ \phi \neq \phi'}} \frac{q^{\dim(\ker(\phi) \cap \ker(\phi'))}}{q^{\dim(\ker(\phi))}} \\
&= \max_{\substack{\phi,\, \phi' \,\in\, \Phi \\ \phi \neq \phi'}} q^{\dim(\ker(\phi) \cap \ker(\phi'))-n+t} \\
&= q^{-d}.
\end{aligned}
$$

It follows that $\dim(\ker(\phi) \cap \ker(\phi')) \leq n - (t + d)$, and the necessity follows.

Conversely, if there is a family $\mathcal{L}$ of subspaces of $V(n, q)$ such that conditions (i) to (iii) are satisfied, then we take $\mathcal{E} = V(n, q)$ and $\mathcal{T} = V(t, q)$. For each subspace $L \in \mathcal{L}$, there exists an $\mathbb{F}_q$-linear transformation ϕ_L from $\mathcal{E}$ to $\mathcal{T}$ such that $L = \ker(\phi_L)$. Let $\Phi = \{\phi_L : L \in \mathcal{L}\}$. Then it is straightforward to verify that $(\Phi, \mathcal{E}, \mathcal{T})$ is an $[n, M, t, d]$-linear A-code. $\qquad\square$

5.6.2 Bounds for Linear A-Codes

In an $[n, M, t, d]$-linear A-code over $\mathbb{F}_q$, given n, t, and d, we would like to have M as large as possible. In this subsection, we derive some upper bounds for M. We denote by $M(n, t, d, q)$ the maximum M for which an $[n, M, t, d]$-linear A-code over $\mathbb{F}_q$ exists.

Let

$$\begin{bmatrix} n \\ k \end{bmatrix}_q = \frac{(q^n - 1)(q^{n-1} - 1) \cdots (q^{n-k+1} - 1)}{(q^k - 1)(q^{k-1} - 1) \cdots (q - 1)}$$

denote the **Gaussian coefficient**. Then the number of k-dimensional subspaces of $V(n, q)$ is $\begin{bmatrix} n \\ k \end{bmatrix}_q$, which gives an upper bound for $M(n, t, d, q)$.

Theorem 5.6.6 *For any integers n, t, d with $n \geq t \geq d$ and prime power q, we have*

$$M(n, t, d, q) \leq \begin{bmatrix} n \\ n - t \end{bmatrix}_q.$$

For $d = 1$, the bound in Theorem 5.6.6 is tight as the following corollary shows.

Corollary 5.6.7 *We have*

$$M(n, t, 1, q) = \begin{bmatrix} n \\ n-t \end{bmatrix}_q.$$

Proof. Let $\mathcal{L}$ be the set of all $(n-t)$-dimensional subspaces of the n-dimensional vector space $V(n, q)$. Then $|\mathcal{L}| = \begin{bmatrix} n \\ n-t \end{bmatrix}_q$. Since, for any $L, L' \in \mathcal{L}$ such that $L \neq L'$, we have $\dim(L \cap L') \leq n - t - 1$, from Theorem 5.6.5, we know that there exists an $[n, \begin{bmatrix} n \\ n-t \end{bmatrix}_q, t, 1]$-linear A-code over $\mathbb{F}_q$. □

If we take $n = 2$ and $t = 1$, then $\begin{bmatrix} 2 \\ 1 \end{bmatrix}_q = q + 1$. We obtain a $[2, q+1, 1, 1]$-linear A-code. In other words, we have a linear A-code $(\mathcal{S}, \mathcal{E}, \mathcal{T})$ with the following parameters

$$|\mathcal{S}| = q + 1, \ |\mathcal{E}| = q^2, \ |\mathcal{T}| = q, \ \text{and} \ P_I = P_S = 1/q.$$

We note that A-codes with these parameters were first constructed by Gilbert, MacWilliams, and Sloane [69]. Their construction uses projective planes and works as follows.

Let q be a prime power. Consider the projective plane $\mathbf{P}^2(\mathbb{F}_q)$ over the field $\mathbb{F}_q$. Fix a line ℓ in $\mathbf{P}^2(\mathbb{F}_q)$. The points on ℓ are regarded as source states, points not lying on ℓ are regarded as the encoding rules (i.e., keys), and the lines different from ℓ are regarded as the messages. To authenticate a source state (a point on ℓ), the transmitter sends to the receiver the entire line that is uniquely determined by the source state and the key (a point not lying on ℓ). This results in an authentication code with $q + 1$ source states, q^2 authentication keys, and q authenticators. The deception probabilities of this code are

$$P_I = P_S = 1/q.$$

On the other hand, choosing different values of t in Corollary 5.6.7 results in linear A-codes with different parameters. The following result improves the bound in Theorem 5.6.6 when $d \geq 2$.

Theorem 5.6.8 *For an $[n, M, t, d]$-linear A-code over $\mathbb{F}_q$, we have*

$$M(n, t, d, q) \leq \frac{\begin{bmatrix} n \\ n-(t+d)+1 \end{bmatrix}_q}{\begin{bmatrix} n-t \\ n-(t+d)+1 \end{bmatrix}_q}.$$

Proof. From Theorem 5.6.5, we know that there is an $[n, M, t, d]$-linear A-code if and only if there is a family $\mathcal{L} = \{V_1, V_2, \ldots, V_M\}$ of subspaces of $V(n, q)$, with $\dim(V_i) = n - t$ and $\dim(V_i \cap V_j) \leq n - (t + d)$, for $i \neq j$. For each $1 \leq i \leq M$, let $\Re_i$ denote the family of subspaces of V_i of dimension $n - (t + d) + 1$. It follows that

$$|\Re_i| = \begin{bmatrix} n - t \\ n - (t + d) + 1 \end{bmatrix}_q.$$

We claim that

$$\Re_i \cap \Re_j = \emptyset, \quad \text{for all } i \neq j.$$

Otherwise, if $C \in \Re_i \cap \Re_j$ is a subspace of dimension $n - (t + d) + 1$, then C is a subspace of both V_i and V_j, which contradicts the assumption that $\dim(V_i \cap V_j) \leq n - (t + d)$. We then have

$$\begin{bmatrix} n \\ n - (t + d) + 1 \end{bmatrix}_q \geq \left| \bigcup_{i=1}^{M} \Re_i \right|$$
$$= M|\Re_i|$$
$$= M \begin{bmatrix} n - t \\ n - (t + d) + 1 \end{bmatrix}_q.$$

The desired result follows immediately. □

For any fixed n and k, as $q \to \infty$, we have

$$\begin{bmatrix} n \\ k \end{bmatrix}_q = \frac{(q^n - 1)(q^{n-1} - 1) \cdots (q^{n-k+1} - 1)}{(q^k - 1)(q^{k-1} - 1) \cdots (q - 1)}$$
$$\approx q^{(n-k)k}.$$

It follows that

$$M(n, t, d, q) \leq \begin{bmatrix} n \\ n - (t + d) + 1 \end{bmatrix}_q \bigg/ \begin{bmatrix} n - t \\ n - (t + d) + 1 \end{bmatrix}_q$$
$$\approx \frac{q^{(n-(t+d)+1)(t+d-1)}}{q^{(n-(t+d)+1)(d-1)}}$$
$$= q^{(n-(t+d)+1)t}. \tag{5.11}$$

In the general theory of A-codes, it is possible, as we have seen earlier, for the size of source states to grow exponentially with the size of the keys, for example, in the construction based on universal hash families in Section 5.4. However, this is not true for linear A-codes. In fact, from Theorem 5.6.8, it is easy to see that $\log_q |\mathcal{S}| \leq n^2 = (\log_q |\mathcal{E}|)^2$, and this bound can be asymptotically achieved. For example, if $(t+d) - 1 \approx t \approx n/2$, then, as we will see in the next subsection, we have a linear A-code with $\log_q M(n, t, d, q) \approx n^2/4$.

5.6.3 Constructions of Linear A-Codes

Rank distance codes were introduced by Gabidulin in [63]. They have been used to construct other authentication schemes such as A^2-codes by Johansson [82] and group A-codes by van Dijk *et al.* [165]. We show that linear A-codes can be constructed from rank distance codes, and such constructions result in linear A-codes that asymptotically meet the bound in Theorem 5.6.8.

Definition 5.6.9 Let $\Lambda = \{A_i\}$ be a set of $m \times t$ matrices over $\mathbb{F}_q$. The distance $d(A, B)$ between two matrices A and B in Λ is defined by $d(A, B) = \text{rank}(A - B)$ and the minimum distance of Λ, denoted by $d(\Lambda)$, is defined as

$$d(\Lambda) = \min_{\substack{A, B \in \Lambda \\ A \neq B}} d(A, B).$$

Let $d = d(\Lambda)$ and $M = |\Lambda|$. We call Λ an $(m \times t, M, d)$-**rank distance code**.

The following theorem establishes a relationship between linear A-codes and rank distance codes.

Theorem 5.6.10 *If there exists an $(m \times t, M, d)$-rank distance code over $\mathbb{F}_q$, then there exists an $[m + t, M, t, d]$-linear A-code over $\mathbb{F}_q$.*

Proof. Let Λ be an $(m \times t, M, d)$-rank distance code. We define a set of $(t+m) \times t$ matrices

$$\Phi = \left\{ \begin{pmatrix} I_t \\ A \end{pmatrix} : A \in \Lambda \right\},$$

where I_t denotes the $t \times t$ identity matrix. For each $\begin{pmatrix} I_t \\ A \end{pmatrix} \in \Phi$, we define

$$\ker \begin{pmatrix} I_t \\ A \end{pmatrix} = \left\{ (\mathbf{e}_1, \mathbf{e}_2) \in \mathbb{F}_q^{t+m} : (\mathbf{e}_1, \mathbf{e}_2) \begin{pmatrix} I_t \\ A \end{pmatrix} = \mathbf{0} \right\},$$

where $\mathbf{e}_1 \in \mathbb{F}_q^t$ and $\mathbf{e}_2 \in \mathbb{F}_q^m$. We consider a set of subspaces of $\mathbb{F}_q^{t+m}$ given by

$$\mathcal{L} = \left\{ \ker \begin{pmatrix} I_t \\ A \end{pmatrix} : \begin{pmatrix} I_t \\ A \end{pmatrix} \in \Phi \right\}.$$

Clearly, $|\mathcal{L}| = M$ and $\dim \left(\ker \begin{pmatrix} I_t \\ A \end{pmatrix} \right) = m$. We show that, for any $A, B \in \Lambda$, $A \neq B$, we have

$$\dim \left(\ker \begin{pmatrix} I_t \\ A \end{pmatrix} \bigcap \ker \begin{pmatrix} I_t \\ B \end{pmatrix} \right) \leq m - d.$$

Indeed,

$$
\left| \ker \begin{pmatrix} I_t \\ A \end{pmatrix} \bigcap \ker \begin{pmatrix} I_t \\ B \end{pmatrix} \right|
$$

$$
= \left| \left\{ (\mathbf{e}_1, \mathbf{e}_2) \in \mathbb{F}_q^{t+m} : (\mathbf{e}_1, \mathbf{e}_2) \begin{pmatrix} I_t \\ A \end{pmatrix} = \mathbf{0}, (\mathbf{e}_1, \mathbf{e}_2) \begin{pmatrix} I_t \\ B \end{pmatrix} = \mathbf{0} \right\} \right|
$$

$$
= \left| \left\{ (-\mathbf{e}_2 A, \mathbf{e}_2) \in \mathbb{F}_q^{t+m} : \mathbf{e}_2 A = \mathbf{e}_2 B \right\} \right|
$$

$$
= \left| \left\{ \mathbf{e}_2 \in \mathbb{F}_q^m : \mathbf{e}_2 (A - B) = \mathbf{0} \right\} \right|
$$

$$
= q^{m - \mathrm{rank}(A-B)}
$$

$$
\leq q^{m-d}.
$$

From Theorem 5.6.5, we know that $(\Phi, \mathbb{F}_q^{t+m}, \mathbb{F}_q^t)$ is an $[m+t, M, t, d]$-linear A-code and the claimed result follows. $\qquad \square$

As shown in [82], in an $(m \times t, M, d)$-rank distance code, we always have $d \leq m - k + 1$, where $k = \log_{q^t} M$. Codes for which the equality holds are called **maximum rank distance codes** (or **MRD codes** for short). Gabidulin [63] showed that MRD codes can be constructed from linearized polynomials.

Recall that a polynomial of the form $F(z) = \sum_{i=0}^{k-1} f_i z^{q^i}$, where $f_i \in \mathbb{F}_{q^t}$, is called a **linearized polynomial** over $\mathbb{F}_{q^t}$. Let k, m, t be integers satisfying $0 < k \leq m \leq t$. By $\mathcal{L}_{k,t}$, we denote the set of all linearized polynomials over $\mathbb{F}_{q^t}$ of degree at most q^{k-1}. Assume that $g_1, g_2, \ldots, g_m$ are specified elements of the field $\mathbb{F}_{q^t}$ which are linearly independent over $\mathbb{F}_q$. For each $F(z) \in \mathcal{L}_{k,t}$, set

$$
\mathbf{c}_{F(z), g_1, \ldots, g_m} = \begin{pmatrix} F(g_1) \\ F(g_2) \\ \vdots \\ F(g_m) \end{pmatrix}.
$$

We associate $\mathbf{c}_{F(z), g_1, \ldots, g_m}$ with an $m \times t$ matrix $A(\mathbf{c}_{F(z), g_1, \ldots, g_m}) = (a_{ij})$, which is obtained by writing $F(g_i)$ (expressed in a fixed basis of $\mathbb{F}_{q^t}$ over $\mathbb{F}_q$) as a row vector with entries $a_{ij} \in \mathbb{F}_q$.

Lemma 5.6.11 ([82]) *The set* $\{A(\mathbf{c}_{F(z), g_1, \ldots, g_m}) : F(z) \in \mathcal{L}_{k,t}\}$ *is an MRD code. In other words,* $\{A(\mathbf{c}_{F(z), g_1, \ldots, g_m}) : F(z) \in \mathcal{L}_{k,t}\}$ *is an* $(m \times t, q^{tk}, m - k + 1)$-rank distance code.

Corollary 5.6.12 *Let* n, t, d *be integers satisfying* $0 < t + d \leq n$ *and let* q *be a prime. Then the above construction from linearized polynomials results in an* $[n, q^{t(n-t-d+1)}, t, d]$-linear A-code.

Proof. Put $k = n - t - d + 1$ and $m = n - t$. Applying Theorem 5.6.10 and Lemma 5.6.11, we obtain the desired result.   □

Note that the parameters given in Corollary 5.6.12 asymptotically meet the bounds in Theorem 5.6.8.

Example 5.6.13 Choosing $n = 2$ and $t = d = 1$, we have a linear A-code $(\mathcal{S}, \mathcal{E}, \mathcal{T})$ with $|\mathcal{S}| = q$, $|\mathcal{E}| = q^2$, and $|\mathcal{T}| = q$, with $P_I = P_S = 1/q$. This code has the same parameters as the A-code $\mathbf{A} = (\mathcal{S}', \mathcal{E}', \mathcal{T}', f')$, where $\mathcal{S}' = \mathbb{F}_q$, $\mathcal{T}' = \mathbb{F}_q$, $\mathcal{E}' = \mathbb{F}_q \times \mathbb{F}_q$, and f' defined as $f'(s, (e_1, e_2)) = e_1 s + e_2$, for all $s \in \mathcal{S}', (e_1, e_2) \in \mathcal{E}'$. It is easy to verify that $\mathbf{A}$ is linear and that $P_I = P_S = 1/q$.

Chapter 6

Frameproof Codes

6.1 Introduction

In order to protect a product such as computer software, for example, a distributor may mark each copy with some codeword and then ships to each user his data marked with that codeword. This marking allows the distributor to detect any unauthorized copy and trace it back to the user. Since a marked object can be traced, the users will be deterred from releasing an unauthorized copy. However, a coalition of users may detect some of the marks, namely, the marks where their copies differ. They can then change these marks arbitrarily. To prevent a group of users from "framing" another user, Boneh and Shaw [27] defined the concept of c-frameproof codes. A c-frameproof code has the property that no coalition of at most c users can frame a user not in the coalition.

In the traitor tracing scheme suggested by Chor, Fiat, and Naor in [41], a sender (such as a distribution center or a company) wishes to broadcast an encrypted message to a set of users who use their individual decoders to decrypt the broadcast message. (More discussion on broadcast encryption may be found later in Chapter 8.) A decoder box consists of N keys, where each key takes on one of q possible values. A decoder box $\mathbf{a}$ can typically be represented as an N-tuple $(a_1, \ldots, a_N)$, where $1 \leq a_i \leq q$ for $1 \leq i \leq N$. A user can redistribute the keys of his decoder box without altering it. If an unauthorized copy of the decoder box is found containing the keys of user u's decoder box, one can accuse u of producing a *pirate* decoder box. However, u could claim that he was framed by a coalition which created a decoder box containing his keys. Thus, it is desirable to construct decoder boxes that satisfy the following property: *no coalition can collude to frame a user not in the coalition.* Codes that satisfy this property are then called **frameproof codes**, or c-**frameproof codes** if the condition is relaxed by limiting the size of the coalition to at most c users.

A second example, described in Fiat and Tassa [58], concerns pay-per-view movies. Suppose a pay-per-view movie is divided into segments, and each segment has possible variations. The possible variations of a segment could have the same "content," but be "marked" in some manner that is not easily detected. A different variation of the movie is broadcast to each subscriber.

A copy of the movie can, therefore, be represented as an N-tuple. A coalition might try to create a pirate copy of the movie by copying segments from the versions broadcast to them, in much the same way as a coalition producing a pirate decoder box in the example described above. The cable company may want to design a scheme that enables the identification of one or more of the members of a coalition that produced a pirated movie.

6.2 Constructions of Frameproof Codes without Algebraic Geometry

We now formally introduce the definition of frameproof codes and some properties. We also discuss some constructions of frameproof codes without using algebraic geometry.

Let S be a set of q elements with $q > 1$ and let N be a positive integer. Define the ith projection, for $1 \leq i \leq N$, as

$$\pi_i : S^N \rightarrow S, \quad (a_1, \ldots, a_N) \mapsto a_i.$$

For a subset $A \subseteq S^N$, we define the **descendants** of A, denoted by $\mathrm{desc}(A)$, to be the set of all $\mathbf{x} \in S^N$ such that, for each $1 \leq i \leq N$, there exists some $\mathbf{a} \in A$ (dependent on i) satisfying $\pi_i(\mathbf{x}) = \pi_i(\mathbf{a})$.

Definition 6.2.1 Let $c \geq 2$ be an integer. A q-ary c-**frameproof code** of length N is a subset $C \subseteq S^N$ such that, for all $A \subseteq C$ with $|A| \leq c$, we have

$$\mathrm{desc}(A) \cap C = A. \tag{6.1}$$

From this definition, it is clear that a q-ary c-frameproof code C is a q-ary c_1-frameproof code for any $2 \leq c_1 \leq c$.

For a finite set S of size q, we denote a q-ary c-frameproof code in S^N of size M by c-FPC(N, M). Since each codeword can be "fingerprinted" into a copy of the distributed documents, we would like the value of M to be as large as possible. This leads to the following definition.

Definition 6.2.2 For a fixed integer $q > 1$ and integers $c \geq 2$ and $N \geq 2$, let $M_q(N, c)$ denote the maximum size of q-ary c-frameproof codes of length N, i.e.,

$$M_q(N, c) \stackrel{\mathrm{def}}{=} \max\{M : \text{there exists a } q\text{-ary } c\text{-FPC}(N, M) \}.$$

In [150], Staddon, Stinson, and Wei proved that $M_q(N, c) \leq c\left(q^{\lceil N/c \rceil} - 1\right)$. This result was improved by Blackburn in [16], where the following was shown.

Theorem 6.2.3 *Let $c \geq 2$ and $N \geq 2$ be integers and let q be a prime power. Then*

$$M_q(N,c) \leq \max \left\{ q^{\lceil N/c \rceil}, r \left(q^{\lceil N/c \rceil} - 1 \right) + (c-r) \left(q^{\lfloor N/c \rfloor} - 1 \right) \right\}, \quad (6.2)$$

where $r = N \bmod c$.

Proof. Let $C \subseteq S^N$ be a q-ary c-frameproof code of cardinality $M = M_q(N,c)$.

If $M \leq q$, then the bound (6.2) is trivially true. Hence, we assume that $M > q$.

For any subset $T \subseteq \{1,2,\dots,N\}$, define U_T by

$$U_T = \{\mathbf{x} \in C : \text{there exists no } \mathbf{y} \in C \setminus \{\mathbf{x}\} \text{ such that } \pi_i(\mathbf{x}) = \pi_i(\mathbf{y}) \text{ for any } i \in T\}.$$

Note that $|U_T| \leq q^{|T|} - 1$ since every codeword $\mathbf{x} \in U_T$ is uniquely identified by the subword $(x_i)_{i \in T}$. Let $T_1, T_2, \dots, T_c \subseteq \{1,2,\dots,N\}$ be disjoint subsets, where $|T_j| = \lceil N/c \rceil$ whenever $1 \leq j \leq r$ and where $|T_j| = \lfloor N/c \rfloor$ whenever $1 + r \leq j \leq c$. Thus, we must have $\cup_{j=1}^c T_j = \{1,2,\dots,N\}$. The bound of the theorem follows if we have $C = \cup_{j=1}^c U_{T_j}$.

Suppose, for a contradiction, that there exists $\mathbf{x} \in C \setminus \cup_{j=1}^c U_{T_j}$. Then there exist $\mathbf{y}_1, \mathbf{y}_2, \dots, \mathbf{y}_c \in C \setminus \{\mathbf{x}\}$ such that $\mathbf{y}_j$ and $\mathbf{x}$ agree in their ith components for all $i \in T_j$. However, in this case, we have $\mathbf{x} \in \mathrm{desc}(\{\mathbf{y}_1, \mathbf{y}_2, \dots, \mathbf{y}_c\})$, which contradicts the c-frameproof property of C. This contradiction proves that $C = \cup_{j=1}^c U_{T_j}$, as desired. $\square$

Remark 6.2.4   (i) For almost all parameter sets, the second term on the right-hand side of (6.2) is the larger of the two.

(ii) An obvious consequence of Theorem 6.2.3 is that, for a q-ary c-frameproof code of length N, the size is at most

$$\delta q^{\lceil N/c \rceil} + O\left(q^{\lceil N/c \rceil - 1} \right),$$

where $\delta = N \bmod c$.

6.2.1   Constructions of Frameproof Codes

In this subsection, we discuss some constructions of frameproof codes.

It has been suggested by many authors that frameproof codes can be constructed from error-correcting codes. We now introduce some such constructions.

Proposition 6.2.5 *The set C of all words of Hamming weight 1 in $\mathbb{Z}_q^N$ forms a c-frameproof code of size $N(q-1)$, for any $c \geq 2$.*

Proof. Let $\mathbf{x} \in C$ be a word of Hamming weight 1 and assume that its ith component is nonzero. Now, any subset P of C with $\mathbf{x} \in \mathrm{desc}(P)$ must contain a codeword $\mathbf{y}$ such that $\pi_i(\mathbf{x}) = \pi_i(\mathbf{y})$. Thus, we must have $\mathbf{x} = \mathbf{y}$ since the Hamming weight of $\mathbf{y}$ is 1. The desired result follows. □

Proposition 6.2.6 *A q-ary (N, M, d)-code C is a q-ary c-FPC(N, M) with $c = \lfloor (N-1)/(N-d) \rfloor$.*

Proof. Let A be a subset of C with $|A| \le c$. Suppose that $\mathrm{desc}(A) \cap C \ne A$. Choose $\mathbf{x} \in (\mathrm{desc}(A) \cap C) \setminus A$. Since there are at most c elements in A, it follows from $\mathbf{x} \in \mathrm{desc}(A)$ that there is a codeword $\mathbf{y} \in A$ such that $\mathbf{y}$ agrees with $\mathbf{x}$ in at least $\lceil N/c \rceil$ positions, i.e., the Hamming weight of $\mathbf{y} - \mathbf{x}$ satisfies

$$\mathrm{wt}(\mathbf{y} - \mathbf{x}) \le N - \left\lceil \frac{N}{c} \right\rceil \le N - \frac{N}{c}.$$

As $\mathbf{x} \ne \mathbf{y}$, we get

$$d \le \mathrm{wt}(\mathbf{y} - \mathbf{x}) \le N - \frac{N}{c},$$

which implies $c \ge N/(N-d)$. This contradicts $c = \lfloor (N-1)/(N-d) \rfloor$. □

Example 6.2.7 Now we consider the construction of frameproof codes in Proposition 6.2.6 by looking at algebraic geometry codes based on projective lines and elliptic curves.

(i) We know that a code from a projective line is an MDS code (cf. Example 2.4.1(i) and Theorem 2.3.24). Let C be a q-ary $[N, k, N - k + 1]$-MDS code. Then, by Proposition 6.2.6, we obtain a q-ary c-FPC(N, q^k) with $c = \lfloor (N-1)/(k-1) \rfloor$.

(ii) A q-ary algebraic geometry code based on an elliptic curve has parameters $[N, k, \ge N - k]$ with $N \le q + 1 + 2\sqrt{q}$. Thus, we obtain a q-ary c-FPC(N, q^k) with $c = \lfloor (N-1)/k \rfloor$.

Proposition 6.2.8 *Let N be an even integer such that $N \ge 4$. Let m be a prime power such that $m \ge N + 1$ and set $q = m^2 + 1$. Then there exists a q-ary 2-frameproof code C of cardinality $2(q-1)^{N/2}(1 - 1/(2\sqrt{q-1}))$.*

Proof. Define F to be the disjoint union $F = \{\infty\} \cup \mathbb{F}_m^2$. Let $\beta_0, \beta_1, \alpha_1, \alpha_2, \dots, \alpha_{N-1}$ be distinct elements of $\mathbb{F}_m$. For polynomials $f, g \in \mathbb{F}_m[x]$, we write $(f, g)(\alpha_i)$ for the element $(f(\alpha_i), g(\alpha_i)) \in F$. Define $C_1 \in F^N$ by

$$C_1 = \{(\infty, (f, g)(\alpha_1), (f, g)(\alpha_2), \dots, (f, g)(\alpha_{N-1}))\},$$

where $f, g \in \mathbb{F}_m[x]$ are such that $\deg(f) = (N/2) - 1$ and $\deg(g) \le (N/2) - 1$. Define $C_2 \in F^N$ by

$$C_2 = \{(t(\beta_0), t(\beta_1)), (s, t)(\alpha_1), (s, t)(\alpha_2), \dots, (s, t)(\alpha_{N-1}))\},$$

where $s, t \in \mathbb{F}_m[x]$ are such that $\deg(s) \leq (N/2) - 2$ and $\deg(t) \leq (N/2)$.

By considering their first components, it is clear that C_1 and C_2 are disjoint. A polynomial of degree at most $(N/2) - 1$ is determined by its values at $N/2$ distinct points, hence the polynomials f and g are uniquely determined by a codeword $\mathbf{x} \in C_1$. There are $m^{N/2} - m^{(N/2)-1}$ choices for f and there are $m^{N/2}$ choices for g, and so $|C_1| = (m^2)^{N/2}(1 - 1/m)$. The polynomial s defining C_2 is determined by $(N/2) - 1$ of the final $N - 1$ components of a codeword $\mathbf{x} \in C_2$. Similarly, the polynomial t is determined by $(N/2) + 1$ of these components. Hence, $|C_2|$ is equal to the number of choices for s and t, and so $|C_2| = m^{(N/2)-1} m^{(N/2)+1} = m^N$. Summing our expressions for $|C_1|$ and $|C_2|$ and using the fact that $m = \sqrt{q-1}$ shows that $|C| = 2(q-1)^{N/2}(1 - 1/(2\sqrt{q-1}))$, as required, where $C = C_1 \cup C_2$.

It remains to show that C is a 2-frameproof code. To this end, we first claim that codewords $\mathbf{x} \in C_1$ and $\mathbf{y} \in C_2$ can agree in at most $(N/2) - 1$ components. The first components of $\mathbf{x}$ and $\mathbf{y}$ are never equal. If $N/2$ of the remaining positions agree, then the definitions of C_1 and C_2 imply that a polynomial f of degree exactly $(N/2) - 1$ and a polynomial s of degree at most $(N/2) - 2$ agree at $N/2$ points. This contradiction establishes our claim.

Let $\mathbf{x} \in C$ and suppose that $P \subset C$ is such that $|P| = 2$ and $\mathbf{x} \in \text{desc}(P)$. We must show that $\mathbf{x} \in P$.

Suppose that $\mathbf{x} \in C_1$. Since $\pi_1(\mathbf{x}) = \infty$, P must contain an element from C_1. Suppose $|P \cap C_1| = 1$. Since a codeword in C_2 agrees with $\mathbf{x}$ in at most $(N/2) - 1$ components, the codeword $\mathbf{y} \in P \cap C_1$ agrees with $\mathbf{x}$ in at least $(N/2) + 1$ positions. Hence, $\mathbf{x}$ and $\mathbf{y}$ agree in at least $N/2$ of the components other than the first. However, a codeword in C_1 is determined by specifying any $N/2$ of its final $N - 1$ components. Hence, $\mathbf{x} = \mathbf{y}$ and our result follows in this case. Now suppose that $|P \cap C_1| = 2$. If a codeword $\mathbf{y} \in P$ agrees with $\mathbf{x}$ in $(N/2) + 1$ positions, then $\mathbf{x} = \mathbf{y}$ as before. Suppose both codewords in P agree with $\mathbf{x}$ in exactly $N/2$ positions. Then one of the codewords, $\mathbf{y} \in P$ say, agrees with $\mathbf{x}$ in $N/2$ of its final $N - 1$ components. However, since any $N/2$ of the last $N - 1$ components determine a codeword in C_1, we have that $\mathbf{x} = \mathbf{y}$. Hence, our result follows when $\mathbf{x} \in C_1$.

Suppose that $\mathbf{x} \in C_2$. By considering the first component of $\mathbf{x}$, the set P must contain a codeword in C_2. If there exists $\mathbf{y} \in P \cap C_2$ that agrees with $\mathbf{x}$ in $(N/2) + 1$ positions, then $\mathbf{x} = \mathbf{y}$ (at least $N/2$ values of s and at least $(N/2) + 1$ values of t are specified, so s and t are determined). In particular, since a codeword in $P \cap C_1$ can agree with $\mathbf{x}$ in at most $(N/2) - 1$ positions, this shows that our result holds in the case when $|P \cap C_1| = 1$. The case remaining is when $P \subseteq C_2$ and each of its codewords agrees with $\mathbf{x}$ in $N/2$ positions. Let $\mathbf{y} \in P$ be the codeword that agrees with $\mathbf{x}$ in the first position. Then $(N/2) - 1$ of the values of s and $(N/2) + 1$ of the values of t are specified by the components where $\mathbf{x}$ and $\mathbf{y}$ agree. Hence, $\mathbf{x} = \mathbf{y}$ in this case. $\square$

6.2.2 Two Characterizations of Binary Frameproof Codes

In this subsection, we concentrate on binary frameproof codes by introducing a combinatorial description of binary frameproof codes and giving some constructions.

For binary frameproof codes, we can introduce an equivalent definition. The advantage of this new definition is to allow an explanation of the practical use of frameproof codes.

For a binary (N, M)-code $C = \{c_1, \ldots, c_M\}$, let $A = \{a_1, \ldots, a_d\}$ be a subset of C. We say $i \in \{1, \ldots, N\}$ is **undetectable** if $\pi_i(a_1) = \cdots = \pi_i(a_d)$. Let $\mathcal{U}(A)$ be the set of undetectable bit positions, then

$$F(A) \overset{\text{def}}{=} \{\mathbf{x} \in \mathbb{F}_2^N : \pi_i(\mathbf{x}) = \pi_i(a_1) \text{ for all } i \in \mathcal{U}(A)\}$$

is called the **feasible set** of A. If $\mathcal{U}(A) = \emptyset$, then we define $F(A) = \mathbb{F}_2^N$.

The feasible set $F(A)$ represents the set of all possible N-tuples that could be produced by the coalition A by comparing the d codewords they jointly hold. If there is a codeword $c_j \in F(A) \setminus A$, then user j could be "framed" if the coalition A produces the N-tuple c_j. The following definition is motivated by the desire for this situation not to occur.

Definition 6.2.9 A binary (N, M)-code is called a c-**frameproof code** C if, for every $A \subseteq C$ such that $|A| \leq c$, we have $F(A) \cap C = A$.

It is easy to verify that this definition coincides with the previous one as the feasible set $F(A)$ is just the set $\text{desc}(A)$.

We now look at some examples.

Example 6.2.10 (i) For any integer $N \geq 1$, there exists a binary N-FPC(N, N). The matrix depicting the code is the $N \times N$ identity matrix (the matrix depicting a code is the one whose rows consist of all the codewords).

(ii) There exists a binary 2-FPC$(3, 4)$. The matrix depicting the code is as follows

$$\begin{pmatrix} 1 & 0 & 0 \\ 0 & 1 & 0 \\ 0 & 0 & 1 \\ 1 & 1 & 1 \end{pmatrix}.$$

We now give a characterization of binary frameproof codes in terms of set systems, defined earlier in Definition 5.2.9.

Since the matrix depicting a binary c-FPC(N, M) is an $M \times N$ $(0, 1)$-matrix, we can view a binary frameproof code as a set system by identifying the depicting matrix with the incidence matrix of this set system. We have the following characterization of frameproof codes as set systems.

Theorem 6.2.11 *There exists a binary c-FPC(N, M) if and only if there exists a set system* $(X, \mathcal{B})$ *such that* $|X| = N$, $|\mathcal{B}| = M$, *and, for any subset of blocks* $B_1, \ldots, B_d$ *with* $d \leq c$, *there does not exist a block* $B \in \mathcal{B}$ *such that*

$$\bigcap_{i=1}^{d} B_i \subseteq B \subseteq \bigcup_{i=1}^{d} B_i.$$

Proof. Assume that we have a binary c-FPC(N, M). Let $\mathbf{c}_1, \ldots, \mathbf{c}_d$ be d codewords with $d \leq c$. Without loss of generality, assume that, in these codewords, the first s bit positions are 0, the next t bit positions are 1, and in every other bit position at least one of the d codewords has the value 1 and at least one has the value 0. (Hence, the undetectable bit positions are the first $s + t$ bit positions.) Then it is not hard to see that the frameproof property is equivalent to saying that any other codeword has at least one 1 in the first s bit positions, or at least one 0 in the next t bit positions. In other words, there does not exist another codeword with 0 in the first s bit positions and 1 in the next t bit positions.

Let $(X, \mathcal{B})$ be the set system, with $X = \{x_1, \ldots, x_N\}$, whose incidence matrix is the matrix depicting the binary c-FPC(N, M) (so $\mathcal{B}$ corresponds to the rows of this matrix). Let $B_1, \ldots, B_d$ be the blocks in the set system corresponding to the d codewords $\mathbf{c}_1, \ldots, \mathbf{c}_d$. Then

$$\bigcap_{i=1}^{d} B_i = \{x_{s+1}, \ldots, x_{s+t}\}$$

and

$$\bigcup_{i=1}^{d} B_i = \{x_{s+1}, \ldots, x_N\}.$$

Hence, the frameproof condition is equivalent to saying that there does not exist a block $B \in \mathcal{B} \setminus \{B_1, \ldots, B_d\}$ such that $\cap_{i=1}^{d} B_i \subseteq B \subseteq \cup_{i=1}^{d} B_i$. $\square$

Next, we give a characterization of 2-frameproof codes in terms of the Hamming distance.

Lemma 6.2.12 *A binary code C with* $|C| \geq 3$ *is a 2-frameproof code if and only if, for any three distinct elements* $\mathbf{x}, \mathbf{y}$, *and* $\mathbf{z}$ *of C, one has the strict triangle inequality*

$$d(\mathbf{x}, \mathbf{z}) < d(\mathbf{x}, \mathbf{y}) + d(\mathbf{y}, \mathbf{z}).$$

Proof. Without loss of generality, we may assume that the undetectable bit positions of $\{\mathbf{x}, \mathbf{z}\}$ are $1, \ldots, r$, then the Hamming distance $d(\mathbf{x}, \mathbf{z})$ satisfies $d(\mathbf{x}, \mathbf{z}) = N - r$. Since the triangle inequality holds for the Hamming distance, we note that showing that the strict triangle inequality holds is equivalent to showing that there is at least one bit position within the first r bit positions where $\mathbf{y}$ is different from $\mathbf{x}$ and $\mathbf{z}$. This is just the condition that the code is

2-frameproof.   □

The following result is an immediate corollary of the previous lemma.

Corollary 6.2.13 *A binary (N, M)-code C is a 2-frameproof code if $2d(C) > d_{\max}(C)$, where $d(C)$ is the minimum Hamming distance of C, while $d_{\max}(C)$ denotes the maximum Hamming distance of C, i.e., $d_{\max}(C) = \max\{d(\mathbf{x}, \mathbf{y}) : \mathbf{x}, \mathbf{y} \in C\}$.*

Theorem 6.2.14 *For any odd prime power $q \geq 31$, there exists a binary 2-FPC$(q, (q^2 - q)/2)$.*

Proof. Let $\chi : \mathbb{F}_{q^2}^* \to \{-1, 1\}$ be the quadratic (Legendre) character, i.e.,

$$\chi(\alpha) = \begin{cases} 1 & \text{if } \alpha \text{ is a square in } \mathbb{F}_{q^2} \\ -1 & \text{otherwise.} \end{cases}$$

Then, for every element $x \in \mathbb{F}_{q^2} \setminus \mathbb{F}_q$, we associate a binary vector $\mathbf{v}_x = (v_{x,\alpha})_{\alpha \in \mathbb{F}_q}$ indexed by the elements of $\mathbb{F}_q$, namely

$$v_{x,\alpha} \stackrel{\text{def}}{=} \frac{1}{2}\left(1 + \chi(x + \alpha)\right).$$

Note that both x and x^q define the same binary vector, i.e., $\mathbf{v}_x = \mathbf{v}_{x^q}$. Hence, the binary code $C = \{\mathbf{v}_x : x \in \mathbb{F}_{q^2} \setminus \mathbb{F}_q\}$ is a $(q, (q^2 - q)/2)$-code with minimum distance $d(C)$ at least $q/2 - 3\sqrt{q}/2$, and maximum distance $d_{\max}(C)$ at most $q/2 + 3\sqrt{q}/2$ (see [150, 158] for the detailed proof). Thus, for $q > 81$, one has $2d(C) > d_{\max}(C)$. In fact, it can be verified by computer that $2d(C) > d_{\max}(C)$ for the codes produced by this construction for all odd prime powers q with $31 \leq q \leq 81$. Applying Corollary 6.2.13, we obtain the desired result.   □

6.2.3   Combinatorial Constructions of Binary Frameproof Codes

We proceed next to give some constructions of frameproof codes from combinatorial objects such as t-designs, packing designs, hash families and orthogonal arrays. All the results on design theory that we require can be found in standard references such as [43].

Recall that t-(v, k, λ) designs have been introduced in Definition 5.2.10. Note that, by simple counting, the number of blocks in a t-$(v, k, 1)$ design is $\binom{v}{t}/\binom{k}{t}$. We use t-designs to construct frameproof codes as described in the following theorem.

Theorem 6.2.15 *If there exists a t-$(N, k, 1)$ design, then there exists a binary c-FPC$(N, \binom{N}{t}/\binom{k}{t})$, where $c = \lfloor (k - 1)/(t - 1) \rfloor$.*

Proof. Let $d \leq c$ and let $B_1, \ldots, B_d$ be d distinct blocks in the given t-$(N, k, 1)$ design $(X, \mathcal{B})$, and let $B \in \mathcal{B} \setminus \{B_1, \ldots, B_d\}$. If $B \subseteq \cup_{i=1}^{d} B_i$, then there exists a B_i, where $1 \leq i \leq d$, such that $|B \cap B_i| \geq t$, given that $d \leq c = \lfloor (k-1)/(t-1) \rfloor$. Since we have a t-design with $\lambda = 1$, it follows that $B = B_i$, a contradiction. Hence, for any $B \in \mathcal{B} \setminus \{B_1, \ldots, B_d\}$, we have that $B \not\subseteq \cup_{i=1}^{d} B_i$. The t-design is a set system satisfying the conditions of Theorem 6.2.11, so the conclusion follows. □

There are many known results on the existence and construction of t-$(v, k, 1)$ designs for $t = 2, 3$. On the other hand, no t-$(v, k, 1)$ design with $v > k > t$ is known to exist for $t \geq 6$. However, known infinite classes of 2- and 3-designs provide some nice infinite classes of frameproof codes. We illustrate with a few samples of typical results that can be obtained.

For $3 \leq k \leq 5$, a 2-$(v, k, 1)$ design exists if and only if $v \equiv 1$ or $k \pmod{(k^2 - k)}$ (see [43]). Hence, we obtain the following result.

Theorem 6.2.16 *There exist frameproof codes as follows:*

(i) *a binary 2-FPC$(N, N(N-1)/6)$ if $N \equiv 1, 3 \pmod 6$;*

(ii) *a binary 3-FPC$(N, N(N-1)/12)$ if $N \equiv 1, 4 \pmod{12}$;*

(iii) *a binary 4-FPC$(N, N(N-1)/20)$ if $N \equiv 1, 5 \pmod{20}$.*

Another type of combinatorial designs which can be used to construct frameproof codes are packing designs. We give the definition as follows.

Definition 6.2.17 A t-(v, k, λ) **packing design** is a set system $(X, \mathcal{B})$, where $|X| = v, |B| = k$ for every $B \in \mathcal{B}$, and every t-subset of X occurs in at most λ blocks in $\mathcal{B}$.

Using the same argument as in the proof of Theorem 6.2.15, we have the following construction for frameproof codes.

Theorem 6.2.18 *If there exists a t-$(N, k, 1)$ packing design with M blocks, then there exists a binary c-FPC(N, M), where $c = \lfloor (k-1)/(t-1) \rfloor$.*

We mentioned previously that no t-$(v, k, 1)$ designs are known to exist if $v > k > t \geq 6$. However, for any t, there are infinite classes of packing designs with a large number of blocks (i.e., close to $\binom{v}{t}/\binom{k}{t}$). These can be obtained from orthogonal arrays (cf. Definition 5.2.7). It is easy to obtain a packing design from an orthogonal array, as shown in the next lemma. Here, we are only interested in orthogonal arrays $\mathsf{OA}_\lambda(t, k, v)$ with $\lambda = 1$, so we simplify the notation to $\mathsf{OA}(t, k, v)$.

Lemma 6.2.19 *If there is an $\mathsf{OA}(t, k, v)$, then there is a t-$(kv, k, 1)$ packing design that contains v^t blocks.*

Proof. Suppose that there is an $\mathsf{OA}(t, k, v)$ with entries from the set $\{0, 1, \ldots, v-1\}$. Define $X = \{(x, y) : \ 0 \le x \le k-1, \ 0 \le y \le v-1\}$. For every row $(y_0, y_1, \ldots, y_{k-1})$ in the orthogonal array, define a block $B = \{(i, y_i) : \ i = 0, 1, \ldots, k-1\}$. Let $\mathcal{B}$ consist of the v^t blocks thus constructed. It is easy to check that $(X, \mathcal{B})$ is a t-$(kv, k, 1)$ packing design. $\square$

The following lemma ([43, Chapter VI.7]) provides infinite classes of orthogonal arrays for any integer t.

Lemma 6.2.20 *If q is a prime power and $t < q$, then there exists an $\mathsf{OA}(t, q+1, q)$, and hence a t-$(q^2 + q, q+1, 1)$ packing design with q^t blocks exists.*

From Theorem 6.2.18 and Lemma 6.2.20, we obtain the following result.

Theorem 6.2.21 *For any prime power q and any integer $t < q$, there exists a $\lfloor q/(t-1) \rfloor$-FPC$(q^2 + q, q^t)$.*

6.3   Asymptotic Bounds and Constructions from Algebraic Geometry

Like asymptotic coding theory, it is natural to look at the asymptotic behavior of c-frameproof codes in the sense that q and c are fixed and the length n tends to infinity. Throughout this section, we assume that $c < q$.

Definition 6.3.1 For fixed integers $q > c \ge 2$, the **rate** of a q-ary c-FPC(n, M) is defined to be the quantity $(\log_q M)/n$. We also define the asymptotic quantity

$$D_q(c) = \limsup_{n \to \infty} \frac{\log_q M_q(n, c)}{n}.$$

It seems that the exact values of $D_q(c)$ are not easy to be determined for any given q and c. Instead, we will obtain some lower bounds for $D_q(c)$. Before looking at lower bounds, we first derive an upper bound for $D_q(c)$ from Theorem 6.2.3.

Theorem 6.3.2 *We have that*

$$D_q(c) \le \frac{1}{c}.$$

Proof. By Theorem 6.2.3, we have

$$M_q(n, c) \le \max \left\{ q^{\lceil n/c \rceil}, r \left(q^{\lceil n/c \rceil} - 1 \right) + (c - r) \left(q^{\lfloor n/c \rfloor} - 1 \right) \right\},$$

where $r \in \{0, 1, \ldots, c-1\}$ is the remainder of n divided by c. Thus, we have

$$M_q(n, c) \leq c q^{\lceil n/c \rceil}.$$

The desired result follows. □

From now on, in this section, we concentrate on lower bounds for $D_q(c)$.

From the relationship between error-correcting codes and frameproof codes given in Proposition 6.2.6, we immediately obtain a lower bound for $D_q(c)$ from the Gilbert-Varshamov bound (see Theorem 2.5.3).

Theorem 6.3.3 *Let $q > c > 1$ be two integers. Then*

$$D_q(c) \geq 1 - H_q\left(1 - \frac{1}{c}\right),$$

where

$$H_q(\delta) = \delta \log_q(q-1) - \delta \log_q \delta - (1-\delta)\log_q(1-\delta)$$

is the q-ary entropy function defined in Theorem 2.5.3.

Proof. Choose a sequence of pairs (n_i, d_i) such that $n_i \to \infty$ as $i \to \infty$, and

$$\frac{n_i - 1}{n_i - d_i} \geq c \quad \text{for all sufficiently large } i \quad \text{and} \quad \lim_{i \to \infty} \frac{n_i}{n_i - d_i} = c.$$

From the proof of the asymptotic Gilbert-Varshamov bound (cf. Theorem 2.5.3), we know that there exists a sequence of q-ary codes $\{C_i\}_{i=1}^{\infty}$ with parameters (n_i, M_i, d_i) such that

$$\lim_{i \to \infty} \frac{d_i}{n_i} = 1 - \frac{1}{c} \quad \text{and} \quad \lim_{i \to \infty} \frac{\log_q M_i}{n_i} \geq 1 - H_q\left(1 - \frac{1}{c}\right).$$

Thus, by Proposition 6.2.6, every C_i is a c-frameproof code, for sufficiently large i. Hence,

$$D_q(c) \geq \lim_{i \to \infty} \frac{\log_q M_i}{n_i} \geq 1 - H_q\left(1 - \frac{1}{c}\right).$$

This completes the proof. □

Remark 6.3.4 The bound in Theorem 6.3.3 is only an existential result as the Gilbert-Varshamov bound is not constructive.

For the rest of this section, we introduce two lower bounds for $D_q(c)$ from algebraic geometry codes. One bound can be obtained by directly applying Proposition 6.2.6 and the Tsfasman-Vlăduţ-Zink bound. However, the second bound employs the group structure of the Jacobians of algebraic curves. Note that the **Jacobian**, denoted by $\mathcal{J}(\mathcal{X}/\mathbb{F}_q)$, of an algebraic curve $\mathcal{X}/\mathbb{F}_q$ is defined to be the quotient group $\mathrm{Div}^0(\mathcal{X})/\mathrm{Princ}(\mathcal{X})$ (see Section 1.4 for the definition of this quotient group).

Theorem 6.3.5 *For a prime power q and an integer $c \geq 2$, we have*

$$D_q(c) \geq \frac{1}{c} - \frac{1}{A(q)},$$

where $A(q)$ is as defined at the end of Chapter 1.

Proof. Let $\delta = 1 - 1/c$. Combining Proposition 6.2.6 with Theorem 2.5.4, and using the same arguments as in the proof of Theorem 6.3.3, we obtain the desired result. □

Remark 6.3.6 (i) The bound in Theorem 6.3.5 is constructive since the algebraic geometry codes in Theorem 2.5.4 are constructive as long as the family of curves is explicit (e.g., that in Example 2.5.6).
(ii) Comparing with the upper bound in Theorem 6.3.2, we find that

$$\frac{1}{c} - \frac{1}{A(q)} \leq D_q(c) \leq \frac{1}{c}.$$

As we have seen from Chapter 1, $1/A(q) \to 0$ as $q \to \infty$, i.e., $D_q(c)$ approaches $1/c$ as $q \to \infty$.

Corollary 6.3.7 (i) *If q is the square of a prime power, then, for any $c \geq 2$,*

$$D_q(c) \geq \frac{1}{c} - \frac{1}{\sqrt{q} - 1}.$$

(ii) *If q is a prime power, then for any $c \geq 2$*

$$D_q(c) \geq \frac{1}{c} - \frac{96}{\log q}.$$

Proof. If q is a square, we have $A(q) = \sqrt{q} - 1$ (see (1.6)).
  If q is any prime power, we have (see [116, Theorem 5.2.9])

$$A(q) \geq \frac{\log q}{96}.$$

The desired result follows from Theorem 6.3.5. □

  The bound in Theorem 6.3.5 can be improved using algebraic geometry codes. However, to this end, we need to modify Goppa's construction of algebraic geometry codes.
  Let $\mathcal{X}$ be an algebraic curve over $\mathbb{F}_q$ with n distinct $\mathbb{F}_q$-rational points $P_1, P_2, \ldots, P_n$. Choose a positive divisor D such that $\mathcal{L}(D - \sum_{i=1}^{n} P_i) = \{0\}$. Let $\nu_{P_i}(D) = v_i \geq 0$ and let t_i be a local parameter at P_i for each i, i.e., $\nu_{P_i}(t_i) = 1$.

Consider the map

$$\phi : \mathcal{L}(D) \longrightarrow \mathbb{F}_q^n, \quad f \mapsto ((t_1^{v_1} f)(P_1), (t_2^{v_2} f)(P_2), \dots, (t_n^{v_n} f)(P_n)).$$

Then the image of ϕ forms a subspace of $\mathbb{F}_q^n$ that is defined as an algebraic geometry code. The image of ϕ is denoted by $C(D; \sum_{i=1}^n P_i)$, or simply, by abuse of notation, $C(D; \mathcal{P})$, where $\mathcal{P} = \{P_1, P_2, \dots, P_n\}$. The map ϕ is an embedding since $\mathcal{L}(D - \sum_{i=1}^n P_i) = \{0\}$, and hence the dimension of $C(D; \mathcal{P})$ is equal to $\ell(D)$.

Remark 6.3.8 Note that the above construction is a modified version of the algebraic geometry codes defined by Goppa (see Chapter 2). The advantage of the above construction is to make it possible to get rid of the condition $\text{Supp}(D) \cap \{P_1, P_2, \dots, P_n\} = \emptyset$. This is crucial for our construction of frameproof codes in this section.

When the condition $\text{Supp}(D) \cap \{P_1, P_2, \dots, P_n\} = \emptyset$ is satisfied, i.e., $v_i = 0$ for all $i = 1, \dots, n$, then the above construction of algebraic geometry codes is consistent with Goppa's construction.

Proposition 6.3.9 *Let $\mathcal{X}/\mathbb{F}_q$ be an algebraic curve of genus g and let $P_1, P_2, \dots, P_n$ be n distinct $\mathbb{F}_q$-rational points of $\mathcal{X}$. Let D be a positive divisor on $\mathcal{X}$ such that $\deg(D) < n$. Let $c \geq 2$ satisfy $\mathcal{L}(cD - \sum_{i=1}^n P_i) = \{0\}$. Then $C(D; \sum_{i=1}^n P_i)$ is a c-FPC$(n, q^{\ell(D)})$.*

Proof. For all $f \in \mathcal{L}(D)$, denote by $\mathbf{c}_f$ the codeword

$$\phi(f) = ((t_1^{v_1} f)(P_1), (t_2^{v_2} f)(P_2), \dots, (t_n^{v_n} f)(P_n)),$$

where $v_i = \nu_{P_i}(D)$, for $1 \leq i \leq n$. Let $A = \{\mathbf{c}_{f_1}, \dots, \mathbf{c}_{f_r}\}$ be a subset of $C \stackrel{\text{def}}{=} C(D; \sum_{i=1}^n P_i)$ with $|A| = r \leq c$. Let $\mathbf{c}_g \in \text{desc}(A) \cap C$ for some $g \in \mathcal{L}(D)$. Then, by the definition of the descendant, for each $1 \leq i \leq n$, we have

$$\prod_{j=1}^r \pi_i(\mathbf{c}_{f_j} - \mathbf{c}_g) = 0,$$

where $\pi_i(\mathbf{c}_{f_j} - \mathbf{c}_g)$ is the ith coordinate of $\mathbf{c}_{f_j} - \mathbf{c}_g$. This implies that

$$\prod_{j=1}^r (t_i^{v_i} f_j - t_i^{v_i} g)(P_i) = 0,$$

i.e.,

$$\nu_{P_i} \left(\prod_{j=1}^r (t_i^{v_i} f_j - t_i^{v_i} g) \right) \geq 1.$$

This is equivalent to

$$\nu_{P_i} \left(\prod_{j=1}^r (f_j - g) \right) \geq -r v_i + 1.$$

Hence,

$$\prod_{j=1}^{r}(f_j - g) \in \mathcal{L}\left(rD - \sum_{i=1}^{n} P_i\right) \subseteq \mathcal{L}\left(cD - \sum_{i=1}^{n} P_i\right) = \{0\}.$$

Thus, the function $\prod_{j=1}^{r}(f_j - g)$ is the zero function. This means that $f_\ell - g = 0$ for some $1 \le \ell \le r$. Hence, $\mathbf{c}_g = \mathbf{c}_{f_\ell} \in A$. $\qquad\square$

Remark 6.3.10 One main feature of the above proposition is that the minimum distance of the algebraic geometry code does not play an important role. However, the above construction has a disadvantage, i.e., it is not explicitly constructive since no explicit construction of the divisor D, for given c, is known. The existence of such a divisor D can, however, be shown, and it is given below.

From Proposition 6.3.9, we know that it is crucial to find a divisor D such that $\mathcal{L}(cD - \sum_{i=1}^{n} P_i) = \{0\}$. We will show some sufficient conditions for the existence of such a divisor D, but first we need to introduce more notations.

For an algebraic curve $\mathcal{X}/\mathbb{F}_q$ of genus g, recall that $\mathcal{J}(\mathcal{X}/\mathbb{F}_q) = \mathrm{Div}^0(\mathcal{X})/\mathrm{Princ}(\mathcal{X})$ denotes the divisor class group of degree zero of $\mathcal{X}/\mathbb{F}_q$. It is a finite abelian group whose order is $L(1)$ (see [151, Theorem 5.1.15]), where $L(t)$ is the L-function of $\mathcal{X}/\mathbb{F}_q$ (see Chapter 1). This order is commonly known as the **divisor class number**, or simply **class number**, of $\mathcal{X}$. Suppose that $\mathcal{X}$ has an $\mathbb{F}_q$-rational point P_0. For a divisor D, we denote by $[D - \deg(D)P_0]$ the class of the degree zero divisor $D - \deg(D)P_0$ in $\mathcal{J}(\mathcal{X}/\mathbb{F}_q)$.

Lemma 6.3.11 *Let $\mathcal{X}/\mathbb{F}_q$ be a smooth projective curve of genus g with an $\mathbb{F}_q$-rational point P_0. Then, for an integer $c \ge 2$ and any fixed integer $m \ge g$, the subgroup of $\mathcal{J}(\mathcal{X}/\mathbb{F}_q)$*

$$\{c[D - mP_0] : \ D > 0, \ \deg(D) = m\}$$

has order at least h/c^{2g}, where h denotes the class number.

Proof. It is a well-known fact that the set

$$\{[D - mP_0] : \ D > 0, \ \deg(D) = m\}$$

is the whole group $\mathcal{J}(\mathcal{X}/\mathbb{F}_q)$ for $m \ge g$ (e.g., see [176, proof of Lemma 2.2]). Therefore,

$$\{c[D - mP_0] : \ D > 0, \ \deg(D) = m\} = c\mathcal{J}(\mathcal{X}/\mathbb{F}_q).$$

Since $\mathcal{J}(\mathcal{X}/\mathbb{F}_q)$ is isomorphic to the group of $\mathbb{F}_q$-rational points on the Jacobian of $\mathcal{X}/\mathbb{F}_q$, the p-rank of $\mathcal{J}(\mathcal{X}/\mathbb{F}_q)$ is at most $2g$ for any prime p (see [110, page 39]). The desired result follows. $\qquad\square$

For $\mathcal{X}/\mathbb{F}_q$ as above, and for any $r \ge 0$, let $\mathcal{A}_r$ denote the set of $\mathbb{F}_q$-rational positive divisors on $\mathcal{X}$ of degree r, and let A_r denote the cardinality of $\mathcal{A}_r$.

Lemma 6.3.12 *Let $\mathcal{X}/\mathbb{F}_q$ be an algebraic curve of genus g with at least one $\mathbb{F}_q$-rational point P_0 and of class number h. Let c, m, n be three integers satisfying $c \geq 2$ and $g \leq m \leq n < cm$, and let G be a fixed positive divisor on $\mathcal{X}$ of degree n. Then there exists a positive divisor D of degree m such that $\mathcal{L}(cD - G) = \{0\}$, provided that $A_{cm-n} < h/c^{2g}$.*

Proof. By Lemma 6.3.11, we have

$$|\{[cH - cmP_0] : \ H > 0, \ \deg(H) = m\}| = |c\mathcal{J}(\mathcal{X}/\mathbb{F}_q)| \geq \frac{h}{c^{2g}}.$$

Moreover,

$$|\{[K + G - cmP_0] : \ K > 0, \ \deg(K) = cm - n\}| \leq |A_{cm-n}| = A_{cm-n}.$$

Thus, $\{[cH - cmP_0] : \ H > 0, \ \deg(H) = m\} \setminus \{[K + G - cmP_0] : \ K > 0, \ \deg(K) = cm - n\}$ is not empty.

Choose an element $[cD - cmP_0]$ from the above nonempty set, for some positive divisor D. We claim that $\mathcal{L}(cD - G) = \{0\}$. Otherwise, there would be a nonzero function $f \in \mathcal{L}(cD - G)$. Therefore, the divisor $\operatorname{div}(f) + cD - G$ is a positive divisor. Put $K = \operatorname{div}(f) + cD - G$. Then $\deg(K) = cm - n$, and cD is equivalent to $K + G$, i.e., $[cD - cmP_0]$ is the same as $[K + G - cmP_0]$. This contradicts the choice of $[cD - cmP_0]$. □

The following lemma can be found in [176, 177].

Lemma 6.3.13 *Let $\mathcal{X}/\mathbb{F}_q$ be a smooth projective curve of genus $g \geq 2$ with class number h. Then the number A_r of $\mathbb{F}_q$-rational positive divisors of $\mathcal{X}/\mathbb{F}_q$ of degree r satisfies*

$$A_r < \frac{(3\sqrt{q} - 1)q^{r+1-g}gh}{(q-1)(\sqrt{q}-1)} \qquad \text{for } 0 \leq r \leq g - 1. \tag{6.3}$$

Proof. It was shown in [115, Lemma 3(ii)] that, in the field $\mathbb{C}(t)$ of rational functions over the complex numbers, we have the identity

$$\sum_{n=0}^{g-2} A_n t^n + \sum_{n=0}^{g-1} q^{g-1-n} A_n t^{2g-2-n} = \frac{L(t) - ht^g}{(1-t)(1-qt)},$$

where $L(t)$ is the L-function of $\mathcal{X}/\mathbb{F}_q$. Letting $t \to 1$, we get

$$\sum_{n=0}^{g-2} A_n + \sum_{n=0}^{g-1} q^{g-1-n} A_n = \lim_{t \to 1} \frac{L(t) - ht^g}{(1-t)(1-qt)}$$

$$= \lim_{t \to 1} \frac{L'(t) - ght^{g-1}}{-(1-qt) - q(1-t)}$$

$$= \frac{L'(1) - gh}{q - 1}.$$

Now $L(t) = \prod_{i=1}^{2g}(1 - \omega_i t)$, with $|\omega_i| = \sqrt{q}$ for all $1 \le i \le 2g$, by Weil's proof of the Riemann Hypothesis for global function fields (see Theorem 1.5.3). By logarithmic differentiation,

$$\frac{L'(t)}{L(t)} = \sum_{i=1}^{2g} \frac{-\omega_i}{1 - \omega_i t}.$$

Putting $t = 1$ and using $L(1) = h$, we obtain

$$|L'(1)| \le h \sum_{i=1}^{2g} \frac{|\omega_i|}{|1 - \omega_i|} \le \frac{2gh\sqrt{q}}{\sqrt{q} - 1},$$

so

$$\left| \frac{L'(1) - gh}{q - 1} \right| \le \frac{(3\sqrt{q} - 1)gh}{(q - 1)(\sqrt{q} - 1)}.$$

Thus, by noting that

$$q^{g-1-r} A_r < \sum_{n=0}^{g-2} A_n + \sum_{n=0}^{g-1} q^{g-1-n} A_n \le \frac{(3\sqrt{q} - 1)gh}{(q - 1)(\sqrt{q} - 1)}$$

for $0 \le r \le g - 1$, the lemma now follows. $\square$

Lemma 6.3.14 *Let $\mathcal{X}/\mathbb{F}_q$ be an algebraic curve of genus g with at least one $\mathbb{F}_q$-rational point. Let c, m, n be three integers satisfying $c \ge 2$, $g \le m \le n < cm$, and*

$$cm - n \le g(1 - 2\log_q c) - 1 - \log_q \frac{(3\sqrt{q} - 1)g}{(q - 1)(\sqrt{q} - 1)}. \tag{6.4}$$

Let G be a fixed positive $\mathbb{F}_q$-rational divisor on $\mathcal{X}$ of degree n. Then there exists a positive divisor D of degree m such that $\mathcal{L}(cD - G) = \{0\}$.

Proof. By rewriting the inequality (6.4), we have

$$\frac{(3\sqrt{q} - 1)gq^{(cm-n)+1-g}h}{(q - 1)(\sqrt{q} - 1)} \le \frac{h}{c^{2g}}.$$

The desired result follows from Lemmas 6.3.12 and 6.3.13. $\square$

Now that we have shown the existence of the positive divisor D with the properties required in Proposition 6.3.9, we are ready to obtain an improvement to Theorem 6.3.5 using algebraic geometry.

Theorem 6.3.15 *For a prime power q and an integer $2 \le c \le A(q)$, we have*

$$D_q(c) \ge \frac{1}{c} - \frac{1}{A(q)} + \frac{(1 - 2\log_q c)}{c} \times \frac{1}{A(q)}. \tag{6.5}$$

Proof. Choose a family of curves $\mathcal{X}_i/\mathbb{F}_q$ with growing genus such that we have $\lim_{i\to\infty} N(\mathcal{X}_i)/g(\mathcal{X}_i) = A(q)$. Put $n_i = N(\mathcal{X}_i)$ and $g_i = g(\mathcal{X}_i)$. Let $\mathbf{P}_i$ be the set of all $\mathbb{F}_q$-rational points of $\mathcal{X}_i$ and put

$$G_i = \sum_{P \in \mathbf{P}_i} P.$$

For any fixed $\varepsilon > 0$, set

$$m_i = \left\lfloor \frac{n_i}{c} + \frac{(1 - 2\log_q c)g_i}{c} - \frac{\varepsilon g_i}{c} \right\rfloor.$$

Then

$$\lim_{i\to\infty} \frac{cm_i - n_i - (1 - 2\log_q c)g_i}{g_i} = -\varepsilon < 0.$$

Therefore, for all sufficiently large i, we have

$$cm_i - n_i \le g_i(1 - 2\log_q c) - 1 - \log_q \frac{(3\sqrt{q} - 1)g_i}{(q-1)(\sqrt{q}-1)}.$$

By Lemma 6.3.14, there exists a divisor D_i of $\mathcal{X}_i$, of degree m_i, such that $\mathcal{L}(cD_i - G_i) = \{0\}$, for each sufficiently large i. Thus, by Proposition 6.3.9, the code $C(D_i; G_i)$ is a c-FPC$(n_i, q^{\ell(D_i)})$. Hence,

$$
\begin{aligned}
D_q(c) &\ge \lim_{i\to\infty} \frac{\log_q q^{\ell(D_i)}}{n_i} \\
&\ge \lim_{i\to\infty} \frac{m_i - g_i + 1}{n_i} \\
&= \frac{1}{c} - \frac{1}{A(q)} + \frac{(1 - 2\log_q c)}{c} \times \frac{1}{A(q)} - \frac{\varepsilon}{cA(q)}.
\end{aligned}
$$

Since the above inequality holds for any $\varepsilon > 0$, we get

$$D_q(c) \ge \frac{1}{c} - \frac{1}{A(q)} + \frac{(1 - 2\log_q c)}{c} \times \frac{1}{A(q)}$$

by letting ε tend to 0. This completes the proof. $\square$

Remark 6.3.16 For $c \le A(q)$, Theorem 6.3.15 improves the lower bound in Theorem 6.3.5 by

$$\frac{(1 - 2\log_q c)}{c} \times \frac{1}{A(q)}.$$

6.4 Improvements to the Asymptotic Bound

In this section, we show several improvements to the asymptotic bound (6.5) in Theorem 6.3.15. The main tools continue to come from algebraic geometry.

First of all, if c is the characteristic of $\mathbb{F}_q$, then Lemma 6.3.11 can be improved.

Lemma 6.4.1 *Let c be the characteristic of $\mathbb{F}_q$. Let $\mathcal{X}/\mathbb{F}_q$ be a smooth projective curve of genus g with an $\mathbb{F}_q$-rational point P_0 and of class number h. Then, for any fixed integer $m \geq g$, the subgroup of $\mathcal{J}(\mathcal{X}/\mathbb{F}_q)$*

$$\{c[D - mP_0] : \ D > 0, \ \deg(D) = m\}$$

has order at least h/c^g.

The proof is exactly the same as that for Lemma 6.3.11, except that we use the fact that the c-rank of $\mathcal{J}(\mathcal{X}/\mathbb{F}_q)$ is at most g (see [110, page 39]).

Subsequently, Theorem 6.3.15 can be modified to the following.

Theorem 6.4.2 *For a prime power q with c being the characteristic of $\mathbb{F}_q$ satisfying $c \leq A(q)$, we have*

$$D_q(c) \geq \frac{1}{c} - \frac{1}{A(q)} + \frac{(1 - \log_q c)}{c} \times \frac{1}{A(q)}. \tag{6.6}$$

The second improvement to Theorem 6.3.15 is based on three papers by Randriam [123, 124, 125].

Let $\mathcal{X}/\mathbb{F}_q$ be a smooth projective curve of genus g with an $\mathbb{F}_q$-rational point P_0. For $r \geq 1$, we denote by W_r the subset

$$\{[D - rP_0] : \ D > 0, \ \deg(D) = r\}$$

of $\mathcal{J}(\mathcal{X}/\mathbb{F}_q)$.

Definition 6.4.3 Let $\mathcal{X}/\mathbb{F}_q$ be an algebraic curve with n distinct $\mathbb{F}_q$-rational points $P_1, \ldots, P_n$ (it is allowed for one of them to be equal to P_0). Let G be the divisor $G = P_1 + P_2 + \cdots + P_n$. For $c \geq 2$, the c-**frameproof Xing number** $x_c \overset{\text{def}}{=} x_c(\mathcal{X}, G)$ is defined to be the largest integer such that

$$c\mathcal{J}(\mathcal{X}/\mathbb{F}_q) - [G - \deg(G)P_0] \overset{\text{def}}{=} \{c[H] - [G - \deg(G)P_0] : \ [H] \in \mathcal{J}(\mathcal{X}/\mathbb{F}_q)\}$$

is not contained in W_{x_c}.

Note that $cJ(X/\mathbb{F}_q) - [G - \deg(G)P_0] \not\subseteq W_{x_c}$ is equivalent to the fact that, for any positive divisor D of degree x_c, there exists a divisor H of degree zero such that the divisor $G - D - \deg(G - D)P_0$ is not equivalent to cH.

Furthermore, we define the following quantities

$$A(X) \stackrel{\text{def}}{=} \frac{|X(\mathbb{F}_q)|}{g(X)}, \quad \delta_c(X) \stackrel{\text{def}}{=} \frac{\log_c |J(\mathbb{F}_q)[c]|}{g(X)} \tag{6.7}$$

and

$$\nu(X, G) \stackrel{\text{def}}{=} \frac{\deg(G)}{g(X)}, \quad \xi_c(X, G) \stackrel{\text{def}}{=} \frac{x_c(X, G)}{g(X)}, \tag{6.8}$$

where $J(\mathbb{F}_q)[c]$ stands for the c-torsion subgroup of $J(X/\mathbb{F}_q)$ and $g(X)$ is the genus of X. Then these quantities satisfy the following conditions.

Proposition 6.4.4 *We have*

(i) $\nu(X, G) \leq A(X)$;

(ii) $\xi_c(X, G) < 1$;

(iii) *If* $c = c_1 c_2$ *with* c_1, c_2 *relatively prime, then*

$$\delta_c(X) = \delta_{c_1}(X) \log_c(c_1) + \delta_{c_2}(X) \log_c(c_2);$$

(iv) $0 \leq \delta_c(X) \leq 2$ *for any* c;

(v) $0 \leq \delta_c(X) \leq 1$ *if* c *is a power of the characteristic of* $\mathbb{F}_q$.

Proof. Part (i) is clear since $\deg(G)$ is at most the number of $\mathbb{F}_q$-rational points on X. Part (ii) follows from the fact that W_g, where $g = g(X)$ is the genus of X, is the whole group $J(X/\mathbb{F}_q)$ (see [176]). Part (iii) follows from the Chinese Remainder Theorem. Parts (iv) and (v) follow from [110, pages 39 and 64]. □

Based on Proposition 6.3.9, we have the following.

Corollary 6.4.5 *Let* $X/\mathbb{F}_q$ *be an algebraic curve of genus* g *with at least* n $\mathbb{F}_q$-*rational points. Suppose* $n \geq g$, *then there exists a* c-$\mathrm{FPC}(n, q^{\lfloor (n+x_c)/c \rfloor - g + 1})$.

Proof. Set $d \stackrel{\text{def}}{=} \lfloor (n + x_c)/c \rfloor$. As $n \geq g > x_c$ and $c \geq 2$, one deduces $d < n$. On the other hand, since $cd - n \leq x_c$, by the definition of x_c, one has that there exists a divisor D_0 of degree zero such that $c[D_0] - [G - nP_0]$ is not contained in W_{cd-n}, where the divisor G is given by $G = \sum_{i=1}^n P_i$. This implies that the divisor $D = D_0 + dP_0$ satisfies $\mathcal{L}(cD - G) = \{0\}$. By Proposition 6.3.9, the algebraic geometry code $C(D; \sum_{i=1}^n P_i)$ defines a frameproof code with the desired parameters. □

Since x_c plays an important role for the parameters of a frameproof code derived from an algebraic geometry code, we give a lower bound on $x_c(X, G)$ (cf. Lemma 6.3.14).

Proposition 6.4.6 *Let $\mathcal{X}/\mathbb{F}_q$ be an algebraic curve of genus g. Set*

$$s(q) = 1 + \log_q \frac{3(\sqrt{q} - 1)}{(q - 1)(\sqrt{q} - 1)}.$$

Then

$$x_c(\mathcal{X}, G) \geq \lfloor g - \log_q |\mathcal{J}(\mathbb{F}_q)[c]| - \log_q g - s(q) \rfloor. \tag{6.9}$$

Proof. Let $\mathcal{J}$ denote the Jacobian $\mathcal{J}(\mathcal{X}/\mathbb{F}_q)$, so that $\mathcal{J}[c]$ denotes the c-torsion subgroup $\mathcal{J}(\mathbb{F}_q)[c]$, and let $h = |\mathcal{J}|$ be the class number. Then $[\mathcal{J} : c\mathcal{J}] = |\mathcal{J}[c]|$. Thus, we have, for G and n as above,

$$\log_q |c\mathcal{J} - [G - nP_0]| = \log_q |c\mathcal{J}| = \log_q h - \log_q |\mathcal{J}[c]|.$$

On the other hand, for $0 \leq r \leq g - 1$, by [177, Lemma 3.9], we have

$$\log_q |W_r| < \log_q h + r - g + \log_q g + s(q).$$

If $c\mathcal{J} - [G - nP_0]$ is contained in W_r, then we have $|c\mathcal{J} - [G - nP_0]| \leq |W_r|$ and hence

$$r > g - \log_q |\mathcal{J}[c]| - \log_q g - s(q),$$

and the desired result follows. $\qquad\square$

We say that a sequence $\{\mathcal{X}_k/\mathbb{F}_q\}$ of curves forms an ∞-sequence if $g(\mathcal{X}_k)$ tends to infinity. From Chapter 1, we know that $A(q)$ is the largest positive number such that there exists an ∞-sequence $\{\mathcal{X}_k/\mathbb{F}_q\}$ of curves satisfying $A(\mathcal{X}_k) \to A(q)$ as $k \to \infty$, where $A(\mathcal{X}_k)$ is as defined in (6.7).

We define two other quantities. Let $\delta_c^-(q)$ be the smallest real number such that there exists an ∞-sequence $\{\mathcal{X}_k/\mathbb{F}_q\}$ of curves satisfying

$$A(\mathcal{X}_k) \to A(q) \text{ and } \delta_c(\mathcal{X}_k) \to \delta_c^-(q) \text{ as } k \to \infty.$$

Let $\xi_c^+(q)$ be the largest real number such that there exist an ∞-sequence $\{\mathcal{X}_k/\mathbb{F}_q\}$ of curves and a family $\{G_k\}$ of divisors on $\{\mathcal{X}_k\}$ satisfying

$$A(\mathcal{X}_k) \to A(q) \text{ and } \xi_c(\mathcal{X}_k, G_k) \to \xi_c^+(q) \text{ as } k \to \infty.$$

Then, by Proposition 6.4.4, we have $0 \leq \delta_c^-(q) \leq 2$ and $0 \leq \xi_c^+(q) \leq 1$. Moreover, we have the following relationship between $\delta_c^-(q)$ and $\xi_c^+(q)$.

Lemma 6.4.7 *We have that*

$$\xi_c^+(q) \geq 1 - \delta_c^-(q) \log_q c.$$

The above lemma is just the asymptotic version of Proposition 6.4.6.

Now we are ready to give an asymptotic version of Corollary 6.4.5.

Theorem 6.4.8 *Let q be a prime power with $A(q) > 1$, then*

(i)

$$D_q(c) \geq \frac{1}{c} - \frac{1}{A(q)} + \frac{\xi_c^+(q)}{cA(q)}; \qquad (6.10)$$

(ii)

$$D_q(c) \geq \frac{1}{c} - \frac{1}{A(q)} + \frac{1 - \delta_c^-(q)\log_q c}{cA(q)}. \qquad (6.11)$$

Proof. Part (ii) can be derived by combining Lemma 6.4.7 and Part (i). Part (i) is actually the asymptotic version of Corollary 6.4.5. To see this, let $\{\mathcal{X}_k/\mathbb{F}_q\}$ be an ∞-sequence of curves and let $\{G_k\}$ be a family of divisors on $\{\mathcal{X}_k\}$ consisting of all the $\mathbb{F}_q$-rational points on $\mathcal{X}_k$, so that $A(\mathcal{X}_k) \to A(q) > 1$ and $\xi_c(\mathcal{X}_k, G_k) \to \xi_c^+(q)$ as $k \to \infty$. Then, for sufficiently large k, we have $g(\mathcal{X}_k) < \deg(G_k)$ and we can apply Corollary 6.4.5 to $(\mathcal{X}_k, G_k)$ to obtain a family of frameproof codes with rate lower bounded by

$$\frac{1}{N(\mathcal{X}_k)}\left(\left\lfloor \frac{N(\mathcal{X}_k) + x_c(\mathcal{X}_k, G_k)}{c} \right\rfloor - g(\mathcal{X}_k) + 1\right) \to \frac{1}{A(q)}\left(\frac{A(q) + \xi_c^+(q)}{c} - 1\right)$$

as $k \to \infty$, where $N(\mathcal{X}_k)$ is the number of $\mathbb{F}_q$-rational points on $\mathcal{X}_k$. This completes the proof. $\qquad \square$

Note that the bound in Theorem 6.3.15 immediately follows from Part (ii) of Theorem 6.4.8 and the fact that $\delta_c^-(q) \leq 2$. We now give the second improvement to Theorem 6.3.15 due to Randriam.

Theorem 6.4.9 *Let q be a prime power with $A(q) > 1$, then*

$$D_q(c) \geq \frac{1}{c} - \frac{1}{A(q)} + \frac{1 - 2\log_q c + \frac{\nu_p(c)}{\nu_p(q)}}{cA(q)}, \qquad (6.12)$$

where p is the characteristic of $\mathbb{F}_q$ and, for any integer x, $\nu_p(x)$ denotes the number a such that p^a exactly divides x.

Proof. Write $c = p^{\nu_p(c)}c'$. Put $c_1 = p^{\nu_p(c)}$ and $c_2 = c'$. By Proposition 6.4.4(iii), we have

$$\delta_c^-(q) \leq \delta_{c_2}^-(q)\log_c c_2 + \delta_{c_1}^-(q)\log_c c_1 \leq 2 - \frac{\nu_p(c)}{\nu_p(q)}\log_c q.$$

The desired result follows. $\qquad \square$

The best possible bound that we can derive from (6.10) is

$$D_q(c) \geq \frac{1}{c} - \frac{1}{A(q)} + \frac{1}{cA(q)} \qquad (6.13)$$

since $\xi_c^+(q) \leq 1$. For the rest of this section, we show that the bound (6.13) can be achieved for $c = 2$ if $A(q) > 4$. The result is given by Randriam in [124]. In fact, we prove in Proposition 6.4.13 that $\xi_2^+(q) = 1$ under the condition that $A(q) > 4$, but first we need some preparatory results.

A result similar to the following lemma can be found in [151, Proposition 1.6.12].

Lemma 6.4.10 *Let $\mathcal{X}/\mathbb{F}_q$ be an algebraic curve of genus g and let D be a divisor on $\mathcal{X}$ with $\deg(D) \leq g - 2$ and $\ell(D) = 0$. Then, for all points $P \in \mathcal{X}(\mathbb{F}_q)$ except for perhaps at most g of them, we have $\ell(D + P) = 0$.*

Proof. Suppose that there are $g + 1$ distinct points $P_1, P_2, \ldots, P_{g+1} \in \mathcal{X}(\mathbb{F}_q)$ such that $\ell(D + P) > 0$. Choose a nonzero function f_i from each space $\mathcal{L}(D + P_i)$. We claim that the functions $f_1, f_2, \ldots, f_{g+1}$ are $\mathbb{F}_q$-linearly independent.

Assume that there exist $\lambda_1, \ldots, \lambda_{g+1} \in \mathbb{F}_q$ such that $\sum_{i=1}^{g+1} \lambda_i f_i = 0$, i.e.,

$$-\lambda_1 f_1 = \sum_{i=2}^{g+1} \lambda_i f_i. \tag{6.14}$$

Since $\mathcal{L}(D) = \{0\}$, the left-hand side of (6.14) is either 0 or has a pole at P_1, while the right-hand side has no pole at P_1. This forces both sides to be equal to 0. This implies that $\lambda_1 = 0$. In the same way, we can show that all other λ_i are equal to 0 as well. Hence, we have a lower bound for the dimension of $\mathcal{L}\left(D + \sum_{i=1}^{g+1} P_i\right)$:

$$\ell\left(D + \sum_{i=1}^{g+1} P_i\right) \geq g + 1. \tag{6.15}$$

Since $\deg(D) \leq g - 2$, we have $\deg(D + \sum_{i=1}^{g+1} P_i) \leq 2g - 1$. Furthermore, by the Clifford Theorem [151, Theorem 1.6.13], we have

$$\ell\left(D + \sum_{i=1}^{g+1} P_i\right) \leq 1 + \frac{1}{2}\deg\left(D + \sum_{i=1}^{g+1} P_i\right) \leq g + \frac{1}{2}. \tag{6.16}$$

The conclusion in (6.16) contradicts (6.15). This completes the proof.   □

The following lemma is crucial for our result.

Lemma 6.4.11 *Let $\mathcal{X}/\mathbb{F}_q$ be an algebraic curve of genus g and let D be a divisor on $\mathcal{X}$ with $\deg(D) \leq g - 3$ and $\ell(D) = 0$. Then, for all points $P \in \mathcal{X}(\mathbb{F}_q)$ except for perhaps at most $4g$ of them, we have $\ell(D + 2P) = 0$.*

Proof. We assume that $|\mathcal{X}(\mathbb{F}_q)| \geq 4g > g$. Otherwise, there is nothing to prove. Without loss of generality, we may assume that $\deg(D) = g - 3$. By the Riemann-Roch Theorem (cf. [144, Chapter II, Theorem 5.4]), we have

$$\ell(D) = \deg(D) - g + 1 + \ell(K - D) = -2 + \ell(K - D)$$

and

$$\ell(D + 2P) = \deg(D + 2P) - g + 1 + \ell(K - D - 2P) = \ell(K - D - 2P),$$

where K is a canonical divisor. If we replace $K - D$ by B, the lemma can be rephrased as follows:

> If B is a divisor on $\mathcal{X}$ with $\deg(B) = g + 1$ and $\ell(B) = 2$, then there exist at most $4g$ points $P \in \mathcal{X}(\mathbb{F}_q)$ such that $\ell(B - 2P) > 0$.

By replacing B with an equivalent divisor, we may assume that B is a positive divisor and $\{1, f\}$ is an $\mathbb{F}_q$-basis of $\mathcal{L}(B)$.

For the rest of the proof, we need the **differential form** of a function. The differential forms of all functions in $\mathbb{F}_q(\mathcal{X})$ form a one-dimensional space. Moreover, if a function t is a local parameter at a point, then the differential form dt is not zero. The reader may refer to [151] for more details on differential forms.

First, we claim that the differential form df of f is not zero. Suppose that $df = 0$, then there exists a function h such that $f = h^p$, where p is the characteristic of $\mathbb{F}_q$. Hence, $\{1, h, f\}$ are linearly independent as they have distinct multiplicities at a certain point. This contradicts $\ell(B) = 2$ since $1, h, f \in \mathcal{L}(B)$.

Let $S \stackrel{\text{def}}{=} \{P \in \mathcal{X}(\mathbb{F}_q) : \ell(B - 2P) > 0\}$. We have to show that $|S| \le 4g$. For any point Q, we consider $\nu_Q(df)$ (note that if t is a local parameter at Q and f can be written as $u(dt)$ for some function u, then $\nu_Q(df)$ is defined to be $\nu_Q(u)$). We are in exactly one of the following four mutually exclusive situations:

(i) $Q \notin S \cup \operatorname{Supp}(B)$. Then $\nu_Q(f) \ge 0$ and $\nu_Q(df) \ge 0$.

(ii) $Q \in \operatorname{Supp}(B) \setminus S$. Then $\nu_Q(f) \ge 0$ and

$$\nu_Q(df) \ge \nu_Q(f) - 1 \ge -\nu_Q(B) - 1 \ge -2\nu_Q(B).$$

(iii) $Q \in S \setminus \operatorname{Supp}(B)$. Then we have $\ell(B) = 2$ and $\ell(B - 2Q) \ge 1$. By the inclusions $\mathcal{L}(B - 2Q) \subseteq \mathcal{L}(B - Q) \subseteq \mathcal{L}(B)$ and the fact that $1 \in \mathcal{L}(B) \setminus \mathcal{L}(B-Q)$, we must have $\ell(B-Q) = \ell(B-2Q) = 1$. Let $\alpha = f(Q)$, then $f - \alpha \in \mathcal{L}(B-Q) = \mathcal{L}(B-2Q)$. Hence, $\nu_Q(f-\alpha) \ge 2$ and, therefore, $\nu_Q(df) = \nu_Q(d(f - \alpha)) \ge 1$.

(iv) $Q \in S \cap \operatorname{Supp}(B)$. Since B is a positive divisor, we have $\nu_Q(B) \ge 1$ and $\nu_Q(f) \ge -\nu_Q(B)$. We claim that it is impossible to have $\nu_Q(B) = 1$ and $\nu_Q(f) = -\nu_Q(B)$ simultaneously. Suppose, to the contrary, that both the equalities hold. Then we must have $f \in \mathcal{L}(B) \setminus \mathcal{L}(B - Q)$ and $1 \in \mathcal{L}(B - Q) \setminus \mathcal{L}(B - 2Q)$. Thus, all the inclusions $\mathcal{L}(B - 2Q) \subseteq \mathcal{L}(B - Q) \subseteq \mathcal{L}(B)$ are strict. Hence, we have $\ell(B) \ge 3$, which is a contradiction.

Finally, since at least one of $\nu_Q(B) \ge 1$ and $\nu_Q(f) \ge -\nu_Q(B)$ is strict, we must have $\nu_Q(df) \ge -2\nu_Q(B) + 1$.

Summing all the inequalities, we obtain

$$2g - 2 = \deg(\mathrm{div}(df)) = \sum_Q \nu_Q(df) \geq -2\deg(B) + |S|.$$

The desired result follows from the above inequality since $\deg(B) = g + 1$. $\square$

Proposition 6.4.12 *Let $X/\mathbb{F}_q$ be an algebraic curve of genus g and let G be a divisor of degree n. If $|X(\mathbb{F}_q)| > 4g$, then there exists a divisor D of degree $\lfloor (n + g - 1)/2 \rfloor$ (or, equivalently, $g - 2 \leq \deg(2D - G) < g$) such that $\ell(2D - G) = 0$.*

Proof. Set $N \stackrel{\mathrm{def}}{=} \lfloor (n+g-1)/2 \rfloor - \lfloor (n-1)/2 \rfloor$. For each $0 \leq i \leq N$, we construct a divisor D_i with

$$\deg(D_i) = i + \left\lfloor \frac{n-1}{2} \right\rfloor \quad \text{and} \quad \ell(2D_i - G) = 0 \tag{6.17}$$

iteratively as follows.

Let D_0 be any divisor of degree $\lfloor (n - 1)/2 \rfloor$. Then $\deg(2D_0 - G) < 0$ and hence $\ell(2D_0 - G) = 0$.

Assume that we have constructed D_i satisfying (6.17) for up to some $i < N$. Then the divisor $A \stackrel{\mathrm{def}}{=} 2D_i - G$ satisfies $-2 \leq \deg(A) < g - 2$ and $\ell(A) = 0$. By Lemma 6.4.11, there exists a point $P \in X(\mathbb{F}_q)$ such that $\ell(A + 2P) = 0$. Now $D_{i+1} \stackrel{\mathrm{def}}{=} D_i + 2P$ satisfies (6.17).

Finally, we set $D = D_N$ and the desired result follows. $\square$

Proposition 6.4.13 *If $A(q) > 4$, then $\xi_2^+(q) = 1$.*

Proof. Let $\{X/\mathbb{F}_q\}$ be an ∞-sequence with $A(X) \to A(q)$. Then, for sufficiently large $g(X)$, one must have $N(X/\mathbb{F}_q) > 4g(X)$. Let $n = N(X/\mathbb{F}_q)$ and let $P_1, P_2, \ldots, P_n$ be n distinct $\mathbb{F}_q$-rational points on X. Put $G = P_1 + P_2 + \cdots + P_n$. By Proposition 6.4.12, there exists a divisor D of degree $\lfloor (n+g-1)/2 \rfloor$, where $g = g(X)$, such that $\ell(2D - G) = 0$, i.e., $2D - G$ is not equivalent to any positive divisor of degree $\deg(2D - G)$. This implies that $2\mathcal{J}(X/\mathbb{F}_q) - [G - \deg(G)P_0]$ is not contained in $W_{\deg(2D - G)}$, where P_0 is a fixed $\mathbb{F}_q$-rational point. By definition, the 2-frameproof Xing number $x_2(X, G)$ is at least $\deg(2D - G) \geq g - 2$. Thus, we have

$$1 \geq \xi_2^+(q) \geq \liminf_{g(X) \to \infty} \xi_2(X, G) = \liminf_{g(X) \to \infty} \frac{x_2(X, G)}{g(X)} \geq 1.$$

This completes the proof. $\square$

Combining Theorem 6.4.8 and Proposition 6.4.13, we immediately have

Theorem 6.4.14 *If q is a prime power with* $A(q) > 4$, *then*

$$D_q(2) \geq \frac{1}{2} - \frac{1}{2A(q)}.$$

Chapter 7

Key Distribution Schemes

Key distribution is one of the major problems in communication and network security. A key predistribution scheme (KPS) is a method by which a trusted authority (TA) distributes secret information among a set of users in such a way that every user in a group in some specified family of privileged subsets is able to compute a common key associated with that group. The key can be used for secure communication among the users in the group or can be used by the TA to send information privately to those users. In addition, certain coalitions of users (called forbidden subsets) outside a privileged subset must not be able to find out any information on the value of the key associated to that subset.

Key predistribution schemes refer to the key distribution methods by which secret keys are assigned to users in the network in advance before the actual communication occurs. The schemes that have been studied can be divided into two classes. In the first class, the schemes are based on the evaluation of multivariate symmetric polynomials over a finite field, first introduced by Blom [23], and generalized by Blundo *et al.* [26]. The construction methods are algebraic in nature and can be formulated using symmetric multilinear functions in linear spaces (see [119, 120]). The schemes in the second class are combinatorial. They are constructed by using **cover-free families**, a method first introduced by Mitchell and Piper [109], later studied by numerous researchers. Cover-free families are combinatorial objects that have interesting links with different subjects such as information theory, group testing, and cryptography.

The main parameter to measure the efficiency of a key predistribution scheme is the **information rate**, which is defined as the ratio between the size of the secret keys and the maximum size of the secret information received by the users.

7.1 Key Predistribution

Given a set $\mathcal{U}$ of users with $|\mathcal{U}| = n$, consider a family $\mathcal{P} \subseteq 2^{\mathcal{U}}$ of **privileged subsets** and a family $\mathcal{F} \subseteq 2^{\mathcal{U}}$ of **forbidden subsets**. In a $(\mathcal{P}, \mathcal{F}, n)$-**key pre-**

distribution scheme, or $(\mathcal{P}, \mathcal{F}, n)$-**KPS** for short, every user $i \in \mathcal{U}$ receives from a **trusted authority** (TA) a secret value (or a **fragment**) u_i such that, for every $P \in \mathcal{P}$, a **common key** from a key space K, say $k_P \in K$, associated with P, can be computed by every user $i \in P$, while the coalitions of users $F \in \mathcal{F}$ with $F \cap P = \emptyset$ do not obtain any information about k_P.

For $1 \leq i \leq n$, let U_i denote the set of all possible secret values that might be distributed to user i by the TA. For any subset of users $X \subseteq \mathcal{U}$, let U_X denote the Cartesian product $U_{i_1} \times \cdots \times U_{i_j}$, where $X = \{i_1, \ldots, i_j\}$ and $i_1 < \cdots < i_j$. We assume that there is a probability distribution on $U_{\mathcal{U}}$, and the TA chooses $u_{\mathcal{U}} \in U_{\mathcal{U}}$ according to this probability distribution. For every $i \in \mathcal{U}$ and $P \in \mathcal{P}$, let $\mathbf{U}_i$ and $\mathbf{K}_P$ denote the random variables corresponding, respectively, to the secret information u_i of user i and the common key k_P.

Definition 7.1.1 We say a key predistribution scheme is a $(\mathcal{P}, \mathcal{F}, n)$-**key predistribution scheme**, or $(\mathcal{P}, \mathcal{F}, n)$-**KPS** for short, if the following conditions are satisfied:

(i) Each user i in any privileged set P can compute k_P, i.e.,

$$H(\mathbf{K}_P \mid \mathbf{U}_i) = 0 \text{ for every } i \in P \in \mathcal{P};$$

(ii) If $F \in \mathcal{F}$ and $P \in \mathcal{P}$ are such that $P \cap F = \emptyset$, then

$$H(\mathbf{K}_P \mid (\mathbf{U}_j)_{j \in F}) = H(\mathbf{K}_P).$$

Observe that all common keys k_P are taken from the key space K. Moreover, we usually assume that all values of k_P are equally probable, i.e., $H(\mathbf{K}_P) = \log |K|$.

We will use the following notations. A $(\mathcal{P}, w, n)$-KPS (respectively, $(\mathcal{P}, \leq w, n)$-KPS), where $1 \leq w \leq n$, denotes a $(\mathcal{P}, \mathcal{F}, n)$-KPS in which the family $\mathcal{F}$ of forbidden subsets consists of all subsets of $\mathcal{U}$ with exactly w users (respectively, at most w users). Note that a $(\mathcal{P}, w, n)$-KPS is automatically a $(\mathcal{P}, \leq w, n)$-KPS. A $(t, \mathcal{F}, n)$-KPS (respectively, a $(\leq t, \mathcal{F}, n)$-KPS), where $2 \leq t \leq n$, denotes a $(\mathcal{P}, \mathcal{F}, n)$-KPS in which the family $\mathcal{P}$ of privileged subsets consists of all subsets of $\mathcal{U}$ with exactly t users (respectively, at most t users). In this chapter, we are interested mainly in (t, w, n)-KPS in which the family $\mathcal{P}$ of privileged subsets consists of all subsets of $\mathcal{U}$ with exactly t users, and the family $\mathcal{F}$ of forbidden subsets consists of all subsets with exactly w users.

We are interested in the efficiency of a KPS, as measured by the amount of secret information that is distributed to each user, as well as the total amount of information distributed to all the users, defined as follows.

Definition 7.1.2 In a $(\mathcal{P}, \mathcal{F}, n)$-KPS,

(i) the **information rate** is defined as

$$\rho = \frac{\log |K|}{\max_{i \in \mathcal{U}} H(\mathbf{U}_i)},$$

i.e., the ratio between the length of the common key and the maximum length of the secret information stored by the users;

(ii) the **total information rate** is defined as

$$\rho_T = \frac{\log |K|}{H(\mathbf{U}_\mathcal{U})},$$

i.e., the ratio between the length of the common keys and the length of the total secret information distributed to all the users.

Example 7.1.3 A trivial $(\mathcal{P}, \leq n, n)$-KPS is constructed by distributing, for every $P \in \mathcal{P}$, a random common key $k_P \in K$, where K is the key space, to all the users in P. In particular, in a trivial $(t, \leq n, n)$-KPS, every user receives as his secret information $\binom{n-1}{t-1}$ elements in K as possible values of the common keys. In this construction, we have

$$\rho = \frac{1}{\binom{n-1}{t-1}}$$

and

$$\rho_T = \frac{1}{\binom{n}{t}}.$$

7.2 Key Predistribution Schemes with Optimal Information Rates

In this section, we review, without proofs, some information-theoretic and combinatorial lower bounds for KPSs that were proved in [26] and [24], as well as some KPSs that attain some of these bounds. KPSs attaining such bounds are said to be optimal.

7.2.1 Bounds for KPSs

First, we consider a lower bound for the size of information held by each user in a (t, w, n)-KPS given in [26].

Theorem 7.2.1 ([26]) *Let $\mathcal{U}$ be a set of n users and let $t, w \leq n - t$ be integers. In a (t, w, n)-KPS, we have*

$$H(\mathbf{U}_i) \geq \binom{t + w - 1}{t - 1} H(\mathbf{K}),$$

for $i = 1, \ldots, n$.

Next, we look at lower bounds for the size of information held by each user in a more general $(\mathcal{P}, \mathcal{F}, n)$-KPS.

Theorem 7.2.2 ([24]) *Let* $\mathcal{U} = \{1, \ldots, n\}$ *be a set of n users.*

(i) *In a* $(\mathcal{P}, \mathcal{F}, n)$-KPS, suppose that, for any $P \in \mathcal{P}$, we have that $\mathcal{U} \backslash P \in \mathcal{F}$. Then, for $i = 1, \ldots, n$, we have

$$H(\mathbf{U}_i) \geq \tau_i H(\mathbf{K}),$$

where $\tau_i = |\{P \in \mathcal{P} : i \in P\}|$.

(ii) *In a* $(\leq n, \mathcal{F}, n)$-KPS, for $i = 1, \ldots, n$, we have

$$H(\mathbf{U}_i) \geq v_i H(\mathbf{K}),$$

where $v_i = |\{F \in \mathcal{F} : i \notin F\}|$.

We are also interested in the number of keys (secret values) that the TA has to generate in order to construct the scheme. Denoting by $\#\Gamma(\mathcal{P}, \mathcal{F}, n)$ the number of keys that the TA has to generate for a $(\mathcal{P}, \mathcal{F}, n)$-KPS Γ (and similarly for $(\mathcal{P}, \leq w, n)$-KPS, $(\leq t, \mathcal{F}, n)$-KPS, (t, w, n)-KPS, etc.), we have the following combinatorial lower bounds.

Theorem 7.2.3 ([24]) *Let* $\mathcal{U}$ *be a set of n users.*

(i) *For any* $(\mathcal{P}, \mathcal{F}, n)$-KPS Γ, suppose that, for any $P \in \mathcal{P}$, we have that $\mathcal{U} \backslash P \in \mathcal{F}$. Then, we have

$$\#\Gamma(\mathcal{P}, \mathcal{F}, n) \geq |\mathcal{P}|.$$

(ii) *For any* $(\leq n, \mathcal{F}, n)$-KPS Γ, we have

$$\#\Gamma(\leq n, \mathcal{F}, n) \geq |\mathcal{F}|.$$

In particular, we have the following corollary.

Corollary 7.2.4 *Let* $\mathcal{U} = \{1, \ldots, n\}$ *be a set of n users.*

(i) *In a* $(\leq n, w, n)$-KPS, for $i = 1, \ldots, n$, we have

$$H(\mathbf{U}_i) \geq \sum_{j=0}^{w} \binom{n-1}{j} H(\mathbf{K}).$$

(ii) *In a* $(\leq n, w, n)$-KPS Γ, we have

$$\#\Gamma(\leq n, w, n) \geq \sum_{j=0}^{w} \binom{n}{j}.$$

7.2.2 The Blom Scheme

In [23], Blom presented a $(2, w, n)$-KPS based on polynomial evaluation over a finite field. Let $q \geq n$ be a prime power. The TA chooses n distinct elements $s_i \in \mathbb{F}_q$, and gives s_i to user i $(1 \leq i \leq n)$. These values are known to all the users. The TA then constructs a random polynomial

$$f(x, y) = \sum_{i=0}^{w} \sum_{j=0}^{w} a_{ij} x^i y^j$$

with coefficients in $\mathbb{F}_q$, such that $a_{ij} = a_{ji}$ for all i, j. In other words, $f(x, y)$ is a random symmetric polynomial in two variables of degree at most w in each variable.

For $1 \leq j \leq n$, the TA computes the polynomial

$$f_j(x) = f(x, s_j) = \sum_{i=0}^{w} b_{ij} x^i,$$

and gives the $w + 1$ values b_{ij} to user j, where these $w + 1$ values constitute the secret information u_j.

The key associated with the pair of users $P = \{i, j\}$ is

$$k_P = f_i(s_j) = f_j(s_i).$$

Theorem 7.2.5 ([23]) *For any $w \geq 1$, there is a $(2, w, n)$-KPS with information rate*

$$\frac{1}{w + 1}$$

and total information rate

$$\frac{1}{\binom{w+2}{2}}.$$

Remark 7.2.6 The original Blom scheme was presented in the setting of MDS (maximum distance separable) codes. Here, we follow the formulation in [26, 155].

Example 7.2.7 Take $n = 3, q = 19$, and $w = 1$, and the public values are $s_1 = 5, s_2 = 8$, and $s_3 = 1$. Now suppose that the TA chooses the symmetric polynomial

$$f(x, y) = 4 + 11(x + y) + 3xy.$$

This gives rise to the following polynomials, which are sent to users $1, 2$, and 3, respectively:

$$f_1(x) = 2 + 7x$$
$$f_2(x) = 16 + 16x$$
$$f_3(x) = 15 + 14x.$$

The three keys determined by this information are: $k_{1,2} = 1, k_{1,3} = 9$, and $k_{3,2} = 13$.

7.2.3 The Blundo *et al.* Scheme

The Blom scheme was generalized by Blundo *et al.* [26] to a (t, w, n)-KPS that is described as follows.

Recall that a polynomial f in t variables is said to be **symmetric** if

$$f(x_1, \ldots, x_t) = f(x_{\sigma(1)}, \ldots, x_{\sigma(t)})$$

for every permutation σ over $\{1, \ldots, t\}$. Let $q \geq n$ be a prime power. The TA chooses n distinct elements $s_i \in \mathbb{F}_q$, and gives s_i to user i $(1 \leq i \leq n)$. These values are known to all the users. The TA chooses uniformly at random a symmetric polynomial $f(x_1, x_2, \ldots, x_t)$ in t variables with coefficients in $\mathbb{F}_q$ and of degree at most w in each variable. Every user $i \in \mathcal{U}$ receives as its secret information u_i the symmetric polynomial in $t - 1$ variables that is obtained from f by fixing the first variable to s_i, i.e.,

$$u_i(x_2, \ldots, x_t) = f(s_i, x_2, \ldots, x_t).$$

The common key corresponding to a privileged subset $P = \{i_1, i_2, \ldots, i_t\}$ is

$$k_P = f(s_{i_1}, s_{i_2}, \ldots, s_{i_t}) \in \mathbb{F}_q.$$

Since the polynomial f is symmetric, this value can be computed by every user in P.

Therefore, the secret information u_i of every user i consists of $\binom{t + w - 1}{t - 1}$ elements in the field $\mathbb{F}_q$, because this is the number of coefficients of a symmetric polynomial in $t - 1$ variables of degree at most w in each variable.

Theorem 7.2.8 ([26]) *For any $t \geq 2$ and $w \geq 1$, there is a (t, w, n)-KPS with information rate*

$$\rho = \frac{1}{\binom{t+w-1}{t-1}}$$

and total information rate

$$\rho_T = \frac{1}{\binom{t+w}{t}}.$$

7.2.4 The Fiat-Naor Scheme

In [57], Fiat and Naor gave a construction for $(\leq n, \mathcal{F}, n)$-KPS. Assume that the set K is an additive abelian group. For every $F \in \mathcal{F}$, a random value $s_F \in K$ is distributed to all the users in $\mathcal{U} \setminus F$. The common key for a set $P \subseteq \mathcal{U}$ is

$$k_P = \sum_{F \in \mathcal{F}, P \cap F = \emptyset} s_F.$$

In particular, in a Fiat-Naor $(\leq n, w, n)$-KPS, the secret information of every user consists of $\sum_{j=0}^{w} \binom{n-1}{j}$ elements in K.

We give an example to illustrate the scheme.

Example 7.2.9 Take $n = 3$, $q = 19$, and $w = 1$, and suppose that the TA chooses the values $s_\emptyset = 7, s_{\{1\}} = 5, s_{\{2\}} = 4$, and $s_{\{3\}} = 7$. Then the eight keys determined by this information are:

$$
\begin{aligned}
k_\emptyset &= s_\emptyset + s_{\{1\}} + s_{\{2\}} + s_{\{3\}} &= 4 \\
k_{\{1\}} &= s_\emptyset + s_{\{2\}} + s_{\{3\}} &= 18 \\
k_{\{2\}} &= s_\emptyset + s_{\{1\}} + s_{\{3\}} &= 0 \\
k_{\{3\}} &= s_\emptyset + s_{\{1\}} + s_{\{2\}} &= 16 \\
k_{\{1,2\}} &= s_\emptyset + s_{\{3\}} &= 14 \\
k_{\{1,3\}} &= s_\emptyset + s_{\{2\}} &= 11 \\
k_{\{2,3\}} &= s_\emptyset + s_{\{1\}} &= 12 \\
k_{\{1,2,3\}} &= s_\emptyset &= 7.
\end{aligned}
$$

Theorem 7.2.10 ([57]) *For any $w \geq 1$, there exists an (n, w, n)-KPS with information rate*

$$
\frac{1}{\sum_{j=0}^{w} \binom{n-1}{j}}
$$

and total information rate

$$
\frac{1}{\sum_{j=0}^{w} \binom{n}{j}}.
$$

Remark 7.2.11 According to the bounds in Section 7.2.1 all the schemes described so far have optimal information rates. The Blom and Blundo *et al.* schemes meet the bound in Theorem 7.2.1, while the Fiat-Naor scheme satisfies the bound in Corollary 7.2.4.

7.3 Linear Key Predistribution Schemes

Padró *et al.* [119, 120] have observed that most of the proposed key predistribution schemes are *linear*, that is, all random variables involved in those schemes are defined by linear mappings. They developed a general framework to study linear KPSs in [119, 120].

Lemma 7.3.1 *Let E, E_0, and E_1 be vector spaces over a finite field $\mathbb{F}_q$. Assume that $\phi_0 : E \to E_0$ and $\phi_1 : E \to E_1$ are linear mappings over $\mathbb{F}_q$, where ϕ_0 is surjective. Let x be a random element chosen from E. Then we have:*

(i) *The value of $x_0 = \phi_0(x)$ can be uniquely determined from $x_1 = \phi_1(x)$ if and only if $\ker \phi_1 \subseteq \ker \phi_0$;*

(ii) *The value of x_1 provides no information about the value of x_0 if and only if $\ker \phi_1 + \ker \phi_0 = E$.*

Proof. The proof is straightforward and we leave it to the reader. $\qquad\square$

Theorem 7.3.2 *Let $\mathcal{U} = \{1, \ldots, n\}$ be the set of users and let $\mathcal{P} \subseteq 2^{\mathcal{U}}$ and $\mathcal{F} \subseteq 2^{\mathcal{U}}$ be the families of privileged subsets and forbidden subsets, respectively. Let E and $E_i \neq \{0\}$, where $i = 0, 1, \ldots, n$, be vector spaces over a finite field $\mathbb{F}_q$. Suppose that there exist a surjective linear mapping $\pi_i : E \to E_i$ for every $i \in \mathcal{U}$ and a surjective linear mapping $\pi_P : E \to E_0$ for every privileged subset $P \in \mathcal{P}$ satisfying:*

(i) $\ker \pi_i \subseteq \ker \pi_P$ *for any $i \in P$;*

(ii) $\cap_{j \in F} (\ker \pi_j) + \ker \pi_P = E$ *for any $F \in \mathcal{F}$ satisfying $F \cap P = \emptyset$.*

Then, there exists a $(\mathcal{P}, \mathcal{F}, n)$-KPS with information rate

$$\rho = \frac{\dim E_0}{\max_{i \in \mathcal{U}} \dim E_i},$$

and total information rate

$$\rho_T = \frac{\dim E_0}{\dim E}.$$

Proof. Let the key space be $K = E_0$. We assume that the vector spaces E, $E_0, E_1, \ldots, E_n$, as well as the mappings π_i and π_P, are publicly known. In the initialization phase, the TA randomly chooses a vector $x \in E$ and sends the vectors $u_i = \pi_i(x) \in E_i$ to every user $i \in \mathcal{U}$ as his or her secret information.

Assume $P \in \mathcal{P}$ is a privileged subset. The key associated with P will be $k_P = \pi_P(x) \in E_0$. Set $\phi_0 = \pi_P$ and $\phi_1 = \pi_i$. Then, applying (i) in Lemma 7.3.1, we know that every user $i \in P$ can compute the key k_P.

On the other hand, for any forbidden subset $F \in \mathcal{F}$ satisfying $F \cap P = \emptyset$, we show that the coalition F obtains no information about k_P. Indeed, to this end, set $\phi_0 = \pi_P$ and $\phi_1 : E \to \prod_{j \in F} E_j$ defined by $\phi_1(x) = (\pi_j(x))_{j \in F}$. Note that $\phi_1(x)$ is the secret information known by the users in F. Since $\ker \phi_1 = \cap_{j \in F} \ker \pi_j$, we have $\ker \phi_0 + \ker \phi_1 = E$ and, applying (ii) in Lemma 7.3.1, we conclude that the users in F cannot obtain any information on $k_P = \phi_0(x)$. $\qquad\square$

Definition 7.3.3 A $(\mathcal{P}, \mathcal{F}, n)$-KPS is called **linear** if it can be constructed according to Theorem 7.3.2.

Example 7.3.4 The Blundo *et al.* scheme is a linear (t, w, n)-KPS. Let E_t (respectively, E_{t-1}) be the vector space of symmetric polynomials in t (respectively, $t - 1$) variables, with coefficients in $\mathbb{F}_q$ and of degree at most w in each variable. For each user $i \in \mathcal{U}$, we consider the surjective linear mapping

$$\pi_i : E_t \to E_{t-1}$$
$$f \mapsto \pi_i(f) = g_i,$$

where $g_i(x_1, \ldots, x_{t-1}) = f(s_i, x_1, \ldots, x_{t-1})$.

For any privileged subset $P = \{i_1, \ldots, i_t\} \subseteq \mathcal{U}$, we consider the linear mapping $\pi_P : E_t \to \mathbb{F}_q$ defined by $\pi_P(f) = f(s_{i_1}, \ldots, s_{i_t})$. Condition (i) in Theorem 7.3.2 is clearly satisfied by these π_i's, while (ii) is true by observing that $\ker \pi_P$ has codimension 1 and that $f(x_1, \ldots, x_t) = \prod_{i=1}^{t} \prod_{j \in F}(x_i - s_j) \in \cap_{j \in F}(\ker \pi_j) \setminus \ker \pi_P$ for any $F \in \mathcal{F}$ satisfying $F \cap P = \emptyset$. These surjective linear mappings define, by Theorem 7.3.2, a linear (t, w, n)-KPS that is obviously equivalent to the Blundo *et al.* (t, w, n)-KPS in Subsection 7.2.3.

Next, we describe a family of linear key predistribution schemes whose specification structure depends on a choice of vectors in a certain vector space. We first need some basic concepts about multilinear functions. Let V be a vector space over a finite field $\mathbb{F}_q$ and let V^t denote the vector space $V \times V \times \cdots \times V$, where there are t copies of V in the product. A mapping $\phi : V^t \to \mathbb{F}_q$ is called a **multilinear function** (or t-**linear function** if we want to specify the number of variables) if, for any $i = 1, 2, \ldots, t$, we have

$$\phi(\mathbf{v}_1, \ldots, \mathbf{v}_i + \mathbf{v}_i', \ldots, \mathbf{v}_t) = \phi(\mathbf{v}_1, \ldots, \mathbf{v}_i, \ldots, \mathbf{v}_t) + \phi(\mathbf{v}_1, \ldots, \mathbf{v}_i', \ldots, \mathbf{v}_t)$$

and

$$\phi(\mathbf{v}_1, \ldots, \lambda \mathbf{v}_i, \ldots, \mathbf{v}_t) = \lambda \phi(\mathbf{v}_1, \ldots, \mathbf{v}_i, \ldots, \mathbf{v}_t) \text{ for any } \lambda \in \mathbb{F}_q.$$

If $\dim V = k$, the t-linear functions $\phi : V^t \to \mathbb{F}_q$ form a vector space $\mathcal{J}^t(V)$ over $\mathbb{F}_q$ of dimension k^t.

A t-linear function $\phi \in \mathcal{J}^t(V)$ is **symmetric** if

$$\phi(\mathbf{v}_1, \ldots, \mathbf{v}_t) = \phi(\mathbf{v}_{\sigma(1)}, \ldots, \mathbf{v}_{\sigma(t)})$$

for any permutation σ over $\{1, \ldots, t\}$ and any $(\mathbf{v}_1, \ldots, \mathbf{v}_t) \in V^t$.

The symmetric t-linear functions form a subspace $\mathcal{S}_t(V) \subset \mathcal{J}^t(V)$. If $\{\mathbf{e}_1, \ldots, \mathbf{e}_k\}$ is a basis of V, a symmetric t-linear function ϕ is uniquely determined by the values $h_{i_1, \ldots, i_t} = \phi(\mathbf{e}_{i_1}, \ldots, \mathbf{e}_{i_t})$, where $1 \leq i_1 \leq \cdots \leq i_t \leq k$. Therefore, $\dim \mathcal{S}_t(V) = \binom{t+k-1}{t}$.

Theorem 7.3.5 ([119]) *Let V be a vector space with $\dim V = k$ over a finite field $\mathbb{F}_q$ and let $\{\mathbf{v}_1, \ldots, \mathbf{v}_n\}$ be a set of vectors in V such that every subset of $w + 1$ vectors is linearly independent. Then, for every t with $2 \leq t \leq q$, there exists a (t, w, n)-KPS with information rate*

$$\rho = \frac{1}{\binom{t+k-2}{t-1}}$$

and total information rate

$$\rho_T = \frac{1}{\binom{t+k-1}{t}}.$$

Proof. Let $V = \mathbb{F}_q^k$. For each user $i \in \mathcal{U}$, consider the surjective linear mapping $\pi_i : \mathcal{S}_t(V) \to \mathcal{S}_{t-1}(V)$ defined by

$$\pi_i \phi(x_1, \ldots, x_t) = \phi(\mathbf{v}_i, x_1, \ldots, x_{t-1}).$$

Let $P = \{i_1, \ldots, i_t\} \in \mathcal{P}$ be a privileged subset. We consider the surjective linear mapping $\pi_P : \mathcal{S}_t(V) \to \mathbb{F}_q$ defined by

$$\pi_P \phi(x_1, \ldots, x_t) = \phi(\mathbf{v}_{i_1}, \ldots, \mathbf{v}_{i_t}).$$

Using elementary linear algebra, we can show that the linear mappings $(\pi_i)_{i \in \mathcal{U}}$ and $(i_P)_{P \in \mathcal{P}}$ satisfy the conditions in Theorem 7.3.2, and hence define a (t, w, n)-KPS whose information rates are derived from the dimensions of $\mathcal{S}_t(V)$ and $\mathcal{S}_{t-1}(V)$. $\qquad\square$

A nice construction of KPS along the line of symmetric linear functions is from error-correcting codes.

Theorem 7.3.6 *Let $C \subseteq \mathbb{F}_q^n$ be an $[n, k]$-linear code such that the dual code $C^\perp$ has minimum distance $d^\perp$. Then, for every t with $2 \le t \le q$, there exists a $(t, d^\perp - 2, n)$-KPS with common keys in $\mathbb{F}_q$, with information rate*

$$\rho = \frac{1}{\binom{t+k-2}{t-1}}$$

and total information rate

$$\rho_T = \frac{1}{\binom{t+k-1}{t}}.$$

Proof. Apply Theorem 7.3.5 to the vectors $\mathbf{v}_1, \ldots, \mathbf{v}_n \in \mathbb{F}_q^k$ corresponding to the columns of a generator matrix of C. Since every $d^\perp - 1$ columns are linearly independent, the result follows immediately. $\qquad\square$

If $C \subseteq \mathbb{F}_q^n$ is a Reed-Solomon code, the construction results in the KPS by Blundo *et al.* [26]. Let $C^\perp$ be the trivial $[n, 1, n]$-code over $\mathbb{F}_q$, spanned by the vector $(1, \ldots, 1) \in \mathbb{F}_q^n$. Then C is an $[n, n-1]$-code over $\mathbb{F}_q$. The $(2, n-2, n)$-KPS obtained from C is similar to the one proposed by Matsumoto and Imai in [103].

Example 7.3.7 If the number n of users is not too large, many known constructions of codes over small fields can be used to construct KPSs. For $q = 2, 3, 4, 5, 7, 8, 9$ and $n \le 256$, the best known codes can be found in the tables in [72]. For example, since there exists a $[256, 224, 9]$-linear code over $\mathbb{F}_2$, we have a $(2, 7, 256)$-KPS over $\mathbb{F}_2$ with the size of the secret information of each user $\binom{t+k-2}{t-1} = 32$. Observe that the size of the secret information of the Blom $(2, 7, 256)$-KPS is at least $(w+1) \log q = 8 \log 256 = 64$ because we have to take $q \ge n$.

Example 7.3.8 Since there exists a $[240, 210, 10]$-linear code over $\mathbb{F}_3$, for $t = 2, 3$, we obtain a $(t, 8, 240)$-KPS over $\mathbb{F}_3$ with the size of the secret information $\binom{t+30-2}{t-1} \log 3$, which is 48 if $t = 2$, and 738 if $t = 3$. In the KPS from the Blundo *et al.* scheme, the size is 72 if $t = 2$, and 357 if $t = 3$. In the second case, the size of the Blundo *et al.* KPS is smaller, but computations must be done in the larger field $\mathbb{F}_{241}$ instead of $\mathbb{F}_3$.

7.4 Key Predistribution Schemes from Algebraic Geometry

We saw in the previous section that, if the number of users is not too large and the common keys are taken from a small field, KPSs can be obtained by using some of the best known codes [72] with suitable parameters, and the size of the secret information for the resulting KPSs can be smaller than that for the schemes with optimal information rates. However, we may be interested in the construction of KPSs for an arbitrarily large number of users, with common secret keys of constant length that are secure against coalitions formed by a constant fraction of the users. Specifically, we want to construct (t, w, n)-KPSs over a fixed base field $\mathbb{F}_q$ such that t is a constant value with $2 \leq t \leq q$ and $w = cn$ for some constant c with $0 < c < 1$.

By the asymptotic Gilbert-Varshamov bound (see Theorem 2.5.3), for every prime power q and for every δ with $0 \leq \delta < (q - 1)/q$, there exists a sequence $\{C_n\}$ of linear codes over $\mathbb{F}_q$ such that C_n has length n, minimum distance $d_n \geq \delta n$, and dimension k_n with $\lim_{n \to \infty} k_n/n = 1 - H_q(\delta)$, where H_q is the q-ary entropy function. In particular, there exist a positive integer n_0 and a constant α with $0 < \alpha < 1$ such that $k_n \geq (1 - \alpha)n$ for all $n \geq n_0$.

Consider $c < \delta < (q - 1)/q$. Then $\delta n \geq cn + 2$ if n is not too small. For these parameters, the dual code $C_n^\perp$ has dimension at most αn and dual minimum distance at least $cn + 2$. Therefore, for every sufficiently large n, there exists a (t, cn, n)-KPS over a fixed base field $\mathbb{F}_q$ in which the size of the secret information of a user is at most $\binom{t+\alpha n-2}{t-1} \log q$. Asymptotically, the size of the key of a user is $O(n^{t-1})$. For a Blundo *et al.* (t, cn, n)-KPS, since the size of the base field depends on the number of users, the size of the key of a user is at least $\binom{t+cn-1}{t-1} \log n$, which asymptotically is $O(n^{t-1} \log n)$. Therefore, the construction based on error-correcting codes is asymptotically better by a factor of $\log n$. We now consider a construction based on algebraic geometry that realizes this coding approach.

Although the proof of the Gilbert-Varshamov bound is non-constructive, constructions of codes exceeding this bound have nonetheless been obtained by using Goppa's algebraic geometry codes [70]. We analyze in the following the parameters of the KPSs that are obtained by using linear codes from algebraic curves.

Let $\mathcal{X}$ be a smooth projective curve over $\mathbb{F}_q$ and let g be its genus. Let $Q, P_1, \ldots, P_n$ be distinct $\mathbb{F}_q$-rational points on $\mathcal{X}$. Consider the divisor $G = mQ$, where $2g - 1 \le m < n$, and the Riemann-Roch space $\mathcal{L}(G) = \{f \in \mathbb{F}_q(\mathcal{X}) : \mathrm{div}(f) + G \ge 0\}$, which is a vector space with dimension $\deg(G) - g + 1 = m - g + 1$. Then, from Section 2.4, we know that

$$C = \{(f(P_1), \ldots, f(P_n)) : f \in \mathcal{L}(G)\} \subseteq \mathbb{F}_q^n$$

is an $[n, k]$-linear code with $k = m - g + 1$ and its dual code has minimum distance $d^\perp \ge \deg(G) - 2g + 2 = m - 2g + 2$.

Theorem 7.4.1 *Let $\mathcal{X}$ be a smooth projective curve over $\mathbb{F}_q$, let g be its genus, and let N be the number of $\mathbb{F}_q$-rational points on $\mathcal{X}$. Consider positive integers t, w, n with $2 \le t \le q$ and $2g + w < n \le N - 1$. Then there exists a (t, w, n)-KPS with information rate*

$$\rho = \frac{1}{\binom{t+w+g-1}{t-1}}$$

and total information rate

$$\rho_T = \frac{1}{\binom{t+w+g}{t}}.$$

Proof. Taking $m = 2g + w$, we obtain a linear code C with dimension $k = g + w + 1$ and dual minimum distance $d^\perp \ge m - 2g + 2 = w + 2$. By Theorem 7.3.6, the code C provides a KPS with the desired information rate and total information rate. $\square$

In Theorem 7.4.1, the case $g = 0$ corresponds to that of Reed-Solomon codes, and hence we obtain the KPS by Blundo *et al.* [26]. By using curves of higher genus, we can obtain efficient KPSs over a constant base field for an arbitrarily large number of users. We analyze in the following the family of KPSs obtained from the family of curves given by Garcia and Stichtenoth in [66].

Let q be a prime power. There exists a family of curves $\{\mathcal{X}_j\}_{j>0}$ defined over $\mathbb{F}_{q^2}$ such that the number of $\mathbb{F}_{q^2}$-rational points on $\mathcal{X}_j$ is $N_j \ge (q-1)q^j$ and the genus of $\mathcal{X}_j$ is $g_j \le q^j$ (cf. Example 2.5.6). By Theorem 7.4.1, for positive integers j, t, w with $2 \le t \le q$ and $2q^j + w < (q-1)q^j - 1$, there exists a (t, w, n)-KPS over the base field $\mathbb{F}_{q^2}$ with $n = (q-1)q^j - 1$, whose information rate satisfies

$$\rho \ge \frac{1}{\binom{t+w+q^j-1}{t-1}}.$$

Note that only KPSs over fields of the form $\mathbb{F}_{q^2}$ are obtained in the construction this way, but there exist other families of algebraic geometry codes that provide similar results for general fields $\mathbb{F}_q$ (see, for example, [163]).

We proceed to compare this family of KPSs with the one by Blundo *et al.* [26]. For simplicity, we consider only the case $t = 2$. In addition, we consider KPSs that are secure against coalitions formed by a constant fraction of the users. By using the curves of Garcia-Stichtenoth, we obtain an infinite family of $(2, w, n)$-KPSs over a fixed base field $\mathbb{F}_{q^2}$ with the size of the user's secret at most $(w + \frac{n}{q-1} + 2)2 \log q$. The size of the user's secret in the Blundo *et al.* $(2, w, n)$-KPS is at least $(w + 1) \log n$. We assume $w = cn - 1$ for some constant c such that $0 < c < 1 - 2/(q - 1)$, where $q > 3$. The upper bound on c guarantees that the condition $w < n - 2q^j$ is satisfied. In this situation, the above KPS improves the size of the secret information in the Blundo *et al.* KPS if

$$j \geq 2 \left(1 + \frac{2}{c(q - 1)} \right).$$

Indeed, in this case, since $n = (q - 1)q^j - 1$, it follows that

$$(w + 1) \log n \geq cnj \log q \geq \left(cn + \frac{2n}{q - 1} \right) 2 \log q \geq \left(cn + \frac{n}{q - 1} + 2 \right) 2 \log q.$$

Remark 7.4.2 In the (t, w, n)-KPS by Blundo *et al.* [26], every coalition $F \subseteq \mathcal{U}$ with $|F| = w + 1$ can compute the secret information of *all* privileged subsets. In the more general construction based on error-correcting codes, the common keys that a coalition $F \subseteq \mathcal{U}$ with $|F| \geq w + 1$ can obtain depend on the users involved in it. In a way, the construction has *ramp* security. From Theorem 7.3.6, we know a coalition $F \subseteq \mathcal{U}$ can obtain the secret key of a privileged subset $P \subseteq \mathcal{U}$ if and only if one of the vectors $\mathbf{v}_i$ with $i \in P$ is a linear combination of the vectors in $\{\mathbf{v}_j : j \in F\}$, where the vector $\mathbf{v}_i$ is the ith column of a generator matrix of the corresponding code C. For instance, in the construction in Theorem 7.4.1, the coalitions with at least $w + 2g + 1$ users can obtain the common keys of all privileged subsets, while the coalitions F with $w + 1 \leq |F| \leq w + 2g$ obtain only partial information of the common keys.

Other constructions of KPSs from algebraic curves can be found in [40].

7.5 Key Predistribution Schemes from Cover-Free Families

In the previous sections, we considered algebraic constructions of KPSs. In the rest of this chapter, we study combinatorial constructions of KPSs from cover-free families. The idea of using cover-free families to construct key predistribution schemes is due to Mitchell and Piper [109], under the name of key distribution pattern. However, the study of cover-free families dates back to the early 1960s. Kautz and Singleton [88] first studied these combinatorial objects under the name of superimposed binary codes. These codes are related to file retrieval, data communication, and magnetic memory. In 1985, Erdős, Frankl, and Füredi [55] studied cover-free families as a generalization of Sperner systems. Furthermore, cover-free families have also been discussed by numerous researchers in other contexts such as information theory, combinatorics, communication, and many other topics in cryptography and information security.

Definition 7.5.1 Let $X = \{x_1, \ldots, x_N\}$ be a set of N elements (points) and let $\mathcal{B} = \{B_1, \ldots, B_T\}$ be a set of T subsets (blocks) of X. Then, for $\mathcal{P}$ and $\mathcal{F}$ collections of subsets of $\{1, \ldots, T\}$, $(X, \mathcal{B})$ is called a $(\mathcal{P}, \mathcal{F}, T)$-**cover-free family** (or $(\mathcal{P}, \mathcal{F}, T)$-**CFF** for short) provided that

$$\bigcap_{i \in P} B_i \not\subseteq \bigcup_{j \in F} B_j$$

for all $P \in \mathcal{P}$ and $F \in \mathcal{F}$ such that $P \cap F = \emptyset$.

We can construct a $(\mathcal{P}, \mathcal{F}, T)$-KPS from a $(\mathcal{P}, \mathcal{F}, T)$-CFF as follows. Assume that a $(\mathcal{P}, \mathcal{F}, T)$-CFF is publicly known. Let the users in the KPS be denoted by $\{1, \ldots, T\}$. For $1 \leq j \leq N$, the TA chooses a random value $s_j \in \mathbb{F}_q$ and gives s_j to every user i for which $x_j \in B_i$. Thus, user i receives $|B_i|$ elements in $\mathbb{F}_q$ as his or her secret information.

The common key k_P for a privileged set P is defined by

$$k_P = \sum_{j : x_j \in \cap_{i \in P} B_i} s_j.$$

Note that each user in P can compute k_P. However, if F is a coalition such that $F \cap P = \emptyset$, then there is at least one element x_j such that $x_j \in \cap_{i \in P} B_i$, but $x_j \notin \cup_{i \in F} B_i$. Hence, F does not know s_j and so has no information about k_P. We have the following theorem.

Theorem 7.5.2 *Suppose $(X, \mathcal{B})$ is a $(\mathcal{P}, \mathcal{F}, T)$-CFF. Then there exists a $(\mathcal{P}, \mathcal{F}, T)$-KPS with information rate*

$$\rho = \frac{1}{\max\{|B_i| : 1 \leq i \leq T\}},$$

and total information rate

$$\rho_T = \frac{1}{N}.$$

We have already seen above that a KPS can be constructed from a cover-free family. However, given an arbitrary $(\mathcal{P}, \mathcal{F}, T)$, it is not easy to find a good $(\mathcal{P}, \mathcal{F}, T)$-CFF that results in an efficient $(\mathcal{P}, \mathcal{F}, T)$-KPS.

In this section, we study the class of CFFs that give rise to (s, t, T)-KPSs, and give some bounds and constructions for this class of CFFs.

Definition 7.5.3 Let X be a set of N elements (points) and let $\mathcal{B}$ be a set of T subsets (blocks) of X. Then $(X, \mathcal{B})$ is called an (s, t)-**cover-free family** provided that, for any s blocks $B_1, \ldots, B_s$ in $\mathcal{B}$ and t other blocks $B'_1, \ldots, B'_t$ in $\mathcal{B}$, one has

$$\bigcap_{i=1}^{s} B_i \not\subseteq \bigcup_{j=1}^{t} B'_j.$$

In other words, no intersection of s blocks is contained in the union of t other blocks. Sometimes, we use the notation (s, t)-CFF(N, T) to denote an (s, t)-cover-free family $(X, \mathcal{B})$ in which $|X| = N$ and $|\mathcal{B}| = T$. We call $(X, \mathcal{B})$ k-**uniform** if $|B| = k$ for all $B \in \mathcal{B}$.

Example 7.5.4 Let
$$X = \{1, 2, 3, 4, 5, 6, 7, 8, 9\}$$

and
$$\mathcal{B} = \{B_1, B_2, B_3, B_4, B_5, B_6, B_7, B_8, B_9, B_{10}, B_{11}, B_{12}\},$$

where

$$
\begin{aligned}
B_1 &= \{4, 5, 6, 7, 8, 9\} & B_2 &= \{2, 3, 5, 6, 8, 9\} \\
B_3 &= \{2, 3, 4, 6, 7, 8\} & B_4 &= \{2, 3, 4, 5, 7, 9\} \\
B_5 &= \{1, 2, 3, 7, 8, 9\} & B_6 &= \{1, 3, 4, 6, 7, 9\} \\
B_7 &= \{1, 3, 4, 5, 8, 9\} & B_8 &= \{1, 3, 5, 6, 7, 8\} \\
B_9 &= \{1, 2, 3, 4, 5, 6\} & B_{10} &= \{1, 2, 4, 5, 7, 8\} \\
B_{11} &= \{1, 2, 5, 6, 7, 9\} & B_{12} &= \{1, 2, 4, 6, 8, 9\}.
\end{aligned}
$$

Then $(X, \mathcal{B})$ is a $(2, 1)$-CFF$(9, 12)$.

Cover-free families have been studied under different names, such as superimposed codes, key distribution patterns, non-adaptive group testing algorithms, etc. For instance, a $(1, t)$-cover-free family is exactly a t-cover-free family studied by Erdős *et al.* [55], and a $(2, t)$-cover-free family was introduced, under the name of **key distribution pattern**, by Mitchell and Piper [109].

In the following, we describe two combinatorial objects equivalent to cover-free families: coverings of hypergraphs [54, 174] and disjunct systems [150, 174].

A **hypergraph** is a pair (V, E), where V is a set of elements called **nodes** (or **vertices**), and E is a collection of nonempty subsets of V called **hyper-edges**.

Let ℓ, u, n be integers such that $0 < \ell < u < n$. Set $[n] = \{1, 2, \ldots, n\}$. Define $P_{n;\ell,u} = \{X \subseteq [n] : \ell \le |X| \le u\}$. Define a hypergraph $G_{n;\ell,u} = (V, E)$ as follows. Let the set of vertices be $V = P_{n;\ell,u}$, and let the set of hyperedges E be the collection of intervals defined as follows:

$$E = \{\{C \subseteq [n] : Y_1 \subseteq C \subseteq Y_2\} \; : \; |Y_1| = \ell, |Y_2| = u, Y_1, Y_2 \subseteq [n]\}.$$

Definition 7.5.5 A **covering** of a hypergraph is a subset S of vertices such that each hyperedge of the hypergraph contains at least one vertex of S.

Example 7.5.6 Let $\ell = 1$, $u = 2$, and $n = 4$. Then the hypergraph $G_{4;1,2}$ has as its set of vertices

$$P_{4;1,2} = \{\{1\}, \{2\}, \{3\}, \{4\}, \{1,2\}, \{1,3\}, \{1,4\}, \{2,3\}, \{2,4\}, \{3,4\}\}$$

and its set of hyperedges

$$E = \left\{ \begin{array}{l} \{\{1\}, \{1,2\}\}, \{\{1\}, \{1,3\}\}, \{\{1\}, \{1,4\}\} \\ \{\{2\}, \{1,2\}\}, \{\{2\}, \{2,3\}\}, \{\{2\}, \{2,4\}\} \\ \{\{3\}, \{1,3\}\}, \{\{3\}, \{2,3\}\}, \{\{3\}, \{3,4\}\} \\ \{\{4\}, \{1,4\}\}, \{\{4\}, \{2,4\}\}, \{\{4\}, \{3,4\}\} \end{array} \right\}.$$

Then, any of the following subsets of $P_{4;1,2}$:

$$\{\{1\}, \{2\}, \{3\}, \{4\}\}$$
$$\{\{1,2\}, \{1,3\}, \{1,4\}, \{2,3\}, \{2,4\}, \{3,4\}\}$$
$$\{\{1\}, \{2\}, \{3\}, \{1,4\}, \{2,4\}, \{3,4\}\}$$

is a covering of $G_{4;1,2}$.

Let $(X, \mathcal{B})$ be a set system, where $X = \{x_1, x_2, \ldots, x_N\}$ and $\mathcal{B} = \{B_1, B_2, \ldots, B_T\}$. Recall that the incidence matrix of $(X, \mathcal{B})$ is the $T \times N$ matrix $A = (a_{ij})$, where

$$a_{ij} = \begin{cases} 1 & \text{if } x_j \in B_i \\ 0 & \text{if } x_j \notin B_i. \end{cases}$$

Conversely, given an incidence matrix, we can define an associated set system in an obvious way.

Definition 7.5.7 A set system $(X, \mathcal{B})$ is an (i, j)-**disjunct system** if, for any $P, Q \subseteq X$ such that $|P| \leq i$, $|Q| \leq j$ and $P \cap Q = \emptyset$, there exists a $B \in \mathcal{B}$ such that $P \subseteq B$ and $Q \cap B = \emptyset$. An (i, j)-disjunct system is denoted as an (i, j)-DS(N, T) if $|X| = N$ and $|\mathcal{B}| = T$.

Example 7.5.8 Let

$$X = \{1, 2, 3, 4, 5, 6, 7, 8, 9, 10, 11, 12\}$$

and

$$\mathcal{B} = \{B_1, B_2, B_3, B_4, B_5, B_6, B_7, B_8, B_9\},$$

where

$$B_1 = \{5, 6, 7, 8, 9, 10, 11, 12\} \quad B_2 = \{2, 3, 4, 5, 9, 10, 11, 12\}$$
$$B_3 = \{2, 3, 4, 5, 6, 7, 8, 9\} \quad B_4 = \{1, 3, 4, 6, 7, 9, 10, 12\}$$
$$B_5 = \{1, 2, 4, 7, 8, 9, 10, 11\} \quad B_6 = \{1, 2, 4, 6, 8, 9, 11, 12\}$$
$$B_7 = \{1, 3, 4, 5, 6, 8, 10, 11\} \quad B_8 = \{1, 2, 3, 5, 7, 8, 10, 12\}$$
$$B_9 = \{1, 2, 4, 5, 6, 7, 11, 12\}.$$

Then $(X, \mathcal{B})$ is a $(2,1)$-DS$(12,9)$. For instance, let $P = \{1, 2\}$ and $Q = \{3\}$. Then we have $B_5 \in \mathcal{B}$ such that $P \subseteq B_5$ and $B_5 \cap Q = \emptyset$.

Theorem 7.5.9 *Let $i, j, T,$ and N be integers with $i, j \leq N$. The following statements are equivalent:*

(i) *There exists a covering of $G_{T;i,T-j}$ of size N.*

(ii) *There exists an (i, j)-DS(T, N).*

(iii) *There exists an (i, j)-CFF(N, T).*

Proof. First, we show that (i) is equivalent to (ii). Observe that S is a covering of $G_{T;i,T-j}$ if and only if, for any $Y_1, Y_2 \subseteq [T]$, $Y_1 \subseteq Y_2$, $|Y_1| = i$, $|Y_2| = T - j$, there exists $C \in S$ such that $Y_1 \subseteq C \subseteq Y_2$. This is equivalent to saying that, for any $Y_1, Y_3 \subseteq [T]$, $|Y_1| = i$, $|Y_3| = T - (T - j) = j$, $Y_1 \cap Y_3 = \emptyset$, there exists $C \in S$ such that $Y_1 \subseteq C$ and $Y_3 \cap C = \emptyset$, which is in turn equivalent to $([T], S)$ being an (i, j)-DS(T, N).

For the equivalence of (ii) and (iii), it is easy to see that A is the incidence matrix of an (i, j)-DS(T, N) if and only if A^T, the transpose of A, is the incidence matrix of an (i, j)-CFF(N, T). $\qquad \square$

7.5.1 Bounds for Cover-Free Families

We now discuss some bounds for cover-free families.

We begin with a trivial construction for (s, t)-CFFs. For any integers $T \geq s > 0$, define $X = \{x_A : A \subseteq [T], |A| = s\}$. For $1 \leq i \leq T$, define $B_i = \{x_A \in X : i \in A\}$ and $\mathcal{B} = \{B_i : i \in [T]\}$. It is easy to see then that $(X, \mathcal{B})$ is an (s, t)-CFF$(\binom{T}{s}, T)$, for any $t \leq T - s$.

We are interested in (s, t)-CFF(N, T)'s with better performance than this trivial construction, i.e., (s, t)-CFF(N, T)'s with $N < \binom{T}{s}$. Note that, given the values of s and t, there is a trade-off between N and T in an (s, t)-CFF(N, T). More precisely, we are interested in (s, t)-CFF(N, T)'s for which T is as large as possible while $s, t,$ and N are given, or equivalently, the value N is as small as possible while $s, t,$ and T are fixed.

Let $N((s, t), T)$ denote the minimum value of N in an (s, t)-CFF(N, T). It is desirable to determine the value of $N((s, t), T)$. Unfortunately, as shown in [174, 29], computing the value of $N((s, t), T)$ turns out to be rather hard.

Theorem 7.5.10 *Given integers $s, t, T,$ and k, the problem of deciding $N((s, t), T) \leq k$ is **NP**-complete.*

Proof. For given integers n, ℓ, and u, let $G_{n;\ell,u}$ be a hypergraph defined above. Set

$$\tau(G_{n;\ell,u}) = \min\{|S| : S \text{ is a covering of } G_{n;\ell,u}\}.$$

It has been proved in [29] that, for any given integers n, ℓ, u, and k, the problem of deciding $\tau(G_{n;\ell,u}) \leq k$ is **NP**-complete. Theorem 7.5.10 now follows from Theorem 7.5.9 immediately.   □

Theorem 7.5.11 *We have the following known bounds on cover-free families:*

(i) ([55]) *In a k-uniform $(1,t)$-CFF(N,T), we have*

$$T \leq \binom{N}{\lceil \frac{k}{t} \rceil} \Big/ \binom{k-1}{\lceil \frac{k}{t} \rceil - 1};$$

(ii) ([50], [130]) *For any $t \geq 2$, given any $(1,t)$-CFF(N,T), we have*

$$N \geq c\frac{t^2}{\log t} \log T,$$

where the constant c is shown to be approximately $1/2$ in D'yachkov and Rykov [50], approximately $1/4$ in Füredi [62], and approximately $1/8$ in Ruszinkó [130];

(iii) ([51]) *For positive integers s, t, and T, we have*

$$N((s,t),T) \geq t(s \log T - \log t - s \log s);$$

(iv) ([54]) *For positive integers s, t, and T, we have*

$$N((s,t),T) \geq \binom{s+t-1}{s} \log(T - t - s + 2);$$

(v) ([54]) *For any $\epsilon > 0$, we have*

$$N((s,t),T) \geq (1-\epsilon)\frac{(s+t-2)^{s+t-2}}{(s-1)^{s-1}(t-1)^{t-1}} \log(T - t - s + 2)$$

for all sufficiently large T;

(vi) ([160]) *For $s, t \geq 1$ and $T \geq s + t > 2$, we have*

$$N((s,t),T) \geq 2c\frac{\binom{s+t}{t}}{\log(s+t)} \log T,$$

where the constant c is the same as in (ii);

(vii) ([160], [98]) *For any integers $s, t \geq 1$ and $T \geq \max\{\lfloor (s+t+1)/2 \rfloor^2, 5\}$,*

$$N((s,t),T) \geq 0.7c\frac{\binom{s+t}{s}(s+t)}{\log \binom{s+t}{s}} \log T,$$

where the constant c is the same as in (ii);

(viii) ([159]) *For positive integers s, t, and T, we have*

$$N((s,t),T) \leq \min\left\{ \left\lceil \frac{(s+t)\log T}{-\log p} \right\rceil, \left\lceil \frac{(s+t-1)\log 2T}{-\log p} \right\rceil \right\},$$

where $p = 1 - s^s t^t/(s+t)^{s+t}$.

The proofs for the bounds in Theorem 7.5.11 are complicated and rather involved. The interested reader may refer to the original papers for the details. We only give a proof for (i), adapted from [174]. We first need a lemma on set systems.

Lemma 7.5.12 ([174]) *Suppose $(X, \mathcal{B})$ is an r-uniform set system, where $|X| = k$. If, for any t blocks $B_1, \ldots, B_t \in \mathcal{B}$, we have $|\cup_{i=1}^t B_i| < k$ and $tr \geq k$, then $|\mathcal{B}| \leq \binom{k-1}{r}$.*

Proof of Theorem 7.5.11(i). When $t = 1$, the theorem is obvious. Therefore, we assume henceforth that $t \geq 2$.

Suppose $(X, \mathcal{B})$ is a k-uniform $(1, t)$-CFF. For $B \in \mathcal{B}$, define

$$\mathbb{N}_r(B) = \{R \subset B : |R| = r, \, \exists B' \neq B, B' \in \mathcal{B} \text{ such that } R \subset B'\},$$

where $r = \lceil \frac{k}{t} \rceil$. We claim that

$$\mathbb{N}_r(B) \leq \binom{k-1}{r}.$$

First, suppose $\mathbb{N}_r(B) < t$. We consider two cases: $t \geq k$ and $t < k$. If $t \geq k$, then $r \leq 1$. It follows that $\mathbb{N}_r(B) < k$ because $(X, \mathcal{B})$ is a $(1, t)$-CFF. Therefore, $\mathbb{N}_r(B) \leq \binom{k-1}{r}$. If $t < k$, then $\mathbb{N}_r(B) < k$. In this case, we also have $\mathbb{N}_r(B) \leq \binom{k-1}{r}$.

In the case $\mathbb{N}_r(B) \geq t$, let $R_1, \ldots, R_t \in \mathbb{N}_r(B)$. Then we have $|\cup_{i=1}^t R_i| \leq k - 1$ by the property of $\mathbb{N}_r(B)$ and that $(X, \mathcal{B})$ is a $(1, t)$-CFF. Since $tr \geq k$, by applying Lemma 7.5.12 to the set system $(B, \mathbb{N}_r(B))$, we have

$$|\mathbb{N}_r(B)| \leq \binom{k-1}{r}.$$

Therefore, for each $B \in \mathcal{B}$, there are at least

$$\binom{k}{r} - \binom{k-1}{r} = \binom{k-1}{r-1}$$

subsets $R \subset B$, which are not contained in any $B' \in \mathcal{B} \setminus \{B\}$. It follows that

$$|\mathcal{B}| \binom{k-1}{r-1} \leq \binom{N}{r}.$$

This completes the proof. □

For some special cases, the bound in Theorem 7.5.11(i) can be attained. For example, Erdős *et al.* [55] gave a probabilistic construction for a $2t$-uniform $(1, t)$-CFF with

$$T = (N^2/(4t - 2)) - o(N^2).$$

However, in general, it remains open whether the bound in Theorem 7.5.11(i) is the best possible for $(1, t)$-CFFs.

Since it is hard to compute the exact value of $N((s, t), T)$ for large values of T, some authors have considered another measurement for the efficiency of CFFs, called the **performance rate**, defined as

$$R(X, \mathcal{B}) = \frac{\log T}{N},$$

for a cover-free-family $(X, \mathcal{B})$. We are interested in the asymptotic behavior of this rate.

Definition 7.5.13 For fixed s and t, we define the **asymptotic rate** of (s, t)-CFFs as

$$R(s, t) = \lim_{T \to \infty} \frac{\log T}{N((s, t), T)}.$$

The following theorem is proved in [94].

Theorem 7.5.14 *For any integers s and t, we have*

$$R(s, t) \leq \min_{0 < x < s} \min_{0 < y < t} \frac{R(s - x, t - y)}{R(s - x, t - y) + (x + y)^{x+y}/(x^x y^y)}.$$

Definition 7.5.15 An (s, t)-CFF(N, T) is said to be **optimal** if $N = N((s, t), T)$.

Although computing the value of $N((s, t), T)$ is hard in general, some cover-free families with small parameters are known to be optimal. There are cases where the trivial solution given at the beginning of this subsection results in optimal solutions. For example, from [54] and [90], we know that, whenever $T \leq s + t + t/s$ or $T \leq \frac{(t+1)s}{s-1} - \sqrt{\frac{36t}{s-1}}$, we have $N((s, t), T) = \binom{T}{s}$, which implies that the trivial solution is an optimal solution. We list some of these optimal families in Table 7.1 (taken from [89, 90]).

7.5.2 Constructions from Error-Correcting Codes

A nice construction for cover-free families is to use error-correcting codes (see [55, 150]).

Let Y be an alphabet of q elements. Recall that a q-ary (n, T, d)-code is a

TABLE 7.1: Optimal (s,t)-CFF(N,T)'s.

$T =$	5	6	7	8	9	10	11-12	16 -20
$N((1,2),T) =$	5	6	7	8	9	9	9	
$N((1,3),T) =$	5	6	7	8	9	10	11-12	16
$N((2,2),T) =$	10	14	14	14	18	18-20	20-22	22-26
$N((2,3),T) =$	10	15	21	24-28	26-30	30	33-45	45-48

set C of T vectors in Y^n such that the Hamming distance between any two distinct codewords in C is at least d. Consider a q-ary (n, T, d)-code C. We write each codeword as $\mathbf{c}_i = (c_{i1}, \ldots, c_{in})$ with $c_{ij} \in Y$, where $1 \leq i \leq T$ and $1 \leq j \leq n$. Set $X = [n] \times Y$ and $\mathcal{B} = \{B_i : 1 \leq i \leq T\}$, where for each $1 \leq i \leq T$, we define $B_i = \{(j, c_{ij}) : 1 \leq j \leq n\}$. It is easy to see that $|X| = nq$, $|\mathcal{B}| = T$ and $|B_i| = n$. For each choice of $i \neq k$, we have $|B_i \cap B_k| = |\{(j, c_{ij}) : 1 \leq j \leq n\} \cap \{(j, c_{kj}) : 1 \leq j \leq n\}| = |\{j : c_{ij} = c_{kj}\}| \leq n - d$.

It is straightforward to show that $(X, \mathcal{B})$ is a $(1, t)$-CFF(nq, T) if the condition $t < \frac{n}{n-d}$ holds. We thus obtain the following theorem.

Theorem 7.5.16 *If there is a q-ary (n, T, d)-code, then there exists a $(1,t)$-CFF(nq, T) provided that $t < \frac{n}{n-d}$.*

Next, we describe the concatenation construction given in [160, 90], which is a powerful method for constructing a larger cover-free family from smaller ones.

Definition 7.5.17 *A matrix $M = (m_{k\ell})_{N \times T}$ with entries from $[q]$ is called an (s,t) **separating matrix** of size (N, T, q) if, for any pair of sets $I, J \subset [T]$ such that $|I| = s$, $|J| = t$ and $I \cap J = \emptyset$, there exists an integer $x \in [N]$ such that the sets $\{m_{xi} : i \in I\}$ and $\{m_{xj} : j \in J\}$ are disjoint.*

Let M be an (s, t) separating matrix of size (N_0, T_0, q) and let $A = (a_{ij})_{N_1 \times q}$ be the incidence matrix of an (s, t)-DS(q, N_1). Denote by $\mathbf{b}_1, \mathbf{b}_2, \ldots, \mathbf{b}_q$ the columns of A. We construct an $N_0 N_1 \times T_0$ matrix $B = M \diamond A$ by substituting the entry i in M by $\mathbf{b}_i$. It can be verified that the resulting matrix B is the incidence matrix of an (s, t)-CFF$(N_0 N_1, T_0)$ (see [157]). From Theorem 7.5.9, we have the following result.

Theorem 7.5.18 *If there exist an (s, t) separating matrix of size (N, T, q) and an (s, t)-CFF(N_0, q), then there exists an (s, t)-CFF(NN_0, T).*

Separating matrices can be constructed from error-correcting codes, hence making another link between cover-free families and error-correcting codes.

Theorem 7.5.19 ([160]) *If there exists a q-ary (N, T, d)-code, then there exists an (s, t) separating matrix of size (N, T, q) provided that*

$$\frac{d}{N} > 1 - \frac{1}{st}.$$

Proof. Let M be an $N \times T$ matrix such that each column is a codeword in a q-ary (N, T, d)-code. Since the minimum distance of the code is d, we know that any two distinct codewords have at most $N - d$ entries in common. It is easy to see that the matrix M is an (s, t) separating matrix if $N > st(N - d)$, proving the desired result. □

7.5.3 Constructions from Designs

Let Y be a set of v elements (points), and let $\mathcal{A} = \{A_1, A_2, \ldots, A_\beta\}$ be a family of k-subsets of Y (blocks). Recall from Definition 5.2.10 that $(Y, \mathcal{A})$ is a t-(v, k, λ) design if every subset of t points occurs in exactly λ blocks. It can be shown by elementary counting that a t-(v, k, λ) design is also a t'-(v, k, λ') design for $1 \leq t' \leq t$, where

$$\lambda' = \frac{\lambda\binom{v-t'}{t-t'}}{\binom{k-t'}{t-t'}}.$$

Theorem 7.5.20 ([155]) *If there exists an $(s+1)$-(n, k, λ) design, then there exists an (s, t)-CFF$(\lambda\binom{n}{s+1}/\binom{k}{s+1}, n)$ provided*

$$t < \frac{n - s}{k - s}.$$

Proof. Let $(Y, \mathcal{A})$ be an $(s + 1)$-(n, k, λ) design, where $Y = \{y_1, y_2, \ldots, y_n\}$ and $\mathcal{A} = \{A_1, A_2, \ldots, A_\beta\}$. We consider the dual $(X, \mathcal{B})$ of $(Y, \mathcal{A})$, defined by $X = \{A_1, A_2, \ldots, A_\beta\}$ and $B_i = \{A_r \in X : y_i \in A_r\}$. We show that $(X, \mathcal{B})$ is an (s, t)-CFF.

For each s-subset Δ of Y, there are exactly $\lambda(n - s)/(k - s)$ elements (blocks) from $\mathcal{A}$ that contain Δ. For any given t-subset $\Lambda \subseteq Y$ satisfying $\Delta \cap \Lambda = \emptyset$, and for each $y \in \Lambda$, there are λ blocks that contain $\Delta \cup \{y\}$. Thus, the number of blocks from $\mathcal{A}$ that contain Δ and at least one member from Λ is at most λt. Since $\lambda t < \lambda(n - s)/(k - s)$, it follows that there exists a block A_i from $\mathcal{A}$ that contains Δ such that $A_i \cap \Lambda = \emptyset$. It is then easy to verify that $(X, \mathcal{B})$ is indeed an (s, t)-CFF$(\lambda\binom{n}{s+1}/\binom{k}{s+1}, n)$. □

Corollary 7.5.21 *An $(s + 1)$-$(n, k, 1)$ design gives rise to an (s, t)-CFF(N, T), where*

$$N = \frac{\binom{n}{s+1}}{\binom{k}{s+1}} = \frac{(n - s)\binom{n}{s}}{(k - s)\binom{k}{s}}, \quad T = n, \quad and \quad t < \frac{n - s}{k - s}.$$

Example 7.5.22 From [109, 155], we know that an inversive plane is a 3-$(q^2 + 1, q + 1, 1)$ design. Such a design is known to exist whenever q is a prime power. Applying Corollary 7.5.21, we know that there exists a $(2, q)$-CFF$(q(q^2 + 1), q^2 + 1)$. Taking $q = 3$, we obtain a $(2, 3)$-CFF$(30, 10)$, which is optimal since $N((2, 3), 10) = 30$ (see [89]). Note that the codewords of weight 4 in the binary extended Hamming $[8, 4, 4]$-code form a 3-$(8, 4, 1)$ design [89]. It follows that there is a $(2, 2)$-CFF$(14, 8)$, which is optimal as well.

Next, we show another construction from a super-simple design. The concept of a super-simple t-design was introduced by Gronau and Mullin [73]. The construction of cover-free families from super-simple designs is proposed by Kim and Lebedev [89].

Definition 7.5.23 A **super-simple** t-(v, k, λ) **design** is a t-(v, k, λ) design with $\lambda > 1$ in which the intersection of any two blocks has at most t elements.

Theorem 7.5.24 ([89]) *A super-simple* s-(n, k, λ) *design gives rise to an* $(s, \lambda - 1)$-CFF$(\lambda \binom{n}{s} / \binom{k}{s}, n)$.

Proof. Let $(Y, \mathcal{A})$ be a super-simple s-(n, k, λ) design, where $Y = \{y_1, y_2, \ldots, y_n\}$ and $\mathcal{A} = \{A_1, A_2, \ldots, A_\beta\}$. As in the proof of Theorem 7.5.20, let $(X, \mathcal{B})$ be the dual of $(Y, \mathcal{A})$, where $X = \{A_1, A_2, \ldots, A_\beta\}$ and $B_i = \{A_r \in X : y_i \in A_r\}$. We show that $(X, \mathcal{B})$ is an $(s, \lambda - 1)$-CFF$(\lambda \binom{n}{s} / \binom{k}{s}, n)$.

For any s points $y_{i_1}, \ldots, y_{i_s} \in Y$, there are exactly λ blocks from $\mathcal{A}$ that contain these s points, i.e.,

$$|B_{i_1} \cap B_{i_2} \cap \cdots \cap B_{i_s}| = \lambda.$$

Consider any other t points $y_{j_1}, \ldots, y_{j_t}$, where $t = \lambda - 1$. Since no two (or more) blocks of a super-simple s-design can have more than s common points, for any ℓ with $1 \leq \ell \leq t$, we have

$$|B_{i_1} \cap B_{i_2} \cap \cdots \cap B_{i_s} \cap B_{j_\ell}| \leq 1.$$

It follows that

$$\left| B_{i_1} \cap B_{i_2} \cap \cdots \cap B_{i_s} \cap \left(\cup_{\ell=1}^t B_{j_\ell} \right) \right| \leq t < \lambda.$$

We then obtain

$$B_{i_1} \cap B_{i_2} \cap \cdots \cap B_{i_s} \not\subseteq \cup_{\ell=1}^t B_{j_\ell},$$

for otherwise, we would have $B_{i_1} \cap B_{i_2} \cap \cdots \cap B_{i_s} \subseteq \cup_{\ell=1}^t B_{j_\ell}$, which implies that

$$\left| B_{i_1} \cap B_{i_2} \cap \cdots \cap B_{i_s} \cap \left(\cup_{\ell=1}^t B_{j_\ell} \right) \right| = |B_{i_1} \cap B_{i_2} \cap \cdots \cap B_{i_s}| = \lambda,$$

a contradiction. This shows that $(X, \mathcal{B})$ is a cover-free family with the desired parameters. $\square$

Note that it is easy to see that an $(s + 1)$-$(n, k, 1)$ design is a super-simple s-$(n, k, (n - s)/(k - s))$ design. Therefore, in this case, Theorem 7.5.24 implies Theorem 7.5.20.

7.5.4 Constructions from Perfect Hash Families

Let n and m be integers such that $2 \leq m \leq n$. Let A be a set of size n and let B be a set of size m. Recall that a function h from A to B is called a hash function. We say a hash function $h : A \to B$ is **perfect** on a subset $X \subseteq A$ if h is injective when restricted to X. Let w be an integer such that $2 \leq w \leq m$ and let $\mathcal{H}$ be a set of hash functions from A to B.

Definition 7.5.25 We say $\mathcal{H}$ is an (n, m, w)-**perfect hash family** if, for any $X \subseteq A$ with $|X| = w$, there exists at least one function $h \in \mathcal{H}$ such that h is perfect on X. We use $\mathrm{PHF}(N; n, m, w)$ to denote an (n, m, w)-perfect hash family with $|\mathcal{H}| = N$.

Example 7.5.26 Let $A = \{1, \ldots, 9\}$ and $B = \{1, 2, 3\}$. Then $\mathcal{H} = \{h_1, h_2, h_3, h_4\}$ given in Table 7.2 is a $\mathrm{PHF}(4; 9, 3, 3)$.

TABLE 7.2: An example of $\mathrm{PHF}(4; 9, 3, 3)$.

	1	2	3	4	5	6	7	8	9
h_1	1	3	2	2	3	2	3	1	1
h_2	1	3	1	3	1	2	2	2	3
h_3	1	2	2	1	3	3	1	2	3
h_4	3	3	2	1	1	3	2	1	2

Perfect hash families, introduced by Mehlhorn [106], originally arose as part of compiler design. They have applications to operating systems, language translation systems, hypertext, hypermedia, file managers, and information retrieval systems (see Czech, Havas, and Majewski [46] for a survey of these results). There are also numerous applications to cryptography, for example, to broadcast encryption [57], secret sharing [17], key distribution patterns, cover-free families, and secure frameproof codes [157].

The next section contains some more detailed discussions on perfect hash families, but for now we give two constructions of cover-free families from perfect hash families.

The first construction is a direct construction from perfect hash families and works only for $(1, t)$-CFFs. Assume that $\mathcal{H}$ is a $\mathrm{PHF}(N; T, m, t+1)$ from A to B. Let $A = \{1, 2, \ldots, T\}$ and $B = \{1, 2, \ldots, m\}$. We define

$$X = \mathcal{H} \times B = \{(h, j) : h \in \mathcal{H}, j \in B\}.$$

For each $1 \leq i \leq T$, we define a subset (block) B_i of X by

$$B_i = \{(h, h(i)) : h \in \mathcal{H}\},$$

and $\mathcal{B} = \{B_i : 1 \leq i \leq T\}$. Then $(X, \mathcal{B})$ is a $(1, t)$-CFF(Nm, T). Indeed, $|X| = Nm$ and $|\mathcal{B}| = T$. For any $t + 1$ blocks $B_{i_1}, \ldots, B_{i_t}, B_j$, since $\mathcal{H}$ is a

PHF($N; T, m, t+1$), there exists a hash function $h \in \mathcal{H}$ such that h restricted to $\{i_1, \ldots, i_t, j\}$ is one-to-one. It follows that $h(i_1), \ldots, h(i_t), h(j)$ are $t+1$ distinct elements in B, which also implies that $(h, h(i_1)), \ldots, (h, h(i_t)), (h, h(j))$ are $t+1$ distinct elements in $B_{i_1}, \ldots, B_{i_t}, B_j$, respectively. Hence, the union of any t blocks in $\mathcal{B}$ cannot cover any remaining block. Thus, we have shown the following result.

Theorem 7.5.27 *If there exists a* PHF($N; T, m, t+1$), *then there exists a* $(1, t)$-CFF(Nm, T).

The second construction from perfect hash families (cf. [155]) provides a method of building a larger cover-free family from smaller ones. It has a similar flavor as the coding construction in Subsection 7.5.2.

The construction works as follows.

Let $(X_0, \mathcal{B}_0)$ be an (s, t)-CFF(N_0, T_0) and let $\mathcal{H} = \{h_1, \ldots, h_N\}$ be a perfect hash family PHF($N; T, T_0, s+t$) from $\{1, \ldots, T\}$ to $\{1, \ldots, T_0\}$. Consider N copies of $(X_0, \mathcal{B}_0)$, denoted by $(X_1, \mathcal{B}_1), \ldots, (X_N, \mathcal{B}_N)$, where X_i and X_j are disjoint sets, i.e., $X_i \cap X_j = \emptyset$, for all $i \neq j$. For each $1 \leq j \leq N$, let $X_j = \{x_1^{(j)}, \ldots, x_{N_0}^{(j)}\}$ and $\mathcal{B}_j = \{B_1^{(j)}, \ldots, B_{T_0}^{(j)}\}$. Clearly, $(X_j, \mathcal{B}_j)$ is an (s, t)-CFF(N_0, T_0). We construct a pair $(X, \mathcal{B})$ with

$$X = X_1 \cup \cdots \cup X_N \quad \text{and} \quad \mathcal{B} = \{B_1, \ldots, B_T\},$$

where $B_i = B_{h_1(i)}^{(1)} \cup \cdots \cup B_{h_N(i)}^{(N)} = \cup_{j=1}^{N} B_{h_j(i)}^{(j)}$, for $1 \leq i \leq T$. In other words, an element of $\mathcal{B}$ is a union of elements of $\mathcal{B}_j$, $1 \leq j \leq N$, chosen through an application of the perfect hash family. We show that $(X, \mathcal{B})$ is an (s, t)-CFF(NN_0, T).

Clearly, $|X| = NN_0$ and $|\mathcal{B}| = T$. For any $s + t$ blocks $B_{i_1}, \ldots, B_{i_s}, B_{j_1}, \ldots, B_{j_t}$, there exists at least one hash function $h_k \in \mathcal{H}$ which is one-to-one on $\{i_1, \cdots, i_s, j_1, \ldots, j_t\}$. For any $1 \leq k \leq N$, since $(X_k, \mathcal{B}_k)$ is an (s, t)-CFF(N_0, T_0), we have

$$\left| \bigcap_{u=1}^{s} B_{i_u} \setminus \bigcup_{v=1}^{t} B_{j_v} \right| \geq \left| \bigcap_{u=1}^{s} B_{h_k(i_u)}^{(k)} \setminus \bigcup_{v=1}^{t} B_{h_k(j_v)}^{(k)} \right|$$

$$\geq 1,$$

proving the desired result. Thus, we have the following result.

Theorem 7.5.28 *Suppose that there exists an* (s, t)-CFF(N_0, T_0) *and a* PHF($N; T, T_0, s+t$). *Then there exists an* (s, t)-CFF(NN_0, T).

7.6   Perfect Hash Families and Algebraic Geometry

We have seen in Subsection 7.5.4 that we can construct cover-free families from perfect hash families. In this section, we show how to construct perfect hash families from algebraic geometry.

Let $\text{PHF}(N; n, m, w)$ denote an (n, m, w)-perfect hash family $\mathcal{H}$ with $|\mathcal{H}| = N$, and let $N(n, m, w)$ denote the minimum N for which a $\text{PHF}(N; n, m, w)$ exists, for given n, m, and w. We are interested in determining these values. In particular, we are interested in the asymptotic behavior of $N(n, m, w)$ as a function of n when m and w are fixed. Bounds for $N(n, m, w)$ have been studied by numerous authors (see, for example, [5, 15, 18, 59, 106, 160]). In particular, it has been proved in [106] that, for fixed m and w, $N(n, m, w)$ is $\Theta(\log n)$. However, this result is non-constructive. It was believed that it would be difficult to give explicit constructions that asymptotically meet this existential result, but explicit constructions turned out to be possible with the use of algebraic geometry (cf. [170]). We first review some bounds for $N(n, m, w)$ without proof.

Theorem 7.6.1 ([59]) *We have that*

$$N(n, m, w) \geq \frac{\binom{n-1}{w-2} m^{w-2} \log(n - w + 2)}{\binom{m-1}{w-2} n^{w-2} \log(m - w + 2)}.$$

As noted in [15], this lower bound is approximately equal to

$$\frac{m^{w-2}}{m(m - 1)(m - 2) \cdots (m - (w - 1))} \frac{\log n}{\log(m - w + 2)}$$

as $n \to \infty$ with w and m fixed.

A weaker bound, due to Mehlhorn [106], is the following:

Theorem 7.6.2 *We have that* $N(n, m, w) \geq \frac{\log n}{\log m}$.

This bound can be met when $w = 2$. In fact, there exists an explicit construction such that $N(n, m, 2) = \lceil \frac{\log n}{\log m} \rceil$, for any integers n and m such that $n \geq m$.

Using an elementary probabilistic argument, the following non-constructive upper bound for $N(n, m, w)$ was proved by Mehlhorn [106].

Theorem 7.6.3 *We have that*

$$N(n, m, w) \leq \left\lceil \frac{\log \binom{n}{w}}{\log(m^w) - \log\left(m^w - w!\binom{m}{w}\right)} \right\rceil.$$

Using straightforward approximations, Theorem 7.6.3 yields the following corollary.

Corollary 7.6.4 ([106]) *We have that* $N(n, m, w) \le \lceil we^{w^2/m} \log n \rceil$.

From Theorem 7.6.2 and Corollary 7.6.4, it follows that, for fixed m and w, $N(n, m, w)$ as a function in n is $\Theta(\log n)$. Efforts have been made to provide explicit constructions which are much more efficient compared to the trivial solutions or which compare well with the asymptotic bound of Corollary 7.6.4. Most known explicit perfect hash families are constructed from error-correcting codes by Alon [2], from resolvable balanced incomplete block designs by Brickell [31], and through various inductive techniques by Atici, Magliveras, Stinson, and Wei [5]. For a good survey of this subject, we refer the reader to Blackburn [15].

In [5], Atici, Magliveras, Stinson, and Wei provided various recursive methods resulting in explicit constructions of $\text{PHF}(N; n, m, w)$ in which N is a polynomial function of $\log n$, for fixed m and w. Stinson, Wei, and Zhu [160] employed combinatorial techniques to generalize and improve the results from [5]. For given m and w, they constructed $\text{PHF}(N; n, m, w)$'s in which N is $O(C^{\log^*(n)} \log n)$, where C is a constant depending only on w, and $\log^*$ is a function from $\mathbb{Z}^+$ to $\mathbb{Z}^+$ recursively defined as follows: $\log^*(1) = 1, \log^*(n) = \log^*(\lceil \log n \rceil) + 1$ for $n > 1$.

Blackburn and Wild introduced and studied **linear perfect hash families** in [18]. They showed that there exist explicit constructions for $\text{PHF}(N; n, m, w)$ with $N = (w - 1) \log n / \log m$. Although these classes of linear perfect hash families are of interest in their own right, their constructions are, however, quite restrictive in general, since they require m to be a prime power and very large compared to w and N.

An important method of constructing perfect hash families is to use error-correcting codes, due to Alon [2].

Theorem 7.6.5 ([2, 5]) *Suppose there is a q-ary (N, K, d)-code. Then there exists a* $\text{PHF}(N; K, q, w)$ *provided that*

$$\frac{d}{N} > 1 - \frac{1}{\binom{w}{2}}.$$

In the rest of this section, we show that, using algebraic geometry, for any fixed integers w and m, we can obtain explicit constructions of $\text{PHF}(N; n, m, w)$'s with $N = C \log n$, where C is a constant depending only on w and m, as n tends to ∞.

Let $\mathcal{X}$ be an algebraic curve defined over $\mathbb{F}_q$ and let $g = g(\mathcal{X})$ be the genus of $\mathcal{X}$. For a divisor G of $\mathcal{X}$, recall that $\mathcal{L}(G)$ denotes the Riemann-Roch space whose dimension is $\ell(G)$.

Let T be a subset of $\mathcal{X}(\mathbb{F}_q)$. Let G be a divisor with $T \cap \text{Supp}(G) = \emptyset$. Each point $P \in T$ can be associated with a map h_P from $\mathcal{L}(G)$ to $\mathbb{F}_q$ defined by

$$h_P(f) = f(P).$$

Lemma 7.6.6 *Let* $\mathcal{H} = \{h_P : P \in T\}$. *If* $\deg(G) \geq 2g + 1$, *then the cardinality of* $\mathcal{H}$ *is equal to* $|T|$.

Proof. It is sufficient to prove that the elements of $\mathcal{H} = \{h_P : P \in T\}$ are pairwise distinct. Assume that $h_P = h_Q$ for P and Q in T, i.e.,

$$h_P(f) = h_Q(f) \tag{7.1}$$

for all $f \in \mathcal{L}(G)$. This is equivalent to the fact that

$$f(P) = f(Q) \tag{7.2}$$

for all $f \in \mathcal{L}(G)$.

Suppose that P is different from Q. As $\deg(G - P) > \deg(G - P - Q) \geq 2g - 1$, we obtain, by the Riemann-Roch Theorem (cf. Theorem 1.4.7),

$$\ell(G - P) = \deg(G) - g \quad \text{and} \quad \ell(G - P - Q) = \deg(G) - g - 1.$$

By the above results on dimensions, we can choose a function u from the set $\mathcal{L}(G - P) \setminus \mathcal{L}(G - P - Q)$. Then it is clear that $u(P) = 0$ and $u(Q) \neq 0$. This contradicts (7.2). Hence, $P = Q$. The proof is complete. $\square$

Theorem 7.6.7 *Let* $\mathcal{X}$ *be an algebraic curve over* $\mathbb{F}_q$ *and let* T *be a set of* $\mathbb{F}_q$-*rational points of* $\mathcal{X}$. *Suppose that* G *is a divisor with* $\deg(G) \geq 2g + 1$ *and* $T \cap \mathrm{Supp}(G) = \emptyset$. *Then there exists a perfect hash family* $\mathrm{PHF}(|T|; q^{\deg(G)-g+1}, q, w)$ *if* $|T| > \deg(G) \times \binom{w}{2}$.

Proof. Let $\mathcal{H}$ be as in Lemma 7.6.6. For a subset X of $\mathcal{L}(G)$ with w elements, consider the set

$$\mathcal{S}_X \overset{\text{def}}{=} \{(u - v)^2 : u \neq v \in X\}.$$

Then $\mathcal{S}_X$ has at most $\binom{w}{2}$ elements and the number of zeros of an element $(u - v)^2$ is equal to the number of zeros of $u - v$, which is at most $\deg(G)$, since $u - v$ is an element of $\mathcal{L}(G)$. Therefore, the number of zeros of all the functions in $\mathcal{S}_X$ is at most

$$\deg(G) \times |\mathcal{S}_X| \leq \deg(G) \times \binom{w}{2}.$$

By the condition $|T| > \deg(G) \times \binom{w}{2}$, we can find a point $R \in T$ such that R is not a zero for any function in $\mathcal{S}_X$.

We claim that the function h_R is one-to-one on the subset X. In fact, suppose u and v are two different elements of X. Then $(u - v)^2 \in \mathcal{S}_X$, thus R is not a zero of $(u - v)^2$, i.e., $u(R) \neq v(R)$. This is equivalent to $h_R(u) \neq h_R(v)$. The proof is complete. $\square$

Theorem 7.6.7 gives a generic construction of perfect hash families based on algebraic curves over finite fields. In the examples below, we apply Theorem 7.6.7 to several special curves to obtain some families with good parameters.

Example 7.6.8 Consider the projective line $\mathcal{X}/\mathbb{F}_q$, where $g = g(\mathcal{X}) = 0$. Let N be an integer between 2 and $q + 1$, and let t, w be two positive integers satisfying $N > t\binom{w}{2}$. Then there exists a subset T of $\mathbb{F}_q$-rational points of $\mathcal{X}$ with $|T| = N$ and a divisor G of degree t with $T \cap \mathrm{Supp}(G) = \emptyset$. Applying Theorem 7.6.7 gives a PHF$(N; q^{t+1}, q, w)$. Taking $N = q + 1$, we obtain a PHF$(q+1; q^t, q, w)$ provided $q+1 > t\binom{w}{2}$. In particular, a very special case is the existence of a PHF$(q + 1; q^2, q, w)$ for $q + 1 > 2\binom{w}{2}$ (also see [5, Corollary 3.2] for this special case).

Example 7.6.9 Let $q = p^u$ for a prime p. Put

$$N_q(1) = \begin{cases} q + \lfloor 2\sqrt{q} \rfloor & \text{if } p \,|\, \lfloor 2\sqrt{q} \rfloor \text{ and } u \geq 3 \text{ odd} \\ q + \lfloor 2\sqrt{q} \rfloor + 1 & \text{otherwise.} \end{cases}$$

Recall from Corollary 3.3.12 that there exists an elliptic curve $\mathcal{X}/\mathbb{F}_q$ with $N_q(1)$ $\mathbb{F}_q$-rational points.

Let N be an integer between 2 and $N_q(1)$, and let t, w be two positive integers with $t \geq 3$ and $N > t\binom{w}{2}$. Then there exists a subset T of $\mathbb{F}_q$-rational points of $\mathcal{X}$ with $|T| = N$ and a divisor G of degree t such that $T \cap \mathrm{Supp}(G) = \emptyset$. Applying Theorem 7.6.7 gives a PHF$(N; q^t, q, w)$ since $g = g(\mathcal{X}) = 1$. In particular, there exists a PHF$(N_q(1); q^t, q, w)$ if $N_q(1) > t\binom{w}{2}$.

Example 7.6.10 Let q be a square prime power. Let $r = \sqrt{q}$. Consider the Hermitian curve $\mathcal{X}/\mathbb{F}_q$ defined by

$$y^r + y = x^{r+1}.$$

Then the number of $\mathbb{F}_q$-rational points of $\mathcal{X}$ is equal to $r^3 + 1 = q\sqrt{q} + 1$ and the genus of $\mathcal{X}$ is $g = \sqrt{q}(\sqrt{q} - 1)/2$. Let N be an integer between $\sqrt{q}(\sqrt{q} - 1) + 2$ and $q\sqrt{q} + 1$, and let t, w be two positive integers with $t \geq \sqrt{q}(\sqrt{q}-1)+1$ and $N > t\binom{w}{2}$. Then there exists a subset T of $\mathbb{F}_q$-rational points of $\mathcal{X}$ with $|T| = N$ and a divisor G of degree t such that $T \cap \mathrm{Supp}(G) = \emptyset$. Applying Theorem 7.6.7 gives a PHF$(N; q^{t+1-\sqrt{q}(\sqrt{q}-1)/2}, q, w)$. In particular, there exists a PHF$(q\sqrt{q} + 1; q^{t+1-\sqrt{q}(\sqrt{q}-1)/2}, q, w)$ if $q\sqrt{q} + 1 > t\binom{w}{2}$.

Next, we give an explicit construction based on the Garcia-Stichtenoth curves in [65, 66]. Let q be a square prime power and put $r = \sqrt{q}$. Consider the sequence of algebraic curves $\{\mathcal{X}_i\}_{i \geq 1}$ whose function fields are $\mathbb{F}_q(x_1, \ldots, x_i)$, with $x_1, x_2, \ldots$ defined by:

$$x_{i+1}^r + x_{i+1} = \frac{x_i^r}{x_i^{r-1} + 1}$$

(see Example 2.5.6). Then the number of $\mathbb{F}_q$-rational points of $\mathcal{X}_i$ is more than $(r^2 - r)r^{i-1}$, and the genus g_i of $\mathcal{X}_i$ is less than r^i, for all $i \geq 1$.

Put

$$N_i = (r^2 - r)r^{i-1} = (\sqrt{q} - 1)q^{i/2}$$
$$t_i = \lfloor (c+1)r^i \rfloor = \lfloor (c+1)q^{i/2} \rfloor$$
$$w = \left\lfloor \sqrt{\tfrac{2}{c+1}} q^{1/4} \right\rfloor,$$

where $c \geq 1$ is a real constant independent of i. Then

$$N_i > t_i \binom{w}{2} \qquad \text{for all } i \geq 1,$$

and, for each $i \geq 1$, there exist a subset T_i of $\mathbb{F}_q$-rational points of $\mathcal{X}_i$ with $|T_i| = N_i$ and a divisor G_i of degree t_i of $\mathcal{X}_i$ such that $T_i \cap \text{Supp}(G_i) = \emptyset$. Applying Theorem 7.6.7 gives a $\text{PHF}(N_i; q^{t_i+1-g_i}, q, w)$ for all $i \geq 1$. Since $N_i = (\sqrt{q} - 1)q^{i/2}$, $w = \left\lfloor \sqrt{\frac{2}{c+1}} q^{1/4} \right\rfloor$, and $t_i + 1 - g_i \geq \lfloor (c+1)r^i \rfloor + 1 - r^i > \lfloor cq^{i/2} \rfloor$, we have:

Theorem 7.6.11 *Let q be a square prime power and let $c \geq 1$ be a real number. Then there exists a*

$$\text{PHF}\left((\sqrt{q} - 1)q^{i/2}; q^{\lfloor cq^{i/2} \rfloor}, q, \left\lfloor \sqrt{\frac{2}{c+1}} q^{1/4} \right\rfloor \right)$$

for each $i \geq 1$. In particular, taking $c = 1$, we obtain a

$$\text{PHF}\left((\sqrt{q} - 1)q^{i/2}; q^{q^{i/2}}, q, \lfloor q^{1/4} \rfloor \right)$$

for all $i \geq 1$.

We need the following lemma, due to Blackburn, Burmester, Desmedt, and Wild [17, 15], for the product construction of perfect hash families.

Lemma 7.6.12 ([15]) *Suppose there exist an explicit $\text{PHF}(N; n, n_0, w)$ and an explicit $\text{PHF}(N_0; n_0, m, w)$. Then there exists an explicit $\text{PHF}(NN_0; n, m, w)$.*

Proof. Assume that $\mathcal{H}_1$ is a $\text{PHF}(N; n, n_0, w)$ from A_1 to B_1 and that $\mathcal{H}_2$ is a $\text{PHF}(N_0; n_0, m, w)$ from A_2 to B_2 such that $B_1 = A_2$. Then it is straightforward to verify that $\mathcal{H} = \{h_2 h_1 : h_1 \in \mathcal{H}_1, h_2 \in \mathcal{H}_2\}$ is a $\text{PHF}(NN_0; n, m, w)$ from A_1 to B_2. $\square$

Combining Theorem 7.6.11 with Lemma 7.6.12, we obtain the following result.

Theorem 7.6.13 *For any positive integers m and w such that $m \geq w$, there exist explicit constructions of $\text{PHF}(N; n, m, w)$ such that $N = C \log n$, where C is a constant independent of n, and n can grow to ∞.*

Proof. Let q be the least square prime power such that $q \geq m$ and $q^{1/4} \geq w$. Since $g(x) = \sqrt{\frac{2}{x+1}} q^{1/4}$ is continuous on $[1, \infty)$, it follows that we may choose $c_0 \in [1, \infty)$ such that $\sqrt{\frac{2}{c_0+1}} q^{1/4} = w$. By Theorem 7.6.11, we know that there is an explicit $\text{PHF}((\sqrt{q}-1)q^{i/2}; q^{\lfloor c_0 q^{i/2} \rfloor}, q, w)$ for all $i \geq 1$. Since $q \geq m$, there exists an explicit $\mathrm{PHF}(N_0; q, m, w)$, where the parameter N_0 can be effectively determined by m. From Lemma 7.6.12, it follows that there exist constructions for

$$\mathrm{PHF}(N_0(\sqrt{q}-1)q^{i/2}; q^{\lfloor c_0 q^{i/2}\rfloor}, m, w)$$

for all $i \geq 1$. We thus obtain $\mathrm{PHF}(N, n, m, w)$ with $N = C \log n$, where

$$C \approx \frac{N_0(\sqrt{q}-1)}{c_0 \log q},$$

in which all the parameters on the right-hand side depend only on m and w, but n can grow to ∞ as $n = q^{\lfloor c_0 q^{i/2}\rfloor}$ for all $i = 1, 2, \ldots$. The desired result follows. $\qquad\square$

We give an example to illustrate the efficiency of the constructions above.

Example 7.6.14 Consider $\mathrm{PHF}(N; n, m, w)$ with $m = w = 3$. Atici, Magliveras, Stinson, and Wei [5] gave an explicit construction of $\mathrm{PHF}(3 \times 4^j; 5^{2^j}, 3, 3)$ for any integer $j \geq 1$. This results in a $\mathrm{PHF}(N; n, 3, 3)$ with $N \approx 0.556(\log n)^2$.

Now we look at the construction above. Let $q = 3^4$. From Theorem 7.6.11 (with $c = 1$), we have an explicit construction of $\mathrm{PHF}(8 \times 3^{2i}; 3^{4 \times 3^{2i}}, 3^4, 3)$ for all $i \geq 1$. We also know that there exist constructions for $\mathrm{PHF}(2k^2 - 2k; 3^k, 3, 3)$ for all $k \geq 2$ (see [5, Corollary 5.2]). Taking $k = 4$, we obtain a $\mathrm{PHF}(24; 3^4, 3, 3)$. Applying Lemma 7.6.12, we have an explicit $\mathrm{PHF}(N; n, 3, 3)$ with $N = 24 \times 8 \times 3^{2i}$, $n = 3^{4 \times 3^{2i}}$ for each $i \geq 1$. It follows that

$$N = \frac{24 \times 8}{4 \times \log 3} \log n \approx 30.285 \log n.$$

Thus, the asymptotic behavior of this construction as $n \to \infty$ is much smaller than that from [5].

Chapter 8

Broadcast Encryption and Multicast Security

Multicast communication allows a sender to efficiently broadcast a message to a specific group of users. Sending a message to a group using unicast requires the sender to send an individual copy to each receiver, while using broadcast results in a copy to be sent to all the users instead of a specific group. These advantages of multicast have made it the preferred communication mode for group communication services such as broadcasting stock quotes, software updates, pay TV, etc.

Securing multicast communication requires access control, that is, ensuring that only authorized users can access the broadcast. In this chapter, we study several techniques for access control when the group is dynamic: at different times, different subgroups of the initial group are authorized to receive the multicast message. This is a common situation in many applications where authorized groups are formed based on the right to access a service.

Broadcast encryption, introduced by Fiat and Naor [57], enables a center to securely transmit the data to a large group of receivers in a way that only a predefined subset of users is able to decrypt the data. Clearly, such a mechanism can be used to send a session key to members of the authorized subset and thus provide a solution to secure multicast. Broadcast encryption systems have been further studied by Blundo *et al.* [25] and Stinson [155], who investigated unconditionally secure models and obtained upper and lower bounds for the communication complexity and key storage of the systems. Luby and Staddon [97] used a combinatorial approach to study the trade-off between the number of keys held by each user and the bandwidth needed for establishing a new session key. Other generalizations of broadcast encryption schemes that address the problem of multicast security can be found in [1, 91, 156].

A different solution to the secure multicast problem can be obtained from logical key hierarchy (LKH) proposed by Wallner *et al.* [168]. In this approach, a logical key tree is used to allocate and update the users' keys. The schemes are primarily designed to support eviction or addition of a single user, and therefore to form an arbitrary group, multiple rounds of the eviction procedure need to be invoked. The LKH approach by Wallner *et al.* has been extended and improved by many other authors (see, for example, [175, 33, 34, 36]).

8.1 One-Time Broadcast Encryption

A **one-time broadcast encryption scheme** (OTBES) consists of two phases. In the first phase, the trusted authority (TA) privately distributes some secret information to every user. In the second phase, the TA selects a privileged subset and broadcasts through a public channel an encrypted common key for that subset. The common key can then be used as a cryptographic key for subsequent communication. Every user in the privileged subset must be able to decrypt the common key by using its secret information, while any forbidden coalition outside the privileged subset obtains no information on the common key. Broadcasting some public information in an OTBES makes it possible, in general, to reduce the amount of secret information that every user receives in the distribution phase.

OTBESs are closely related to key predistribution schemes. We will use many of the notations from Chapter 7. Given a set $\mathcal{U}$ of users with $|\mathcal{U}| = n$, we consider a family $\mathcal{P} \subseteq 2^{\mathcal{U}}$ of privileged subsets and a family $\mathcal{F} \subseteq 2^{\mathcal{U}}$ of forbidden subsets.

A $(\mathcal{P}, \mathcal{F}, n)$-**one-time broadcast encryption scheme** (or $(\mathcal{P}, \mathcal{F}, n)$-**OTBES** for short) consists of two phases. In the first one, the key predistribution phase, the TA privately distributes to every user $i \in \mathcal{U}$ some secret $u_i \in U_i$, where U_i denotes the set of all possible secret values that might be distributed to the user i by the TA. In the second one, the broadcast phase, for a selected privileged subset $P \in \mathcal{P}$ and a **secret message** (or a **common key**) $m_P \in M_P$, where M_P denotes the set of all possible secret messages that might be broadcast to the users in P, the TA publicly broadcasts a **broadcast message** $b_P \in B_P$ that is an encryption of m_P, where B_P denotes the set of all possible messages associated with the subset P of users that might be broadcast. Every user $i \in P$ can compute m_P from its secret u_i and the broadcast message b_P, while, even after seeing the broadcast message, the users in a forbidden subset $F \in \mathcal{F}$ with $F \cap P = \emptyset$ obtain no information about m_P.

As in Chapter 7, we can formally state these properties by using entropies.

(i) Without knowing the broadcast, no subset of users has any information about m_P, even if given all the secret information $u_{\mathcal{U}}$:

$$H(\mathbf{M}_P \mid (\mathbf{U}_i)_{i \in \mathcal{U}}) = H(\mathbf{M}_P)$$

for every $P \in \mathcal{P}$. In other words, the secret message m_P is independent of the secrets distributed in the key predistribution phase.

(ii) The message for the privileged subset P is uniquely determined by the broadcast and the secret information of the user:

$$H(\mathbf{M}_P \mid \mathbf{B}_P, \mathbf{U}_i) = 0$$

for every $i \in P$.

(iii) After receiving the broadcast, no forbidden subset F disjoint from P has any information on m_P: if $F \in \mathcal{F}$ is such that $P \cap F = \emptyset$, then

$$H(\mathbf{M}_P \mid \mathbf{B}_P, (\mathbf{U}_j)_{j \in F}) = H(\mathbf{M}_P).$$

As with KPSs, we use the following notation. If $\mathcal{P}$ consists of all the t-subsets of $\mathcal{U}$, then we write $(t, \mathcal{F}, n)$-OTBES for such a $(\mathcal{P}, \mathcal{F}, n)$-OTBES. Similarly, if $\mathcal{P}$ consists of all the subsets of $\mathcal{U}$ of size at most t, then we write $(\leq t, \mathcal{F}, n)$-OTBES. We also use an analogous notation for $\mathcal{F}$. For example, a $(\leq n, 1, n)$-OTBES is a one-time broadcast encryption scheme in which a secret key can be broadcast to any privileged subset of users such that no individual user outside the privileged subset has any information on the broadcast message. Note that the family of forbidden subsets is monotone decreasing: in any $(\mathcal{P}, \mathcal{F}, n)$-OTBES, if $F \in \mathcal{F}$ and $F' \subseteq F$, then $F' \in \mathcal{F}$. Hence, a $(\mathcal{P}, w, n)$-OTBES is the same as a $(\mathcal{P}, \leq w, n)$-OTBES.

We often measure the efficiency of OTBESs by the actual bit-length of the secrets stored by the users and the bit-length of the broadcast message. The amount of secret information that is distributed to each user is measured by the information rate as in the case of KPSs. We assume that, for all $P \in \mathcal{P}$, $M_P = M$ for some M, i.e., the sets of possible common keys for all the privileged subsets $P \in \mathcal{P}$ are the same.

The **information rate** of an OTBES is defined as

$$\rho = \frac{\log|M|}{\max_{i \in \mathcal{U}} H(\mathbf{U}_i)},$$

i.e., the ratio between the length of the common key and the maximum length of the secret information stored by the users.

We also have to take into account the length of the broadcast message. Hence, the **broadcast information rate** of an OTBES is

$$\rho_B = \frac{\log|M|}{\max_{P \in \mathcal{P}} H(\mathbf{B}_P)}.$$

Clearly, we have that $\rho \leq 1$ and $\rho_B \leq 1$. We say that an OTBES has **optimal information rate** (respectively, **optimal broadcast information rate**) if $\rho = 1$ (respectively, $\rho_B = 1$).

There is a trade-off between the amount of secret information held by each user and the length of the broadcast. More precisely, to increase ρ_B, ρ must be decreased, and *vice versa*. To analyze this trade-off, we sometimes look at the **total information rate**, which is defined as

$$\rho_T = \frac{\log|M|}{\max_{P \in \mathcal{P}} H(\mathbf{U}_\mathcal{U}, \mathbf{B}_P)}.$$

8.1.1 Two Trivial Constructions

It is not possible to optimize both the information rate and the broadcast information rate of an OTBES. However, there are two simple ways to construct an OTBES with either optimal information rate or optimal broadcast information rate. We describe such schemes below.

We first give a scheme with optimal broadcast information rate.

> **Key Predistribution:** Suppose there is a key predistribution $(\mathcal{P}, \mathcal{F}, n)$-KPS with key space $\mathbb{F}_q$. The TA gives the user i a secret key u_i in accordance with the $(\mathcal{P}, \mathcal{F}, n)$-KPS.
>
> **Broadcast:** For a privileged set $P \in \mathcal{P}$, the TA broadcasts
>
> $$b_P = m_P + k_P,$$
>
> where $m_P \in \mathbb{F}_q$ is the message the TA wants to broadcast, and $k_P \in \mathbb{F}_q$ is the common key associated with the set P. In other words, the message m_P is encrypted by using the secret common key k_P corresponding to the set P for the underlying KPS.
>
> **Decryption:** Every user from P knows the secret key k_P and can compute $m_P = b_P - k_P$.

In this construction, the lengths of both the broadcast message and the secret key are the same, therefore it gives rise to a scheme with optimal broadcast information rate. The following theorem is immediate.

Theorem 8.1.1 ([155]) *If there exists a $(\mathcal{P}, \mathcal{F}, n)$-KPS having information rate σ and total information rate τ, then there is a $(\mathcal{P}, \mathcal{F}, n)$-OTBES having information rate σ, broadcast information rate 1, and total information rate $\tau/(\tau + 1)$.*

At the other extreme, it is also trivial to construct a $(\mathcal{P}, \mathcal{F}, n)$-OTBES having optimal information rate. The scheme works as follows.

> **Key Predistribution:** For $1 \leq i \leq n$, the TA chooses a random element $u_i \in \mathbb{F}_q$ and gives it to the user i.
>
> **Broadcast:** For a privileged set $P \subseteq \mathcal{U}$, the TA broadcasts
>
> $$b_P = (b_i = u_i + m_P : i \in P),$$
>
> where $m_P \in \mathbb{F}_q$ is the secret message that the TA wants to broadcast to the users in P.
>
> **Decryption:** Every user i from P knows the secret key u_i, and can then compute $m_P = b_i - u_i$.

This construction has the same length for the common key and the secret value stored by each user, so it results in a scheme with optimal information rate.

Theorem 8.1.2 ([155]) *For any integers t, w, n such that $2 \leq t \leq n$ and $1 \leq w \leq n$, there is a $(\leq t, \leq w, n)$-OTBES having information rate 1, broadcast information rate $1/n$, and total information rate $1/(n+1)$.*

8.1.2 The Blundo-Frota Mattos-Stinson Scheme

We describe in the following a family of (t, w, n)-OTBESs proposed by Blundo, Frota Mattos, and Stinson [25], which is an improvement of the scheme by Beimel and Chor [10].

Consider a prime power $q \geq n$ and a positive integer ℓ that is a divisor of a given integer t. Let $P \subseteq \mathcal{U}$ be a set of users with $|P| = t$. Consider the collection of subsets (called blocks) of P with ℓ users. Since ℓ divides t, these $\binom{t}{\ell}$ blocks can be partitioned into $r = \binom{t-1}{\ell-1}$ **parallel classes**, where a parallel class is a collection of blocks that partition the point set P (see, for example, [166]). Each of these classes consists of t/ℓ blocks that form a partition of P. We denote these classes by $\mathcal{C}_1, \ldots, \mathcal{C}_r$ and the blocks in $\mathcal{C}_i$ are denoted by $B_{i,j}$, where $j = 1, \ldots, t/\ell$.

> **Key Predistribution:** The TA distributes keys to the users in $\mathcal{U}$ according to the Blundo *et al.* $(\ell, t + w - \ell, n)$-KPS over $\mathbb{F}_q$ (see Section 7.2.3). Therefore, for every set $Q \subseteq \mathcal{U}$ with $|Q| = \ell$, there is a common key $k_Q \in \mathbb{F}_q$ that can be computed by the users in Q.

> **Broadcast:** To encrypt a secret message $m_P = (m_1, \ldots, m_r) \in \mathbb{F}_q^r$, where $r = \binom{t-1}{\ell-1}$, addressed to the users in a set P with $|P| = t$, the TA computes $b_{i,j} = k_{B_{i,j}} + m_i$ and broadcasts the message
> $$(b_{i,j})_{1 \leq i \leq r, 1 \leq j \leq t/\ell}.$$

> **Decryption:** Every user h in P uses all his keys, denoted by $k_{B_{i,h_i}}$, $1 \leq i \leq r$, to compute $m_i = b_{i,h_i} - k_{B_{i,h_i}}$ for all $1 \leq i \leq r$.

Theorem 8.1.3 ([25]) *Suppose t and ℓ are integers, where $\ell \geq 2$ divides t. Then there is a $(t, \leq w, n)$-OTBES having information rate*

$$\frac{\binom{t-1}{\ell-1}}{\binom{t+w-1}{\ell-1}},$$

broadcast information rate ℓ/t, and total information rate

$$\frac{\binom{t-1}{\ell-1}}{\binom{t+w}{\ell} + \binom{t-1}{\ell-1}}.$$

Proof. For each user $h \in P$, we show that h can decrypt the broadcast. For each i ($1 \le i \le r$), there is a block $B_{i,h_i} \in \mathcal{C}_i$ such that $h \in B_{i,h_i}$. Thus, h can compute all the keys $k_{B_{i,h_i}} \in \mathcal{C}_i$ such that $h \in B_{i,h_i}$. Hence, h can compute all the keys $k_{B_{i,h_i}}, 1 \le i \le r$, and then compute $m_i = b_{i,h_i} - k_{B_{i,h_i}}$, for all $1 \le i \le r$. Therefore, each user in P can recover the message m_P.

Next, we show that a coalition of w users disjoint from a privileged set P has no information about m_P after observation of the broadcast. Indeed, it can be proved that, if $|P| = t$, the vector $(k_Q : Q \subseteq P, |Q| = \ell)$ is uniformly distributed in $\mathbb{F}_q^{\binom{t}{\ell}}$. Then the $\binom{t}{\ell}$ keys k_Q's appear to be independent random elements in $\mathbb{F}_q$. Each of these keys is used to encrypt one element in $\mathbb{F}_q$, so these keys function as a one-time pad.

It is easy to see that the bit-length of the secret information of each user is $\binom{t+w-1}{\ell-1} \log q$, while the length of the broadcast message is $\binom{t}{\ell} \log q$. Therefore, the claimed information rates follow. $\square$

We illustrate the Blundo *et al.* OTBES with an example (see [25]).

Example 8.1.4 Let $\mathcal{U} = \{1, \ldots, n\}$. Assume $t = 2$ and $\ell = 2$. Let $P = \{1, 2, 3, 4, 5, 6\}$. In the key predistribution phase, the TA constructs a random symmetric polynomial in variables x and y with coefficients in $\mathbb{F}_q$, in which each of the degrees of x and y is at most w:

$$f(x, y) = \sum_{i=0}^{w} \sum_{j=0}^{w} a_{ij} x^i y^j, \quad \text{where } a_{ij} = a_{ji} \text{ for all } 0 \le i, j \le w.$$

For each $1 \le i \le n$, the TA computes a polynomial $g_i(x) = f(x, y_i)$ and gives $g_i(x)$ to the user i as his secret information, where the values $y_i \in \mathbb{F}_q$ ($1 \le i \le n$) are publicly known.

Next, the $\binom{6}{2} = 15$ pairs of users in P can be partitioned into $r = \binom{5}{1} = 5$ disjoint parallel classes, as follows:

$$\mathcal{C}_1 = \{\{5, 6\}, \{1, 4\}, \{2, 3\}\} \quad \mathcal{C}_2 = \{\{1, 5\}, \{2, 6\}, \{3, 4\}\}$$
$$\mathcal{C}_3 = \{\{2, 5\}, \{1, 3\}, \{4, 6\}\} \quad \mathcal{C}_4 = \{\{3, 5\}, \{1, 6\}, \{2, 4\}\}$$
$$\mathcal{C}_5 = \{\{4, 5\}, \{3, 6\}, \{1, 2\}\}.$$

Suppose the TA wants to broadcast $m_P = (m_1, m_2, m_3, m_4, m_5) \in \mathbb{F}_q^5$ to the users in P. The TA broadcasts b_P, where b_P consists of the following 15 values:

$$
\begin{array}{lll}
b_1 = k_{5,6} + m_1 & b_2 = k_{1,4} + m_1 & b_3 = k_{2,3} + m_1 \\
b_4 = k_{1,5} + m_2 & b_5 = k_{2,6} + m_2 & b_6 = k_{3,4} + m_2 \\
b_7 = k_{2,5} + m_3 & b_8 = k_{1,3} + m_3 & b_9 = k_{4,6} + m_3 \\
b_{10} = k_{3,5} + m_4 & b_{11} = k_{1,6} + m_4 & b_{12} = k_{2,4} + m_4 \\
b_{13} = k_{4,5} + m_5 & b_{14} = k_{3,6} + m_5 & b_{15} = k_{1,2} + m_5,
\end{array}
$$

where $k_{i,j} = f(y_i, y_j) = g_i(y_j) = g_j(y_i) \in \mathbb{F}_q$.

Now, we see, for instance, how user 4 can decrypt the broadcast to recover m_P. Since user 4 knows five keys $k_{1,4}$, $k_{3,4}$, $k_{4,6}$, $k_{2,4}$, and $k_{4,5}$, he can then calculate

$$\begin{aligned}
m_1 &= b_2 - k_{1,4} &&= b_2 - f(y_1, y_4) \\
m_2 &= b_6 - k_{3,4} &&= b_6 - f(y_3, y_4) \\
m_3 &= b_9 - k_{4,6} &&= b_9 - f(y_6, y_4) \\
m_4 &= b_{12} - k_{2,4} &&= b_{12} - f(y_2, y_4) \\
m_5 &= b_{13} - k_{4,5} &&= b_{13} - f(y_5, y_4).
\end{aligned}$$

8.1.3   KIO Construction

In [155], Stinson gave a general construction of $(\mathcal{P}, \mathcal{F}, n)$-OTBESs that combines a key predistribution scheme (KPS) and an ideal secret sharing scheme (ISSS).

Key Predistribution: The TA chooses v independent $(\mathcal{P}_i, \mathcal{F}_i, n)$-KPSs, $1 \le i \le v$, and distributes keys to the users in $\mathcal{U}$ in accordance with these KPSs.

Broadcast: To encrypt a secret message m_P for a privileged set $P \in \mathcal{P}$, the TA splits m_P into v shares $s_1, \ldots, s_v$ using an ideal (t, v)-threshold scheme. For $1 \le i \le v$, the TA encrypts the share s_i with a key k_i from the $(\mathcal{P}_i, \mathcal{F}_i, n)$-KPS, in a way that the following conditions are satisfied:

  (i) Every user $j \in P$ can compute at least t of the keys $k_1, \ldots, k_v$ (hence user j has t shares of m_P and can reconstruct m_P);

 (ii) Any coalition $F \in \mathcal{F}$ such that $F \cap P = \emptyset$ can compute at most $t - 1$ of the keys $k_1, \ldots, k_v$ (hence F can obtain at most $t - 1$ shares from $s_1, \ldots, s_v$, and therefore F cannot obtain any information about m_P).

Decryption: Every user j in P uses t keys to decrypt the t shares of the (t, v)-threshold scheme, and then reconstructs m_P.

Although the idea of the KIO construction is quite simple, it is not straightforward to see how to choose appropriate $(\mathcal{P}_i, \mathcal{F}_i, n)$-KPSs such that the two conditions in the broadcast phase are satisfied, so that the required $(\mathcal{P}, \mathcal{F}, n)$-OTBES can be obtained. To solve this problem, we next give an implementation of the general KIO construction from the Fiat-Naor KPS.

Suppose that $\mathcal{B} = \{B_1, \ldots, B_v\}$ is a family of subsets of $\mathcal{U}$, where $\mathcal{B}$ is public knowledge. Let $\alpha \ge 0$ be an integer. For each B_j $(1 \le j \le v)$, suppose a Fiat-Naor $(\le |B_j|, \le \alpha, |B_j|)$-KPS is constructed with respect to the user set B_j. The secret keys associated with the jth scheme are denoted by $k_{j,C}$, for any $C \subseteq B_j$ and $|C| \le \alpha$. Note that, according to the Fiat-Naor KPS, the value $k_{j,C}$ is given to every user in $B_j \setminus C$.

Let $\mathcal{F} \subseteq 2^{\mathcal{U}}$. Assume that the following two conditions are satisfied:

(i) For every $i \in \mathcal{U}$,
$$|\{B_j \in \mathcal{B} : i \in B_j\}| \geq t;$$

(ii) For each $F \in \mathcal{F}$,
$$|\{B_j \in \mathcal{B} : |F \cap B_j| \geq \alpha + 1\}| \leq t - 1.$$

We can then construct a $(\leq n, \mathcal{F}, n)$-OTBES as follows. Let $P \subseteq \mathcal{U}$. To broadcast a message m_P to the users in P, the TA applies the following algorithm:

(i) For each $1 \leq j \leq v$, the TA computes a share $s_j \in \mathbb{F}_q$ from a (t, v)-threshold scheme with respect to the secret m_P.

(ii) For each $B_j \in \mathcal{B}$, the TA computes the key k_j for the set $P \cap B_j$ in the Fiat-Naor $(\leq |B_j|, \leq \alpha, |B_j|)$-KPS implemented on the user set B_j:

$$k_j = \sum_{C \subseteq B_j : C \cap P = \emptyset, |C| \leq \alpha} k_{j,C}.$$

(iii) The TA broadcasts

$$b_P = (b_j \stackrel{\text{def}}{=} s_j + k_j : 1 \leq j \leq v).$$

We show that the above algorithm for the broadcast indeed results in a $(\leq n, \mathcal{F}, n)$-OTBES. Consider a user $i \in P$. We let $A_i = \{j : i \in B_j\}$. Then, the user i can compute k_j for every $j \in A_i$. It follows that, for each $j \in A_i$, i can compute $s_j = b_j - k_j$. Since, by condition (i) above, $|A_i| \geq t$, it follows that the user i has at least t shares of the (t, v)-threshold scheme, therefore the user i can recover the secret m_P.

On the other hand, let $F \in \mathcal{F}$ with $F \cap P = \emptyset$. Let $A_F = \{j : |F \cap B_j| \geq \alpha + 1\}$. Then the coalition F can compute k_j, and so s_j, for every $j \in A_F$. However, it has no information on k_j, $j \notin A_F$, and hence no information about s_j, $j \notin A_F$. By condition (ii) above, $|A_F| \leq t - 1$, hence F has at most $t - 1$ shares of the (t, v)-threshold scheme. Therefore, F has no information about the secret m_P.

Next, we give another implementation of the KIO construction for $(\leq n, \leq w, n)$-OTBESs, due to Stinson [155], from **BIBDs (balanced incomplete block designs)**.

Let Y be a set of n elements (called *points*), and let $\mathcal{A} = \{A_1, \ldots, A_\beta\}$ be a family of k-subsets of Y (called *blocks*). We say $(Y, \mathcal{A})$ is an $(n, \beta, r, k, \lambda)$-**BIBD** if each point is contained in exactly r blocks and any pair of points occurs in exactly λ blocks. It is a well-known fact that, in a $(n, \beta, r, k, \lambda)$-BIBD, we have the following two identities: $nr = \beta k$ and $\lambda(n - 1) = r(k - 1)$. For other results on BIBDs and design theory, we refer the reader to Beth, Jungnickel, and Lenz [13].

Theorem 8.1.5 ([155]) *Suppose there is an* $(n, \beta, r, k, \lambda)$-*BIBD such that* $r > \lambda\binom{w}{2}$. *Then there exists a* $(\leq n, \leq w, n)$-*OTBES having information rate* $1/(rk)$, *broadcast information rate* $1/\beta$, *and total information rate*

$$\frac{1}{\lambda\binom{w}{2} + 1 + \beta(k+1)}.$$

Proof. Let $\mathcal{U}$ be the set of users, and let $\mathcal{B}$ be a family of subsets of $\mathcal{U}$. Suppose $(\mathcal{U}, \mathcal{B})$ is an $(n, \beta, r, k, \lambda)$-BIBD such that $r > \lambda\binom{w}{2}$. Note that, for any w-subset $F \subseteq \mathcal{U}$, F intersects at most $\lambda\binom{w}{2}$ blocks in at least two points. We choose $\alpha = 1$ and $t = \lambda\binom{w}{2} + 1$ for a (t, β)-threshold scheme in the KIO construction, then it is easy to check that the two conditions for a $(\leq n, \leq w, n)$-OTBES in the KIO construction are satisfied. Thus, we obtain a $(\leq n, \leq w, n)$-OTBES.

Next, we calculate the values of the associated information rates. We use the Shamir threshold scheme for the required $(\lambda\binom{w}{2} + 1, \beta)$-threshold scheme, which is based on a polynomial of degree at most $\lambda\binom{w}{2}$ over a finite field $\mathbb{F}_q$. The following entropies are straightforward to compute:

$$\begin{aligned}
H(\mathbf{B}_P) &= \beta \log q \\
H(\mathbf{M}_P) &= \log q \\
H(\mathbf{U}_i) &= rk \log q \\
H(\mathbf{U}_{\mathcal{U}}) &= \beta(k+1) \log q \\
H(\mathbf{B}_P \mid \mathbf{U}_{\mathcal{U}}) &= \left(\lambda\binom{w}{2} + 1\right) \log q \\
H(\mathbf{U}_{\mathcal{U}}, \mathbf{B}_P) &= \left(\lambda\binom{w}{2} + 1 + \beta(k+1)\right) \log q.
\end{aligned}$$

Therefore, the information rate is

$$\rho = \frac{\log q}{\max_{i \in \mathcal{U}} H(\mathbf{U}_i)} = \frac{\log q}{rk \log q} = \frac{1}{rk};$$

the broadcast information rate is

$$\rho_B = \frac{\log q}{\max_{P \in \mathcal{P}} H(\mathbf{B}_P)} = \frac{\log q}{\beta \log q} = \frac{1}{\beta};$$

and the total information rate is

$$\rho_T = \frac{\log q}{\max_{P \in \mathcal{P}} H(\mathbf{U}_{\mathcal{U}}, \mathbf{B}_P)} = \frac{1}{\lambda\binom{w}{2} + 1 + \beta(k+1)}.$$

$\square$

Example 8.1.6 ([155]) We construct a $(\leq 7, \leq 2, 7)$-OTBES from a $(7, 7, 3, 3, 1)$-BIBD. The blocks of the BIBD are as follows:

$$\begin{aligned}
B_1 &= \{1, 2, 4\} & B_2 &= \{2, 3, 5\} \\
B_3 &= \{3, 4, 6\} & B_4 &= \{4, 5, 7\} \\
B_5 &= \{1, 5, 6\} & B_6 &= \{2, 6, 7\} \\
B_7 &= \{1, 3, 7\}.
\end{aligned}$$

A total of seven Fiat-Naor $(\leq 3, \leq 1, 3)$-KPSs, denoted by Σ_i $(1 \leq i \leq 7)$, are constructed. Each Σ_i has four keys, and each user receives three out of these four keys. The keys are distributed in accordance to the following rules: Suppose $B_i = \{a, b, c\}$, then a key $k_{i,\emptyset}$ is given to a, b, and c, a key $k_{i,a}$ is given to b and c, a key $k_{i,b}$ is given to a and c, and a key $k_{i,c}$ is given to a and b.

It follows that a total of nine keys are given to each user, as listed in Table 8.1.

TABLE 8.1: Example of seven Fiat-Naor $(\leq 3, \leq 1, 3)$-KPSs.

1	2	3	4	5	6	7
$k_{1,\emptyset}$	$k_{1,\emptyset}$	$k_{2,\emptyset}$	$k_{1,\emptyset}$	$k_{2,\emptyset}$	$k_{3,\emptyset}$	$k_{4,\emptyset}$
$k_{1,2}$	$k_{1,1}$	$k_{2,2}$	$k_{1,1}$	$k_{2,2}$	$k_{3,3}$	$k_{4,4}$
$k_{1,4}$	$k_{1,4}$	$k_{2,5}$	$k_{1,2}$	$k_{2,3}$	$k_{3,4}$	$k_{4,5}$
$k_{5,\emptyset}$	$k_{2,\emptyset}$	$k_{3,\emptyset}$	$k_{3,\emptyset}$	$k_{4,\emptyset}$	$k_{5,\emptyset}$	$k_{6,\emptyset}$
$k_{5,5}$	$k_{2,3}$	$k_{3,4}$	$k_{3,3}$	$k_{4,4}$	$k_{5,1}$	$k_{6,2}$
$k_{5,6}$	$k_{2,5}$	$k_{3,6}$	$k_{3,6}$	$k_{4,7}$	$k_{5,5}$	$k_{6,6}$
$k_{7,\emptyset}$	$k_{6,\emptyset}$	$k_{7,\emptyset}$	$k_{4,\emptyset}$	$k_{5,\emptyset}$	$k_{6,\emptyset}$	$k_{7,\emptyset}$
$k_{7,3}$	$k_{6,6}$	$k_{7,1}$	$k_{4,5}$	$k_{5,1}$	$k_{6,2}$	$k_{7,1}$
$k_{7,7}$	$k_{6,7}$	$k_{7,7}$	$k_{4,7}$	$k_{5,6}$	$k_{6,7}$	$k_{7,3}$

Now, suppose that the TA wants to broadcast a message m_P to a set $P = \{2, 3, 5\}$. The seven keys that will be used in the Fiat-Naor $(\leq 3, \leq 1, 3)$-KPSs are computed as follows:

$$\begin{aligned}
k_1 &= k_{1,\emptyset} + k_{1,1} + k_{1,4} \\
k_2 &= k_{2,\emptyset} \\
k_3 &= k_{3,\emptyset} + k_{3,6} + k_{3,4} \\
k_4 &= k_{4,\emptyset} + k_{4,4} + k_{4,7} \\
k_5 &= k_{5,\emptyset} + k_{5,1} + k_{5,6} \\
k_6 &= k_{6,\emptyset} + k_{6,6} + k_{6,7} \\
k_7 &= k_{7,\emptyset} + k_{7,1} + k_{7,7}.
\end{aligned}$$

Suppose that a Shamir $(2, 7)$-threshold scheme is implemented on the seven blocks and that the public value associated with the block B_j is $x_j \in \mathbb{F}_q$ $(1 \leq j \leq 7)$. The TA chooses a random linear polynomial $f(x) = m_P + ax$ for some $a \in \mathbb{F}_q$. Then the TA computes $s_j = m_P + ax_j$ and broadcasts $b_j = k_j + s_j$ $(1 \leq j \leq 7)$. Therefore, the broadcast b_P consists of the following seven values:

$$\begin{aligned}
b_1 &= k_{1,\emptyset} + k_{1,1} + k_{1,4} + ax_1 + m_P \\
b_2 &= k_{2,\emptyset} + ax_2 + m_P \\
b_3 &= k_{3,\emptyset} + k_{3,6} + k_{3,4} + ax_3 + m_P \\
b_4 &= k_{4,\emptyset} + k_{4,4} + k_{4,7} + ax_4 + m_P \\
b_5 &= k_{5,\emptyset} + k_{5,1} + k_{5,6} + ax_5 + m_P \\
b_6 &= k_{6,\emptyset} + k_{6,6} + k_{6,7} + ax_6 + m_P \\
b_7 &= k_{7,\emptyset} + k_{7,1} + k_{7,7} + ax_7 + m_P.
\end{aligned}$$

8.1.4 The Fiat-Naor OTBES

In this subsection, we present a construction of OTBESs, due to Fiat and Naor [57], based on perfect hash families. This is one of the first constructions of OTBESs.

Suppose $\mathcal{H}$ is a $PHF(N; n, m, w)$ from $A = \{1, \ldots, n\}$ to $B = \{1, \ldots, m\}$. Let $\mathcal{U} = A = \{1, \ldots, n\}$. For each element $(h, j) \in \mathcal{H} \times B$, we associate a block

$$B_{h,j} = \{i : h(i) = j\},$$

and let

$$\mathcal{B} = \{B_{h,j} : (h, j) \in \mathcal{H} \times B\}.$$

For each pair $(h, j) \in \mathcal{H} \times B$, suppose a Fiat-Naor $(\leq |B_{h,j}|, \leq 1, |B_{h,j}|)$-KPS with respect to the user set $B_{h,j}$ is implemented in the key distribution phase. The secret values associated with the (h, j)th KPS are denoted by $k_{(h,j),C}$, where $C \subseteq B_{h,j}$, and $|C| \leq 1$. The value $k_{(h,j),C}$ is given to every user in $B_{h,j} \setminus C$.

Now we construct a $(\leq n, \leq w, n)$-OTBES as follows. Let $P \subseteq \mathcal{U}$. The TA broadcasts a message $m_P \in \mathbb{F}_q$ to P using the following algorithm:

(i) For each $h \in \mathcal{H}$, the TA computes a share $s_h \in \mathbb{F}_q$ of an (N, N)-threshold scheme corresponding to the secret m_P, i.e.,

$$m_P = \sum_{h \in \mathcal{H}} s_h.$$

(ii) For each $B_{h,j} \in \mathcal{B}$, the TA computes the key $k_{h,j}$ corresponding to the set $P \cap B_{h,j}$ in the Fiat-Naor $(\leq |B_{h,j}|, \leq 1, |B_{h,j}|)$-KPS with respect to the user set $B_{h,j}$:

$$k_{h,j} = \sum_{C \subseteq B_{h,j}\,:\,C \cap P = \emptyset, |C| \leq 1} k_{(h,j),C}.$$

(iii) For each $B_{h,j} \in \mathcal{B}$, the TA computes

$$b_{h,j} = s_h + k_{h,j}.$$

(iv) The TA then broadcasts

$$b_P = (b_{h,j} : B_{h,j} \in \mathcal{B}).$$

Note that it may be the case that $B_{h,j}$ is empty for some h and j. When this is the case, we exclude $B_{h,j}$ from $\mathcal{B}$. It follows that $|\mathcal{B}| \leq Nm$. We have the following result.

Theorem 8.1.7 ([57, 155]) *Suppose that there is a $PHF(N; n, m, w)$. Then there is a $(\leq n, \leq w, n)$-OTBES having information rate at least $1/(nN)$, broadcast information rate at least $1/(mN)$, and total information rate at least $1/((n + m + 1)N)$.*

Proof. Suppose $F, P \subseteq \mathcal{U}$ with $|F| \leq w$ and $P \cap F = \emptyset$. Let i be a user in P. For each $h \in \mathcal{H}$, we have $i \in B_{h,h(i)}$. It follows that the user i can compute $k_{h,h(i)}$, and so $s_h = b_{h,h(i)} - k_{h,h(i)}$, for each $h \in \mathcal{H}$. Therefore, i can compute the message $m_P = \sum_{h \in \mathcal{H}} s_h$.

Next, we show that the coalition F has no information about m_P. Assume $F = \{i_1, \ldots, i_\ell\}$. Since $|F| \leq w$, there exists a hash function $h^* \in \mathcal{H}$ such that h^* is one-to-one on F. It follows that

$$|F \cap B_{h^*,j}| \leq 1,$$

for any $j \in B$. This implies that F cannot compute $k_{h^*,j}$ for any $j \in B$, hence F has no information about s_{h^*}, and thus no information about m_P.

We have the following entropies of the resulting scheme:

$$H(\mathbf{B}_P) \leq mN \log q$$
$$H(\mathbf{U}_i) \leq nN \log q$$
$$H(\mathbf{U}_\mathcal{U}) \leq (n+m)N \log q$$
$$H(\mathbf{B}_P \mid \mathbf{U}_\mathcal{U}) = N \log q$$
$$H(\mathbf{U}_\mathcal{U}, \mathbf{B}_P) \leq (n+m+1)N \log q.$$

Thus, the claimed information rates for the scheme follow immediately. □

A class of PHFs of particular interest is when $w = 2$. In this case, we know that there exists an explicit construction such that $N(n, m, 2) = \lceil \log n / \log m \rceil$ (cf. Theorem 7.6.2).

Corollary 8.1.8 *Suppose $m \geq 2$ is an integer and n is an integral power of 2. Then there is an $(\leq n, \leq 2, n)$-OTBES having information rate $\log m / (n \log n)$ and broadcast information rate $\log m / (m \log n)$ (see Theorem 7.6.2).*

We give an example to illustrate the above construction method.

Example 8.1.9 Suppose $n = 5$ and $m = w = 2$. Since

$$\left\lceil \frac{\log 5}{\log 2} \right\rceil = 3,$$

we have a PHF$(3; 5, 2, 2)$, say $\mathcal{H} = \{f, g, h\}$ from $\{1, 2, 3, 4, 5\}$ to $\{1, 2\}$, given in Table 8.2.

TABLE 8.2: Example of a PHF$(3; 5, 2, 2)$.

	1	2	3	4	5
f	1	1	1	2	2
g	1	1	2	1	2
h	1	2	1	1	2

We then have the following collection of blocks:

$$B_1 = B_{f,1} = \{1,2,3\} \quad B_2 = B_{f,2} = \{4,5\}$$
$$B_3 = B_{g,1} = \{1,2,4\} \quad B_4 = B_{g,2} = \{3,5\}$$
$$B_5 = B_{h,1} = \{1,3,4\} \quad B_6 = B_{h,2} = \{2,5\}.$$

For each block B_i, we set up a Fiat-Naor KPS for the user set B_i and distribute the keys to the users according to Table 8.3.

TABLE 8.3: Example of the Fiat-Naor KPS for Example 8.1.9.

1	2	3	4	5
$k_{1,\emptyset}$	$k_{1,\emptyset}$	$k_{1,\emptyset}$	$k_{2,\emptyset}$	$k_{2,\emptyset}$
$k_{1,2}$	$k_{1,1}$	$k_{1,1}$	$k_{2,5}$	$k_{2,4}$
$k_{1,3}$	$k_{1,3}$	$k_{1,2}$		
$k_{3,\emptyset}$	$k_{3,\emptyset}$	$k_{4,\emptyset}$	$k_{3,\emptyset}$	$k_{4,\emptyset}$
$k_{3,2}$	$k_{3,1}$	$k_{4,5}$	$k_{3,1}$	$k_{4,3}$
$k_{3,4}$	$k_{3,4}$		$k_{3,2}$	
$k_{5,\emptyset}$	$k_{6,\emptyset}$	$k_{5,\emptyset}$	$k_{5,\emptyset}$	$k_{6,\emptyset}$
$k_{5,3}$	$k_{6,5}$	$k_{5,1}$	$k_{5,1}$	$k_{6,2}$
$k_{5,4}$		$k_{5,4}$	$k_{5,3}$	

Suppose the privileged set is $P = \{2,4,5\}$. The following keys will be used in the six Fiat-Naor KPSs:

$$k_1 = k_{1,\emptyset} + k_{1,1} + k_{1,3}$$
$$k_2 = k_{2,\emptyset}$$
$$k_3 = k_{3,\emptyset} + k_{3,1}$$
$$k_4 = k_{4,\emptyset} + k_{4,3}$$
$$k_5 = k_{5,\emptyset} + k_{5,1} + k_{5,3}$$
$$k_6 = k_{6,\emptyset}.$$

Now, to broadcast the message m_P, the TA will split m_P into three shares using a $(3,3)$-threshold scheme, i.e., the TA chooses s_1, s_2, and s_3 such that $m_P = s_1 + s_2 + s_3$. The broadcast b_P consists of the following six values:

$$b_1 = k_1 + s_1$$
$$b_2 = k_2 + s_1$$
$$b_3 = k_3 + s_2$$
$$b_4 = k_4 + s_2$$
$$b_5 = k_5 + s_3$$
$$b_6 = k_6 + s_3.$$

8.2   Multicast Re-Keying Schemes

The main challenge in securing multicast applications has been to control access to ensure that only authorized users can access the broadcast. Key management schemes allocate keys to users such that it is possible for the group controller to form new groups by sending multicast messages over public channels. The authorized group is not fixed but changes over time. Hence, efficient methods of re-keying are of central importance.

In a multicast re-keying scheme, during the initialization phase, the TA allocates a (common) group key and some auxiliary keys to each user through a secure channel. The auxiliary keys will be used during the key update phase and are allocated such that unauthorized users cannot find the updated key.

The group is dynamic, i.e., at different times, different subgroups of the initial group $\mathcal{U}$ are authorized to receive the multicast message. To allow users of a specified subgroup to decrypt a broadcast message, the TA uses a public (insecure) channel to multicast one or more messages that allow the establishment of a new session key. The smallest change to $\mathcal{U}$ is by the eviction of a single user or the joining of a new user. In both cases, the session key must be changed: in the former case, so that the evicted user cannot access future communication, while in the latter, a new session key must be established so that the new user cannot access the old communication. To add a new user, a new session key will be encrypted with the old session key and the secret key of the new user. Hence, everyone, including the new user, will share the new key. However, evicting a user is not straightforward.

The efficiency of multicast re-keying schemes is measured in terms of communication complexity, key storage, and computation complexity. These measure, respectively, the required bandwidth for the re-keying message, the amount of secure storage for the TA and users, and the amount of computation by each. We focus on the first two measures.

Let $\mathcal{U} = \{1, \ldots, n\}$ denote the set of all users, called the **multicast group**, and let TA denote the group center. We assume that TA $\notin \mathcal{U}$. During the initialization, each user of $\mathcal{U}$ obtains a shared session key $k^{\mathcal{U}}$ and some auxiliary keys from the TA. We may assume that each user joining the multicast group has an authenticated secure unicast channel with the TA in the initialization (in practice, this may be obtained by using a public key cryptosystem). After the initialization phase and throughout the life of the system, the only means of communication among the users and the TA is a multicast channel that will be observed by all the users.

A **re-keying scheme** specifies an algorithm by which the TA may update the session key for the following two operations: (i) the removal of the users $\mathcal{U} \setminus P$ from the multicast group $\mathcal{U}$ and the updating of the session key k^P for a subgroup $P \subseteq \mathcal{U}$; (ii) the addition of the users of a subgroup P to the multicast group $\mathcal{U}$, where $P \cap \mathcal{U} = \emptyset$, and the generation of a new session key

$k^{P \cup \mathcal{U}}$ known to all users in $P \cup \mathcal{U}$. We will only be concerned with re-keying protocols for user removal, since, as we noted earlier, the joining of new users is much simpler. We also assume there are at most w users who are corrupted by the adversary.

A re-keying scheme enables the TA to give a new session key k^P to a set P of users by encrypting it in such a way that the users in P have enough key information to decrypt it while the other users in $\mathcal{U} \setminus P$ cannot do so. We model the encryption with a publicly available black-box pair E, D, such that, given a key k and a message m, E outputs a random ciphertext $c = E_k(m)$; and, given a ciphertext c and a key k, D outputs the plaintext m. This guarantees that any user holding k will be able to decrypt and that any coalition of users that cannot compute k gains no information from the ciphertext.

Formally, we define a re-keying scheme as follows.

Definition 8.2.1 A **re-keying scheme** consists of the following three phases:

(i) **Key Initialization**: The TA generates a set of auxiliary keys $\mathcal{K}$, and securely distributes the auxiliary keys $\mathcal{K}(i) \subseteq \mathcal{K}$ to the user i, for $1 \leq i \leq n$.

(ii) **Broadcast**: To re-key the multicast group $P \subseteq \mathcal{U}$ (i.e., to evict the users in $\mathcal{U} \setminus P$), the TA generates a new session key k^P and broadcasts the encrypted messages $\{E_{k'}(k^P) : k' \in \mathcal{K}'\}$, where $E_k(\cdot)$ is a symmetric cipher and $\mathcal{K}'$ is the set of keys derived from $\mathcal{K}$ through certain appropriate operations.

(iii) **Re-Keying**: Each user i in P can compute at least one element k' in $\mathcal{K}'$ through certain appropriate operations on his auxiliary keys $\mathcal{K}(i)$, in order to decrypt the message $E_{k'}(k^P)$, while any other user from $\mathcal{U} \setminus P$ is unable to obtain any k' in $\mathcal{K}'$ and so cannot decrypt the broadcast data.

A re-keying scheme is w-**resilient** if any coalition of up to w users from $\mathcal{U} \setminus P$ is unable to compute any keys in $\mathcal{K}'$.

8.2.1 Re-Keying Schemes from Cover-Free Families

A general framework for multicast re-keying schemes can be described through the combinatorial structure of its key allocation. Observe that, in order to thwart the collusion attack by a group F of up to w malicious users, for each user i, there must exist at least one key in $\mathcal{K}(i)$ which is unknown to any other w users. Otherwise, $\mathcal{K}(i)$ is covered by the keys of the group of users F, and so any multicast message that is decryptable by i is also decryptable by the coalition of F. This property is exactly captured by the combinatorial structure of cover-free families that we have studied in Chapter 7. Multicast re-keying schemes based on cover-free families were originally proposed by Kumar *et al.* in [91]. They are also called **blacklisting schemes**.

Recall that a $(1, w)$-CFF(v, n) (cf. Definition 7.5.3) is a set system $(X, \mathcal{B})$ in which no block is a subset of the union of w other blocks. Assume that the auxiliary keys of the users and the system are allocated in accordance with a $(1, w)$-CFF(v, n), i.e., $\mathcal{K} = X$ and $B_i = \mathcal{K}(i)$, $i \in \mathcal{U}$. Then this results in a re-keying scheme with n users and a total of v keys that can revoke up to w users.

Theorem 8.2.2 *If there exists a $(1, w)$-CFF(v, n), then there exists a w-resilient re-keying scheme in which the number of auxiliary keys of the TA is v, the number of keys for each user i is $|B_i|$, and the number of broadcast transmissions to remove up to w users is at most v.*

Proof. Denote the set of v keys in the system by $X = \{k_1, \ldots, k_v\}$. Suppose that F is a subset of users who are to be revoked. Assume $|F| = w$. The TA chooses a new key $k^{\mathcal{U} \setminus F}$. Then, for every

$$k_j \in X \setminus \cup_{i \in F} \mathcal{K}(i),$$

the TA computes $b_j = E_{k_j}(k^{\mathcal{U} \setminus F})$ and broadcasts b_j. Due to the properties of a cover-free family, it is easy to see that no user in F can compute $k^{\mathcal{U} \setminus F}$, even if all the users in F pool all their keys together. On the other hand, every user $h \notin F$ has at least one key from $X \setminus \cup_{i \in F} \mathcal{K}(i)$. Therefore, h can decrypt at least one of the ciphertexts b_j's and can compute $k^{\mathcal{U} \setminus F}$. □

Example 8.2.3 Let $X = \{1, 2, 3, 4, 5, 6, 7\}$ and $\mathcal{B} = \{B_1, B_2, B_3, B_4, B_5, B_6, B_7\}$, where

$$\begin{array}{lll} B_1 = \{2, 3, 5\} & B_2 = \{3, 4, 6\} & B_3 = \{4, 5, 7\} \\ B_4 = \{1, 5, 6\} & B_5 = \{2, 6, 7\} & B_6 = \{1, 3, 7\} \\ B_7 = \{1, 2, 4\}. & & \end{array}$$

Then $(X, \mathcal{B})$ is a $(1, 2)$-CFF$(7, 7)$. We construct a re-keying scheme based on this cover-free family. The TA generates seven keys, $k_1, \ldots, k_7$, and distributes the keys to each user in accordance with the underlying cover-free family. Hence, user 1 receives the keys k_2, k_3, and k_5, user 2 gets the keys k_3, k_4, k_6, and so on.

The scheme is 2-resilient. Suppose that users 3 and 6 are to be revoked. The new key k' is encrypted using all the keys not in $\{k_4, k_5, k_7\} \cup \{k_1, k_3, k_7\}$. This means that the broadcast consists of $E_{k_2}(k')$ and $E_{k_6}(k')$. Obviously, users 3 and 6 are unable to get the new key k'. However, users 1, 5, and 7 know k_2 and can therefore obtain k' from the first ciphertext, while users 2 and 4 can recover k' from the second ciphertext.

8.2.2   Re-Keying Schemes from Secret Sharing

Anzai, Matsuzaki, and Matsumoto in [3], and independently Naor and Pinkas in [111], proposed a multicast re-keying scheme based on the Shamir threshold scheme. The basic idea of their re-keying scheme is as follows. The TA uses a Shamir $(w+1, n)$-threshold scheme to construct n shares, $s_1, \ldots, s_n$, for the group key k. Every user j is given the share s_j. To revoke a set of w users $F = \{i_1, \ldots, i_w\}$, the TA simply broadcasts the w shares of revoked users, $s_{i_1}, \ldots, s_{i_w}$. After the broadcast, each user not in F has $w + 1$ shares and can reconstruct the secret. It is easy to see that, if the number of revoked users is less than w, the TA can add some *dump* shares and broadcasts w shares (shares from the revoked users, plus the dump shares). The above basic scheme only allows a single revocation. However, the scheme can be modified for multiple revocations, through combining with the Diffie-Hellman key exchange protocol, and its security is based on the **decisional Diffie-Hellman problem**. We describe the modified scheme as follows.

Assume that p and q are large primes such that $q \mid p-1$, and g is a generator of the multiplicative group $\mathbb{Z}_p^*$, in which the decisional Diffie-Hellman problem is intractable.

Key Initialization: The TA chooses a secret key $k \in \mathbb{Z}_q$ and constructs the polynomial

$$F(x) = k + \sum_{j=1}^{w} a_j x^j \bmod q,$$

where a_j are random values in $\mathbb{Z}_q$. The TA generates shares $s_{x_i} = F(x_i)$ for $1 \le i \le n$, where $x_1, \ldots, x_n \in \mathbb{Z}_q$ are the public x-coordinates of the users, and distributes s_i to user i, for $1 \le i \le n$, through secure channels.

Broadcast: To revoke w users, say $F = \{i_1, \ldots, i_w\}$, the TA chooses a random value $r \in \mathbb{Z}_q$ and broadcasts g^r, together with $y_{i_1} = g^{rs_{i_1}}, \ldots, y_{i_w} = g^{rs_{i_w}}$.

Re-Keying: A non-revoked user, say i_{w+1}, can compute his own exponentiated share

$$y_{i_{w+1}} = g^{rs_{i_{w+1}}} = (g^r)^{s_{i_{w+1}}}.$$

Then the user i_{w+1} can compute the new common key $k = g^{rk}$ by

$$g^{rk} = \prod_{j=1}^{w+1} y_{i_j}^{b_j} \bmod p,$$

where

$$b_j = \prod_{1 \le \ell \le w+1, \ell \neq j} \frac{x_{i_\ell}}{x_{i_\ell} - x_{i_j}} \bmod q.$$

Theorem 8.2.4 ([111]) *The construction above gives rise to a w-resilient re-keying scheme in which the secret key of each user is a single element of $\mathbb{Z}_q$, and the broadcast message of each revocation is of length $O(\log w)$. Moreover,* *the scheme can be used for virtually an unlimited number of re-keyings, of up to w users per revocation.*

Proof. The proof is based on the decisional Diffie-Hellman problem. For the sake of clarity, we show the case when $w = 1$.

Assume that the scheme is insecure with parameter $w = 1$ and can be broken by a user $v \in \mathcal{U}$. This user can be simulated by an algorithm D' which receives the following inputs: a value $s_{x_v} = F(x_v)$ of the linear polynomial $F(x)$, polynomially many tuples $\langle g^{r_i}, g^{r_i F(x_v)}, g^{r_i F(0)} \rangle$ generated with random choices of r_i's, and a pair $(g^r, g^{r F(x_v)})$. If the scheme is insecure, then D' can distinguish between $g^{r F(0)}$ and a random value.

We construct an algorithm D that uses D' to solve the decisional Diffie-Hellman problem. The algorithm D is given inputs g^a, g^b, and a value C which is either g^{ab} or random. Let D generate inputs to D' (with $F(0) = b$ and $r = a$): it generates a random key $\langle x_v, F(x_v) \rangle$ and gives to D'; it then generates random r_i's and gives the tuples $\langle g^{r_i}, g^{r_i F(x_v)}, g^{r_i b} \rangle$ to D'; it also gives the pair (g^a, C) to D'. Let the outputs of D be the same answer that D' outputs. Then D's success probability of solving the decisional Diffie-Hellman problem is the same as D''s probability of breaking the re-keying scheme. $\square$

8.2.3 Logical Key Hierarchy Schemes

Logical key hierarchy (LKH) schemes were first proposed by Wallner *et al.* in [168], and independently by Wong *et al.* in [175]. The n users of the multicast group are associated with the leaves of a q-ary tree; each node and leaf of the tree is assigned a key. A user receives all the keys in the path from his associated node to the root. Hence, the key corresponding to the root node is shared by all the users.

Figure 8.1 is an LKH scheme based on a binary tree for eight users. Each user holds the keys on the path from its leaf node to the root. The key shared by all the group users is k_0. To evict user u_1, the TA chooses a key k'_0, encrypts it with $k_{2.2}, k_{1.2}$, and $k_{0.2}$ and broadcasts $E_{k_{2.2}}(k'_0), E_{k_{1.2}}(k'_0)$, and $E_{k_{0.2}}(k'_0)$. All users except u_1 will have at least one of the keys used for encryption and so can obtain the new group key k'_0. Each user stores $O(\log n)$ keys and the TA has to store $O(n)$ keys. The communication complexity for each key update is $O(\log n)$. A number of variations of this scheme, with the aim of lowering communication cost, have also been proposed. For example, the scheme proposed by Canetti *et al.* [34] uses hash functions to improve communication complexity while maintaining the key storage of the Wallner *et al.* scheme.

In [132], Safavi-Naini *et al.* gave a variant of the Wallner *et al.* scheme, called the w-**resilient LKH**, which we describe below.

Consider a full q-ary tree F of height ℓ. To each leaf, we associate a q-ary vector of length ℓ corresponding to the set of keys from that leaf to the root

FIGURE 8.1: An example of an LKH scheme with eight users.

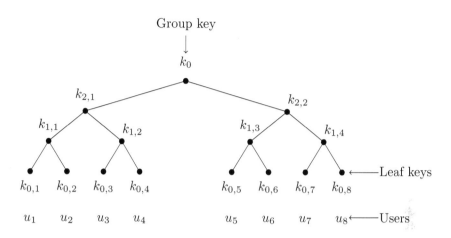

of the tree. Thus, a leaf can be uniquely identified with a vector of length ℓ. For each level i, $1 \leq i \leq \ell$, we choose q values $\{a_1^i, a_2^i, \ldots, a_q^i\}$ and use them to label the q^i nodes of the ith level of the tree such that the q nodes with the same parent are labeled with distinct values.

Definition 8.2.5 Let L be a subset of leaves of a full q-ary tree $\mathcal{T}$. Let $\mathcal{T}(L)$ be a subtree of $\mathcal{T}$ with L as its set of leaves. The subtree $\mathcal{T}(L)$ is called a w-**resilient LKH** if, for any $w+1$ leaves $\{L_{j_1}, \ldots, L_{j_w}, L_j\} \subseteq L$, there exists i such that the ith component of the vector attached to L_j is different from that of L_{j_s}, for all $1 \leq s \leq w$.

When $|L| = n$, we call $\mathcal{T}(L)$ an (n, ℓ, w) q-**ary resilient LKH**. Figure 8.2 shows an example of a $(9, 3, 2)$ ternary resilient LKH.

We describe two multicast re-keying schemes, called the **OR** and the **AND** re-keying schemes, built on the key initialization scheme from the q-ary resilient LKH.

8.2.3.1 Key Initialization Scheme from Resilient LKH

Suppose that $\mathcal{T}(L)$ is an (n, ℓ, w) q-ary resilient LKH with $L = \{L_1, \ldots, L_n\}$ as the the set of leaves. We associate each leaf with a user in the multicast group. The scheme proceeds as follows:

(i) For each $1 \leq i \leq \ell$, the TA independently chooses q auxiliary keys $\{k_1^i, \ldots, k_q^i\}$ and assigns them to the nodes in the ith level of $\mathcal{T}(L)$ such that each node receives a key in accordance with its labeled value, i.e., if a node is labeled as a_s^i, the key assigned to that node is k_s^i.

FIGURE 8.2: An example of a 2-resilient LKH from a full ternary tree.

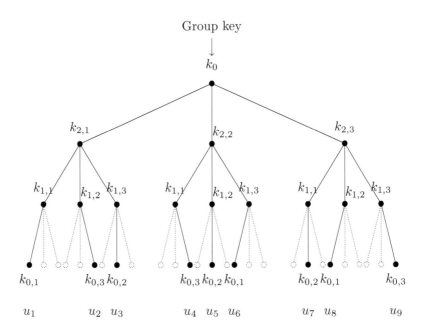

(ii) Each user is given the keys on the nodes along the path from his leaf to the root.

The tree $\mathcal{T}(L)$ is of height ℓ and each level has q keys, requiring the TA to generate $q\ell$ node keys. There are ℓ nodes from each leaf to the root, so each user receives exactly ℓ node keys. Since the tree is w-resilient, for each user j, there exists i such that, in the ith level of the tree, the labeled value of the node associated with the user j is different from the labeled values of the nodes associated with any set of users $\{j_1, \ldots, j_w\}$ different from j, so the key assigned to the user j at level i is different from the keys assigned to $\{j_1, \ldots, j_w\}$.

Theorem 8.2.6 *In the above key initialization scheme, the numbers of auxiliary keys for the TA and each user are $q\ell$ and ℓ, respectively. Moreover, for any $w + 1$ users $\{j_1, \ldots, j_w, j\}$, we have,*

$$|\mathcal{K}(j) \setminus \cup_{s=1}^{w} \mathcal{K}(j_s)| \geq 1.$$

In other words, the family of the key sets for the users and the set of the TA's keys forms a $(1, w)$-CFF$(q\ell, n)$.

Next, we give two re-keying schemes based on the above key initialization scheme, called the **OR broadcast scheme** and the **AND broadcast scheme**.

8.2.3.2   The OR Broadcast Scheme

`Broadcast`: To form a new multicast group $P \subseteq \mathcal{U}$, the TA randomly chooses a session key k^P, encrypts it with those keys not belonging to the users in $\mathcal{U} \setminus P$, and broadcasts it to all the users. In other words, the TA broadcasts

$$\{E_k(k^P) \ : \ k \in \mathcal{K}, k \notin \mathcal{K}(\mathcal{U} \setminus P)\},$$

where $\mathcal{K}(\mathcal{U} \setminus P) = \cup_{i \in \mathcal{U} \setminus P} \mathcal{K}(i)$.

`Re-Keying`: Each user $j \in P$ uses one of his keys $k \in \mathcal{K}(j)$ to decrypt $E_k(k^P)$ and obtain the session key k^P.

The key initialization is based on a q-ary w-resilient LKH. The users outside P do not possess the keys that are used to encrypt k^P in the broadcast, and are therefore unable to decrypt the encrypted k^P, even if they collude. On the other hand, any user in P has at least one key to decrypt. Indeed, by assumption, there are at most w revoked users; we may assume that $\mathcal{U} \setminus P = \{i_1, \dots, i_s\}$ and $s \leq w$. For each $j \in P$, from Theorem 8.2.6, we know that there exists a key $k_j \in \mathcal{K}(j) \setminus \cup_{t=1}^{s} \mathcal{K}(i_t)$. Thus, by the OR broadcast scheme, user j can use this key k_j to decrypt one of the broadcast messages $E_{k_j}(k^P)$ and obtain the new session key. From Theorem 8.2.2, it is obvious that the numbers of keys for each user and the TA are ℓ and $q\ell$, respectively. The number of broadcast ciphertexts is $|\mathcal{K}| - |\mathcal{K}(\mathcal{U} \setminus P)| \leq q\ell - \ell = (q-1)\ell$. We obtain the following result.

Theorem 8.2.7 *If there exists an (n, ℓ, w) q-ary resilient LKH, then there exists a w-resilient re-keying OR scheme in which the numbers of auxiliary keys for each user and for the TA are ℓ and $q\ell$, respectively, and the number of broadcast transmissions to revoke up to w users is not greater than $(q-1)\ell$.*

We note that for the OR scheme based on a w-resilient LKH, the number of revoked users is at most w. This means that the scheme can be used to form any subgroup of size $n - i$, for any $1 \leq i \leq w$.

8.2.3.3   The AND Broadcast Scheme

`Broadcast`: To form a new multicast group $P \subseteq \mathcal{U}$, for each $j \in P$, the TA computes

$$\widehat{k(j)} = \sum_{k \in \mathcal{K}(j)} k,$$

where we assume all the keys are from a finite field or an abelian group. The TA then randomly chooses a new session key k^P and broadcasts

$$\{E_{\widehat{k(j)}}(k^P) \ : \ j \in P\}.$$

Re-Keying: Each user j in P computes $\widehat{k(j)}$ and decrypts $E_{\widehat{k(j)}}(k^P)$ to obtain the new session key k^P.

Again, the key initialization is based on the q-ary resilient LKH. Each user $j \in P$ can compute the key $\widehat{k(j)} = \sum_{k \in \mathcal{K}(j)} k$ and decrypt $E_{\widehat{k(j)}}(k^P)$ to obtain the new session key. We can show that the coalition of any w users $i_1, \ldots, i_w$ outside P has no information to compute k^P. Indeed, for each user $j \in P$, from Theorem 8.2.2, we know that there exists at least one key in $\mathcal{K}(j)$ that is unknown to the coalition of users $i_1, \ldots, i_w$, so it is unable to compute $\widehat{k(j)}$.

Theorem 8.2.8 *If there exists an (n, ℓ, w) q-ary resilient LKH, then there exists a w-resilient re-keying AND scheme. To form a new multicast group $P \subseteq \mathcal{U}$, it requires each user to store ℓ keys and the TA to store $q\ell$ keys. The size of the transmissions for the encrypted new session key is $|P|$.*

We note that the AND scheme can be used to revoke any number of users and so form multicast groups of any size. Compared to the OR scheme, which can form a new multicast group with at least $n - w$ users (i.e., revoke at most w users), the AND scheme is more flexible in terms of the size of the multicast groups.

8.3   Re-Keying Schemes with Dynamic Group Controllers

All the re-keying schemes in the previous section have a single, fixed group controller, the TA, who decides and manages the key updates. In many applications, such as dynamic conferences, users in the group may wish to transmit data to a subgroup of users. If the single group controller model is used, then either all communication from the users goes through the group controller, who then disseminates to the designated group through multicast, or numerous session keys need to be established. These solutions have problems such as single point of failure, communication overhead for the group controller, and communication delay.

In contrast to the previous solution in which the controller is fixed, we consider the scenario where the group controller is dynamic, that is, after the system initialization, each user can behave like a TA and establish a new subgroup. This is achieved by a single broadcast from a user to the group, which can only be decrypted by the users of the target subgroup.

A trivial solution to the problem of dynamic controllers is to take the scheme of a single controller as a building block and associate a single controller scheme to each user such that the corresponding user is the TA and the rest of the group are the receivers. The obvious drawback of this solution is that the key storage of each user is prohibitively large, namely, $n - 1$ times the storage of a user plus the storage of the TA in a single controller scheme, which is linear in n. For large groups, this is very inefficient.

In the model of multicast re-keying with dynamic controllers, we assume that there is a TA who initializes the system by assigning auxiliary keys to the group. After the initialization phase, the only means of communication among the group users is via multicast channels on which any user in the group may broadcast messages that will be received by all the users in the group. The goal is to enable any user to securely establish a common secret (e.g., a session key) among the users of a designated subgroup.

Formally, a **dynamic controller re-keying scheme** consists of three phases:

(i) `Key Initialization`: This is the phase during which the TA generates and securely distributes the auxiliary keys to each user of the group.

(ii) `Broadcast`: Any user of the group (or a subset of the group) broadcasts an encrypted secret which is only decryptable by the users from a specified target group.

(iii) `Re-Keying`: The authorized users are able to decrypt the encrypted secret while unauthorized users are not.

We also call the scheme w-**resilient** if any collusion of up to w unauthorized users cannot find the secret.

Re-keying schemes with dynamic group controllers were introduced by Safavi-Naini and Wang in [131]. We describe their two schemes based on perfect hash families.

In this section, we denote the set of users of the group by $\mathcal{U} = \{u_1, \ldots, u_n\}$.

8.3.1 The OR Re-Keying Scheme with Dynamic Controller

Assume that, in the key initialization, the users $u_1, \ldots, u_n$ share a common key $k^{\mathcal{U}}$, and each user u_i is secretly given a set of auxiliary keys $\mathcal{K}(u_i) \subseteq \mathcal{K}$, where $\mathcal{K}$ denotes the set of all auxiliary keys. Later, the user u_i wants to update the secret for a new group $\mathcal{U} \setminus F$ for some $F \subseteq \mathcal{U}$. In other words, a subset F is to be revoked from the group.

> `Key Initialization`: Let $\mathcal{H} = \{h_1, \ldots, h_N\}$ be a PHF$(N; n, m, w + 2)$ from $\{1, 2, \ldots, n\}$ to $\{1, 2, \ldots, m\}$. The TA generates a group key $k^{\mathcal{U}}$ and a set of auxiliary keys $\mathcal{K}$, consisting of N symmetric $m \times m$ matrices with auxiliary keys as entries,
>
> $$G^1 = (k_{i,j}^1)_{1 \leq i,j \leq m}, \ldots, G^N = (k_{i,j}^N)_{1 \leq i,j \leq m},$$
>
> where $k_{i,j}^\ell = k_{j,i}^\ell$ for all ℓ, i, j. To each user u_i, the TA securely sends the group key $k^{\mathcal{U}}$ and the set of auxiliary keys $\mathcal{K}(u_i)$, consisting of the

$h_\ell(i)$th row of the matrix G^ℓ, for all $1 \le \ell \le N$, i.e.,

$$\mathcal{K}(u_i) = \{(k^1_{h_1(i),1}, \ldots, k^1_{h_1(i),m}), \ldots, (k^N_{h_N(i),1}, \ldots, k^N_{h_N(i),m})\}.$$

Broadcast: Suppose that a user u_i wants to establish a secret for a group $\mathcal{U} \setminus F$. He randomly chooses a secret $k^{\mathcal{U}\setminus F}$ and encrypts it with all his auxiliary keys except those belonging to F, and broadcasts the encrypted messages. In other words, the user u_i broadcasts

$$\{E_k(k^{\mathcal{U}\setminus F}) : k \in \mathcal{K}(u_i), k \notin \mathcal{K}(F)\},$$

where $\mathcal{K}(F) = \cup_{u \in F} \mathcal{K}(u)$.

Decryption: Each user u_j in $\mathcal{U} \setminus F$ uses one of his auxiliary keys to decrypt the encrypted secret in the broadcast.

Theorem 8.3.1 *If there exists a* $\mathrm{PHF}(N; n, m, w + 2)$*, then there exists a* w*-resilient re-keying scheme with dynamic controller in which the number of auxiliary keys for each user is* Nm*. To multicast a secret* $k^{\mathcal{U}\setminus F}$*, for a group* $\mathcal{U} \setminus F$*, where* $|F| \le w$*, the number of transmissions for the ciphertexts is less than* mN*.*

Proof. We apply the above OR re-keying scheme with dynamic controller, keeping the same notation. Assume u_i wants to establish a common key for a group $\mathcal{U} \setminus F$, where $|F| \le w$. Clearly, the users in F, even if they collude, do not possess the keys used to encrypt $k^{\mathcal{U}\setminus F}$ in the broadcast and are hence unable to compute the secret $k^{\mathcal{U}\setminus F}$. We are left to show that any user u_j in $\mathcal{U} \setminus F$ has at least one key to decrypt one of the broadcast ciphertexts, and hence obtain $k^{\mathcal{U}\setminus F}$. Assume $F = \{u_{i_1}, \ldots, u_{i_w}\}$. Let $X = \{i_1, \ldots, i_w, i, j\}$, where $i, j \notin F$. Since $\mathcal{H}$ is a $\mathrm{PHF}(N; n, m, w + 2)$, we know there exists $h_\alpha \in \mathcal{H}$ such that h_α is perfect on X, which means that $h_\alpha(i_1), \ldots, h_\alpha(i_w), h_\alpha(i), h_\alpha(j)$ are all distinct. It follows that $k^\alpha_{h_\alpha(i),h_\alpha(j)} \in \mathcal{K}(u_i) \setminus \mathcal{K}(F)$. The user u_j holds the keys of the $h_\alpha(j)$th row of the matrix G^α and so possesses the key $k^\alpha_{h_\alpha(j),h_\alpha(i)}$. Since the matrix G^α is symmetric, it follows that $k^\alpha_{h_\alpha(i),h_\alpha(j)} = k^\alpha_{h_\alpha(j),h_\alpha(i)}$, and so u_i and u_j share a common key which is not in $\mathcal{K}(\mathcal{U} \setminus F)$, and this key can be used by u_j to decrypt the ciphertext in the broadcast. The various parameters of the schemes are obvious. $\square$

From Section 7.6 (more specifically, Theorem 7.6.13) we know that there exist explicit constructions for $\mathrm{PHF}(N; n, m, w)$ with $N = O(\log n)$. The following corollary is immediate.

Corollary 8.3.2 *For a given* w*, there exists a* w*-resilient re-keying scheme with dynamic controller in which the number of auxiliary keys for each user is* $O(\log n)$*. To establish a common key for a multicast group to exclude up to* w *users, the number of transmissions for the ciphertexts is* $O(\log n)$ *as well.*

Next, we show how to improve the communication efficiency of the above OR scheme using erasure codes. The basic idea is to first encode the secret with an erasure code, then apply the OR re-keying scheme (slightly modified) to the new codeword.

We begin with the definition of an erasure code.

Definition 8.3.3 An $[n, t, m]$-**erasure code** over a finite field $\mathbb{F}_q$ is a polynomial-time function $C : \mathbb{F}_q^t \to \mathbb{F}_q^n$ for which there exists a polynomial-time function $D : \widehat{\mathbb{F}_q}^n \to \mathbb{F}_q^t$, where $\widehat{\mathbb{F}_q} = \mathbb{F}_q \cup \{\perp\}$, such that: for all $\mathbf{v} \in \mathbb{F}_q^t$, if $\mathbf{u} \in \widehat{\mathbb{F}_q}^n$ is such that $\mathbf{u}$ agrees with $C(\mathbf{v})$ in at least m places, and is $\perp$ elsewhere, then $D(\mathbf{u}) = \mathbf{v}$.

Given an $[n, t, m]$-erasure code over $\mathbb{F}_q$, one can encode a message $\mathbf{v} \in \mathbb{F}_q^t$ to obtain a codeword $C(\mathbf{v}) \in \mathbb{F}_q^n$. The message $\mathbf{v}$ can be reconstructed even if up to $n - m$ positions of $C(\mathbf{v})$ are damaged or erased. Erasure codes can be constructed using error-correcting codes, such as the Reed-Solomon codes.

Given a message vector $\mathbf{v} = (v_0, v_1, \ldots, v_{t-1}) \in \mathbb{F}_q^t$, we construct the polynomial $p_{\mathbf{v}}(x) = v_0 + v_1 x + \cdots + v_{t-1} x^{t-1}$. Let $x_1, x_2, \ldots, x_n$ be n distinct elements in $\mathbb{F}_q$. The encoding is defined by $C(\mathbf{v}) = (p_{\mathbf{v}}(x_1), p_{\mathbf{v}}(x_2), \ldots, p_{\mathbf{v}}(x_n))$, and the decoding D uses t pairs $(x_i, p_{\mathbf{v}}(x_i))$ to interpolate the polynomial and reconstruct the coefficients of $p_{\mathbf{v}}(x)$ to obtain the source message $\mathbf{v}$.

In order to apply erasure codes to improve the communication complexity, we extend the definition of PHFs to α-PHFs.

Definition 8.3.4 Let $\mathcal{H}$ be a PHF$(N; n, m, w)$ from A to B. We call $\mathcal{H}$ an α-PHF$(N; n, m, w)$ if, for any subset $X \subseteq A$ with $|X| = w$, there exist at least α functions from $\mathcal{H}$ such that they are all perfect on X.

We note that a 1-PHF$(N; n, m, w)$ is exactly a PHF$(N; n, m, w)$ and that most techniques for constructing PHFs can be generalized to α-PHFs in a straightforward manner. For example, in Chapter 7 (cf. Theorem 7.6.5), we have seen that a q-ary (N, n, d)-code results in a PHF$(N; n, q, w)$ provided that $d > (\binom{w}{2} - 1)N/\binom{w}{2}$. It is easy to see that the hash family is actually an α-PHF if $d \geq ((\binom{w}{2} - 1) + \alpha)N/\binom{w}{2}$.

The basic OR re-keying scheme can now be modified as follows. Assume that the auxiliary keys of the users are elements of a finite field $\mathbb{F}_q$. Let $\mathcal{H}$ be an α-PHF$(N; n, m, w)$ and let the auxiliary keys for the users be allocated using the basic OR re-keying scheme. The only difference in the modified scheme is in the broadcast transmissions. To revoke a subset F of users, the sender (TA) first divides the secret $k^{\mathcal{U}\backslash F}$ into t pieces $k^{\mathcal{U}\backslash F} = (k_1^{\mathcal{U}\backslash F}, \ldots, k_t^{\mathcal{U}\backslash F})$, and then encodes $k^{\mathcal{U}\backslash F}$ using an $[Nm, t, \alpha]$-erasure code to obtain a codeword $C(k^{\mathcal{U}\backslash F}) = (c_1, \ldots, c_{Nm})$. Next, the sender (TA) uses all the auxiliary keys that do not belong to the users of F to encrypt the corresponding components of $C(k^{\mathcal{U}\backslash F})$ and broadcasts the encrypted messages to all the users. In other words, the sender (TA) broadcasts

$$\{E_{k_i}(c_i) : k_i \in \mathcal{K}, k_i \notin \mathcal{K}(F)\}.$$

Since each non-revoked user has at least α keys that can decrypt α messages of $\{E_{k_i}(c_i) : k_i \in \mathcal{K}, k_i \notin \mathcal{K}(F)\}$, he can decrypt the α messages and then apply the erasure code to obtain the secret, while a user in F, as in the basic OR scheme, cannot find the secret.

In an $[n, t, m]$-erasure code over $\mathbb{F}_q$, the length of the codeword $C(\mathbf{v})$ is $n \log q$ bits, whereas the length of the source message $\mathbf{v}$ is $t \log q$ bits. Hence, the rate, which indicates the extra bandwidth, is n/t. We note that the basic OR scheme uses an $[n, 1, 1]$-erasure code in the construction. In general, we expect t to be as large as possible to minimize the extra bandwidth.

8.3.2 The AND Re-Keying Scheme with Dynamic Controller

In the OR re-keying scheme with dynamic controller, the efficiency relies on the size of the perfect hash family $\mathcal{H}$. We would like $|\mathcal{H}|$ to be as small as possible (e.g., $O(\log n)$). However, the value of $|\mathcal{H}|$ can be substantially smaller than n only if w is much smaller than n. Therefore, efficiency can be achieved only when the size of the revoked users is small. In other words, the OR re-keying schemes are efficient only for large multicast groups.

We present an AND dynamic controller scheme to cater for the revocation of a large number of users, thus making it suitable for establishing common secrets for small multicast groups. The scheme works as follows.

> **Key Initialization:** This phase is the same as in the OR re-keying scheme with dynamic controller.

> **Broadcast:** Assume that a user u_i wants to multicast a secret to the group $\mathcal{U} \setminus F$ with $|F| > w$. For each $u_j \in \mathcal{U} \setminus F$, and w users $u_{i_1}, \ldots, u_{i_w}$ in F, let
>
> $$H[u_i \rightarrow u_j; u_{i_1}, \ldots, u_{i_w}] = \{h \in \mathcal{H} : h \text{ is perfect on } \{i, j, i_1, \ldots, i_w\}\}$$
>
> and
>
> $$\mathcal{H}[u_i \rightarrow u_j; F] = \bigcup_{u_{i_1}, \ldots, u_{i_w}} H[u_i \rightarrow u_j; u_{i_1}, \ldots, u_{i_w}],$$
>
> where $u_{i_1}, \ldots, u_{i_w}$ run through all combinations of w distinct users in F. To multicast a secret for a group $\mathcal{U} \setminus F$, the user u_i chooses a session key $k^{\mathcal{U} \setminus F}$ and broadcasts
>
> $$\{E_{k(\widehat{u_i \rightarrow u_j})}(k^{\mathcal{U} \setminus F}) : u_j \in \mathcal{U} \setminus F\},$$
>
> with $k(\widehat{u_i \rightarrow u_j})$ defined by
>
> $$k(\widehat{u_i \rightarrow u_j}) = \sum_{h_\ell \in \mathcal{H}[u_i \rightarrow u_j; F]} k^\ell_{h_\ell(i), h_\ell(j)},$$
>
> where we assume all the auxiliary keys are elements of a finite field or an abelian group.

Decryption: Each user u_j in $\mathcal{U} \setminus F$ computes $k(\widehat{u_i \to u_j})$ and decrypts $E_{k(\widehat{u_i \to u_j})}(k^{\mathcal{U} \setminus F})$.

Theorem 8.3.5 *If there exists a* PHF$(N; n, m, w + 2)$, *then there exists a w-resilient re-keying scheme with dynamic controller with n users such that the number of auxiliary keys for each user is Nm. To multicast a secret for a group $\mathcal{U} \setminus F$, the number of transmissions for the ciphertexts is $|\mathcal{U} \setminus F|$.*

Proof. Every user $u_j \in \mathcal{U} \setminus F$ knows the key $k^\ell_{h_\ell(j), h_\ell(i)}$. Since the matrix G^ℓ is symmetric, we have $k^\ell_{h_\ell(j), h_\ell(i)} = k^\ell_{h_\ell(i), h_\ell(j)}$. It follows that u_j can compute $k(\widehat{u_i \to u_j})$ and decrypt $E_{k(\widehat{u_i \to u_j})}(k^{\mathcal{U} \setminus F})$. We show that the scheme is w-resilient.

Assume that w users $u_{i_1}, \ldots, u_{i_w}$ from F collude. They succeed in computing $k^{\mathcal{U} \setminus F}$ only if they can compute $k(\widehat{u_i \to u_j})$ for some j. However, this is not possible. Indeed, since for each j there exists a function $h_\ell \in \mathcal{H}[u_i \to u_j; F]$ such that h_ℓ is perfect on $\{i, j, i_1, \ldots, i_w\}$, it follows that the key $k^\ell_{h_\ell(i), h_\ell(j)}$ is unknown to $u_1, \ldots, u_w$, so they are unable to compute $k(\widehat{u_i \to u_j})$. Various parameters of the scheme can be easily found in terms of the parameters of the underlying PHF$(N; n, m, w + 2)$. $\quad\square$

8.4 Some Applications from Algebraic Geometry

In this final section, we point out some connections between the constructions for OTBESs/multicast re-keying schemes with algebraic geometry.

8.4.1 OTBESs over Constant Size Fields

We have seen in Section 8.1 that OTBESs are closely related to KPSs. In Chapter 7, we showed how to construct KPSs over constant size fields for an arbitrarily large number of users by using algebraic geometry codes. We present in the following an application of the results of Chapter 7 to the construction of OTBESs over constant size fields.

In the family of (t, w, n)-OTBESs proposed by Blundo *et al.* in Subsection 8.1.2, the key predistribution phase is done according to the Blundo *et al.* $(\ell, t + w - \ell, n)$-KPS over $\mathbb{F}_q$, so it requires $q \geq n$. By using instead KPSs from algebraic geometry codes, for instance, the ones constructed in Section 7.4, we can modify the OTBESs from [25] to obtain OTBESs over constant size fields.

Stinson and Wei [158] proposed a construction of OTBESs that combines

KPSs, set systems (such as combinatorial designs), and ramp secret sharing schemes, improving the efficiency from the KIO construction. Since they use Shamir-like ramp schemes, the size of the secret keys, and hence the size of the secrets given to each user, depend on the parameters of the set system. As we have shown in Section 4.9, ramp secret sharing schemes over constant size fields can be constructed from algebraic curves. Those results can be used to improve the efficiency of OTBESs as well.

8.4.2 Improving the Fiat-Naor OTBES

The Fiat-Naor OTBES in Subsection 8.1.4 is essentially a combination of a Fiat-Naor KPS and a perfect hash family, and its performance is determined by the underlying PHF. As before, let $N(n, m, w)$ denote the minimum N for which a PHF$(N; n, m, w)$ exists. Various bounds on $N(n, m, w)$ given in Section 7.6 can be used to derive the complexity of the corresponding OTBES from Theorem 8.1.7. For example, applying Corollary 7.6.4, we have the following result.

Corollary 8.4.1 *Suppose n, m, and w are integers and $n \geq m \geq w$. There exists an $(\leq n, \leq w, n)$-OTBES having information rate*

$$\frac{1}{n \lceil w e^{w^2/m} \log n \rceil}$$

and broadcast information rate

$$\frac{1}{m \lceil w e^{w^2/m} \log n \rceil}.$$

However, most upper bounds on $N(n, m, w)$ are non-constructive, so the results on OTBESs from Corollary 8.4.1 are also non-constructive. In Section 7.6, we showed how to apply algebraic curves to obtain explicit constructions of PHFs with good parameters. From Theorem 8.1.7, this in turn results in explicit constructions of the Fiat-Naor OTBES with good performance.

8.4.3 Improving the Blacklisting Scheme

The blacklisting scheme proposed by Kumar *et al.* in [91], discussed in Subsection 8.2.1, is based on a cover-free family. The performance of the resulting re-keying scheme is completely determined by the parameters of the underlying $(1, w)$-CFF(v, n). Using the techniques developed in Chapter 7, we know that there exist $(1, w)$-CFF(v, n)'s in which $v = O(\log n)$. For example, in [55], Erdős, Frankl, and Füredi showed that, for a given n, there exists a $(1, w)$-CFF(v, n) with $v = O(w^2 \log n)$. However, the result is again non-constructive. Explicit constructions that asymptotically achieve the lower bound can be obtained through using algebraic curves as shown in Section 7.6.

Corollary 8.4.2 *For any n, there exists an explicit construction for $(1, w)$-CFF(v, n) in which $v = O(w^4 \log n)$.*

Proof. From Theorem 7.5.27, we know that a PHF$(N; n, m, t + 1)$ explicitly yields a $(1, t)$-CFF(Nm, n). From Theorem 7.6.11, we have that, for any square prime power q, there exist explicit constructions of PHF $\left((\sqrt{q} - 1)q^{i/2}; q^{q^{i/2}}, q, \lfloor q^{1/4} \rfloor\right)$ for all $i \geq 1$. Taking $w = q^{1/4}$, the result follows. $\qquad\square$

Example 8.4.3 From Example 7.6.14, we know that there is an explicit construction of a PHF$(N; n, 3, 3)$ with $N \approx 30 \log n$. From Theorem 7.5.27, we obtain an explicit construction of $(1, 2)$-CFF(ℓ, v) with $\ell \approx 90 \log v$. Erdős, Frankl, and Füredi showed in [55] that, for any $(1, 2)$-CFF(ℓ, v), one has $\ell \geq \lceil 3.1 \log v \rceil$. Dyer, Fenner, Frieze, and Thomason gave in [51] a probabilistic construction with $\ell = \lceil 13 \log v \rceil$. Thus, the explicit construction from algebraic curves increases the parameter ℓ by around seven times from the best known probabilistic construction.

8.4.4 Construction of w-Resilient LKHs

The efficiency of the multicast OR and AND re-keying schemes in Subsection 8.2.3 relies on the parameters of the (n, ℓ, w) q-ary resilient LKH. Next, we give a construction of resilient LKHs from error-correcting codes.

Let C be the set of codewords in a q-ary (ℓ, n, d)-code. We write each element of C as $\mathbf{c}_i = (c_{i1}, \ldots, c_{i\ell})$ with $c_{ij} \in \{1, \ldots, q\}$, where $1 \leq i \leq n$ and $1 \leq j \leq \ell$. To each codeword $\mathbf{c}_i \in C$, we associate a leaf L_i in the full q-ary tree $\mathcal{T}$ in a natural way, i.e., the node in the jth level of the tree, along the path from L_i to the root, is labeled by c_{ij}. Let $L = \{L_i : \mathbf{c}_i \in C\}$, and let $\mathcal{T}(L)$ be the subtree of $\mathcal{T}$ consisting of the set of leaves L.

For each leaf L_i, we denote by V_i the set of labeled values of the nodes along the path from L_i to the root. Then, clearly, we have $|V_i| = \ell$ and $|V_i \cap V_j| \leq \ell - d$ for any distinct leaves L_i and L_j. Now, for any $w + 1$ distinct leaves $L_{i_1}, \ldots, L_{i_w}, L_j$, we have

$$\left| V_j \setminus \bigcup_{s=1}^{w} V_{i_s} \right| \geq |V_j| - \sum_{s=1}^{w} |V_j \cap V_{i_s}| \geq \ell - w(\ell - d) \geq 1,$$

provided $w \leq \frac{\ell-1}{\ell-d}$. Thus, we obtain the following result.

Theorem 8.4.4 *If there exists a q-ary (ℓ, n, d)-code, then there exists an $(n, \ell, \lfloor \frac{\ell-1}{\ell-d} \rfloor)$ q-ary resilient LKH.*

One straightforward application of Theorem 8.4.4 is to use Reed-Solomon codes. Since a Reed-Solomon code is an $(\ell, q^t, \ell - t + 1)$-code for any integer $t < \ell - 1$, from Theorem 8.4.4, we have the following corollary.

Corollary 8.4.5 *Let q be a prime power. For any integers $\ell, t,$ and w such that $t < \ell - 1$, there exists an $(q^t, \ell, \lfloor \frac{\ell-1}{t-1} \rfloor)$ q-ary resilient LKH.*

The construction based on Reed-Solomon codes is still restrictive because it requires $\ell \leq q$. In order to relax the condition "$\ell \leq q$," one can use algebraic geometry codes. Thus, from the results in Chapter 2, we can restate our result as follows.

Corollary 8.4.6 *For a given integer $w \geq 2$ and a prime power q, there exists a sequence of (ℓ, n, w) q-ary resilient LKHs in which $\ell = O(\log n)$.*

Combining Theorems 8.2.7, 8.2.8, and Corollary 8.4.6, we have

Theorem 8.4.7 *Let $\mathcal{U} = \{u_1, \ldots, u_n\}$ and let $P \subseteq \mathcal{U}$ be a multicast group.*

1. *For $|P| \geq n - w$, there exists a w-resilient multicast re-keying scheme such that the numbers of keys of each user and the TA are both $O(\log n)$, and the number of broadcast transmissions for the encrypted session key is $O(\log n)$.*

2. *For $1 \leq |P| \leq n-1$, there exists a w-resilient multicast re-keying scheme such that the numbers of keys of each user and the TA are both $O(\log n)$, and the number of broadcast transmissions for the encrypted session key is $|P|$.*

Chapter 9

Sequences

9.1 Introduction

Stream ciphers form an important class of encryption algorithms. They encrypt individual characters (usually binary bits) of a plaintext, one at a time, using an encryption transformation which varies with time. Stream ciphers are generally faster than block ciphers in hardware, and have less complex hardware circuitry.

A stream cipher can be either a symmetric (i.e., private-key) or public-key cryptosystem. Since most stream ciphers use the same keys for encryption and decryption, they are symmetric. In practical implementations, a stream cipher will be a bit-based cryptosystem. The plaintext, the ciphertext, and the key are all bit strings of the same length, but this length can be arbitrary (as opposed to a block cipher where the lengths are fixed). Encryption proceeds by taking the plaintext string and bit-wise XORing it with the key string (or, in other words, adding the two strings bit by bit in the finite field $\mathbb{F}_2$). It is clear that the plaintext is recovered by bit-wise XORing the ciphertext string and the key string. Thus, in a bit-based stream cipher, the encryption and decryption algorithms are identical, which has the practical advantage that the same hardware can be used for both operations.

From the theoretical point of view, it does not make any difference whether we consider stream ciphers over $\mathbb{F}_2$ or over an arbitrary finite field $\mathbb{F}_q$. Therefore, we consider the general case where the plaintext and the ciphertext are strings (or, in other words, finite sequences) of elements of $\mathbb{F}_q$, and the encryption (respectively, decryption) proceeds by term-wise addition (respectively, subtraction) of the same key string of elements of $\mathbb{F}_q$. The key string is commonly called the **keystream** and known only to authorized users. It is convenient from now on to speak of strings (respectively, sequences) over $\mathbb{F}_q$ when we mean strings (respectively, sequences) of elements of $\mathbb{F}_q$.

In an ideal situation, the keystream would be a "truly random" string over $\mathbb{F}_q$. In this case, the stream cipher would be perfectly secure since the ciphertexts will carry absolutely no information, and so there will be no basis for an attack on the cryptosystem. In practice, sources of true randomness are hard to come by, so keystreams are taken to be pseudorandom strings that are obtained from certain secret seed data by some (perhaps even publicly available) algorithm. A central issue in the security analysis of stream ciphers is then the quality assessment of these pseudorandom keystreams. In other words, we need to know how close a given keystream is to true randomness. We focus here on the complexity-theoretic aspects of this assessment, where global function fields have been known to play some methodological role. There are also statistical techniques for this assessment, but they will not be studied in this chapter.

9.2 Linear Feedback Shift Register Sequences

A **linear feedback shift register** (**LFSR**) is a special kind of electronic switching circuit handling information in the form of elements of $\mathbb{F}_q$. This circuit contains constant multipliers, adders, and delay elements. LFSRs are very popular because they are extremely easy to implement in hardware and they produce sequences with good statistical properties. Algorithmically, an LFSR is described by a linear recurrence relation defined below.

Definition 9.2.1 Let k be a positive integer. A k**th-order LFSR** over $\mathbb{F}_q$ is a pair $< f(T), k >$, where $f(T) = 1 - \sum_{i=1}^{k} c_i T^i \in \mathbb{F}_q[T]$ (note that $c_k = 0$ is allowed). An infinite sequence $\mathbf{s} = s_0 s_1 s_2 \ldots$ of elements of $\mathbb{F}_q$ is said to satisfy a kth-order linear recurrence relation over $\mathbb{F}_q$ if there exists a kth-order LFSR $< f(T) = 1 - \sum_{i=1}^{k} c_i T^i, k >$ over $\mathbb{F}_q$ such that

$$s_j = c_1 s_{j-1} + c_2 s_{j-2} + \cdots + c_k s_{j-k}, \tag{9.1}$$

for all $j \geq k$, and the sequence is called a k**th-order linear recurring sequence**, or a k**th-order LFSR sequence**, over $\mathbb{F}_q$.

The linear recurrence relation (9.1) and the initial terms $s_0, s_1, \ldots, s_{k-1}$ determine the rest of the sequence $s_k, s_{k+1}, \ldots$ uniquely. We denote this sequence by $< f(T), k; \mathbf{s}_0 >$, where $\mathbf{s}_0$ is the **seed** $s_0, s_1, \ldots, s_{k-1}$. The vector $\mathbf{s}_n = (s_n, s_{n+1}, \ldots, s_{n+k-1})$ is called the n**th-state vector** and the seed $\mathbf{s}_0 = (s_0, s_1, \ldots, s_{k-1})$ is also called the **initial state vector**.

Remark 9.2.2 A linear recurring sequence satisfies many other linear recurrence relations apart from the one defining it. For instance, consider the binary sequence $1011011011011011 \ldots$. It is defined by $s_j = s_{j-1} + s_{j-2}$ with the seed 10, but it also satisfies the linear recurrence relation $s_j = s_{j-3}$ with the seed 101.

Definition 9.2.3 Let $\mathbf{s} = s_0 s_1 s_2 \ldots$ be an arbitrary sequence over $\mathbb{F}_q$. If there exist integers $r > 0$ and $n_0 \geq 0$ such that $s_{n+r} = s_n$ for all $n \geq n_0$, then the sequence is called **ultimately periodic** and the number r is called a **period**

of the sequence. The smallest among all possible periods of the sequence **s** is called its **least period**. An ultimately periodic sequence $s_0 s_1 s_2 \ldots$ with period N is called **(purely) periodic** if $s_{n+N} = s_n$ holds for all $n \geq 0$. If **s** is ultimately periodic with period N, then the least nonnegative integer n_0 such that $s_{n+N} = s_n$ for all $n \geq n_0$ is called the **preperiod** of **s**.

Example 9.2.4   (i) The sequence generated by $s_j = s_{j-1} + s_{j-2}$ $(j \geq 3)$ with the seed 111, namely $1110110110110110\ldots$, is an ultimately periodic sequence with least period 3. The preperiod is 1.

(ii) The binary sequence in Remark 9.2.2 is periodic with period 3.

Proposition 9.2.5 *A period of an ultimately periodic sequence* **s** *is divisible by the least period of* **s**.

Proof. Let r and N be a period and the least period, respectively, of **s**. Then, by definition, there exist $n \geq 0$ and $m \geq 0$ such that $s_{i+r} = s_i$ for all $i \geq n$ and $s_{j+N} = s_j$ for all $j \geq m$. Hence, $s_{i+r} = s_i$ and $s_{i+N} = s_i$ for all $i \geq \max\{n, m\}$. Let $r = t \cdot N + u$ with $0 \leq u \leq N - 1$ and $t \geq 1$. Then, for any $i \geq \max\{n, m\}$, we have

$$s_i = s_{i+r} = s_{i+t \cdot N + u} = s_{i+(t-1) \cdot N + u} = \cdots = s_{i+u}.$$

This implies that $u = 0$ since N is the least period. $\square$

To each infinite sequence **s** over $\mathbb{F}_q$, we can associate a generating function, which is a power series

$$s(x) = \sum_{i=0}^{\infty} s_i x^i$$

with an indeterminate x.

Theorem 9.2.6 *An infinite sequence* $\mathbf{s} = s_0 s_1 s_2 \ldots$ *over* $\mathbb{F}_q$ *is ultimately periodic if and only if its generating function is rational.*

Proof. If **s** is ultimately periodic, then there exist integers $r > 0$ and $n \geq 0$ such that $s_{i+r} = s_i$ for all $i \geq n$. Thus, the generating function of **s** is

$$
\begin{aligned}
s(x) &= \sum_{j=0}^{\infty} s_j x^j \\
&= \sum_{j=0}^{n-1} s_j x^j + \sum_{j=n}^{\infty} s_j x^j \\
&= \sum_{j=0}^{n-1} s_j x^j + \sum_{a=0}^{\infty} \sum_{b=0}^{r-1} s_{n+b+ar} x^{n+b+ar} \\
&= \sum_{j=0}^{n-1} s_j x^j + \sum_{b=0}^{r-1} \sum_{a=0}^{\infty} s_{n+b} x^{n+b+ar} \\
&= \sum_{j=0}^{n-1} s_j x^j + \sum_{b=0}^{r-1} s_{n+b} x^{n+b} \frac{1}{1-x^r},
\end{aligned}
$$

which is rational.

Now assume that the generating function is a rational function $f(x)/g(x)$ with $f(x), g(x) \in \mathbb{F}_q[x]$. Write the rational function into $f(x)/g(x) = h(x) + x^t u(x)/g(x)$ with $h(x), u(x) \in \mathbb{F}_q[x]$, $\gcd(x, g(x)) = 1$, and $\deg(u(x)) + t < \deg(g(x))$. Then we can find an integer $n > 0$ such that $g(x)|(x^n - 1)$ (see [95]). Write $x^n - 1 = g(x)v(x)$ for some $v(x) \in \mathbb{F}_q[x]$ and $x^t u(x)v(x) = x^t \sum_{i=0}^{n-1} a_i x^i$ for some $a_i \in \mathbb{F}_q$, $0 \leq i \leq n-1$. Hence,

$$
\begin{aligned}
f(x)/g(x) &= h(x) + x^t u(x)/g(x) \\
&= h(x) + x^t u(x)v(x)/(x^n - 1) \\
&= h(x) + x^t \left(\sum_{i=0}^{n-1} a_i x^i \right) \left(-\sum_{j=0}^{\infty} x^{jn} \right) \\
&= h(x) - x^t \sum_{j=0}^{\infty} \left(\sum_{i=0}^{n-1} a_i x^i \right) x^{jn}.
\end{aligned}
$$

Hence, n is a period of the sequence. $\qquad\square$

9.3 Constructions of Almost Perfect Sequences

Complexity measures for sequences are important in the system-theoretic approach to stream cipher design. Sequences that are suitable as keystreams should satisfy certain criteria. One requirement is that it should be computationally very hard to replicate such sequences. From this point of view, the linear complexity of sequences is introduced: the notion of linear complexity is based on the shortest LFSR generating a given sequence and this sequence can be computed efficiently by the Berlekamp-Massey algorithm. Another requirement is that the sequence must belong to a large class of sequences exhibiting similar behavior (in some suitable sense).

In this section, we discuss the notions of linear complexity, perfect and almost perfect sequences, as well as some constructions of such sequences using algebraic curves.

9.3.1 Linear Complexity

There are several complexity measures for sequences (see [128, 114]). In this chapter, we concentrate mainly on the linear complexity of sequences defined as follows.

Definition 9.3.1 The **linear complexity** of an infinite sequence $\mathbf{s}$, denoted by $\ell(\mathbf{s})$, is defined as follows:

(i) if $\mathbf{s}$ is the zero sequence $\mathbf{0}$, then $\ell(\mathbf{s}) = 0$;

(ii) if no LFSR generates $\mathbf{s}$, then $\ell(\mathbf{s}) = \infty$;

(iii) otherwise, $\ell(\mathbf{s})$ is the length of a shortest LFSR that generates $\mathbf{s}$.

In practice, it should be hard to replicate the key, which is a sequence, in a stream cipher. It is obvious that the linear complexity of this sequence must be large enough so that it is computationally infeasible to exhaustively search those LFSRs that can generate this key.

Although the key in a stream cipher is infinite, only a finite part of this key is used in practice as the plaintext is always of finite length. Therefore, to recover a plaintext, it is sufficient to figure out the finite length of the key which is used for the encryption for this plaintext. Thus, we need to define the linear complexity of sequences of finite length.

Definition 9.3.2 The **linear complexity** of a finite sequence $\mathbf{s}$ of length n, denoted by $\ell(\mathbf{s})$, is the length of a shortest LFSR that generates an infinite sequence having $\mathbf{s}$ as the first n terms.

The **Berlekamp-Massey algorithm** provides an efficient way for determining the linear complexity of a sequence of finite length n. The algorithm takes n iterations, with the mth iteration computing the linear complexity of the first m terms of the sequence. In fact, the algorithm tells not only the linear complexity but also the linear recurrence that the sequence obeys.

We refer the reader to [49] for the proof of correctness of the Berlekamp-Massey algorithm in Figure 9.1. Here we make some remarks.

FIGURE 9.1: The Berlekamp-Massey algorithm.

Objective: Given a finite sequence $\mathsf{s} = s_0 s_1 \ldots s_{n-1}$, find a shortest LFSR $< f(T), k >$ that generates s.

Step 1: Assume that $m \geq 0$ is the smallest index such that $s_m \neq 0$, i.e., $s_0 = s_1 = \cdots = s_{m-1} = 0$ and $s_m \neq 0$; then set the initial values

$$d_0 = d_1 = \cdots = d_{m-1} = 0, \quad d_m = s_m,$$

$$f_1(T) = f_2(T) = \cdots = f_m(T) = 1, f_{m+1}(T) = 1 - d_m T^{m+1},$$

and

$$k_1 = k_2 = \cdots = k_m = 0.$$

Step 2: Assume that, for $i = 1, 2, \ldots, N$, we have constructed the polynomial $f_i(T)$ with least degree k_i that generates the first i terms of s. Let $f_N(T) = 1 + c_{N1}T + \cdots + c_{Nk_N}T^{k_N}$ and compute

$$d_N = s_N + c_{N1}s_{N-1} + \cdots + c_{Nk_N}s_{N-k_N}.$$

Case 1: If $d_N = 0$, then set $f_{N+1}(T) = f_N(T)$ and $k_{N+1} = k_N$;
Case 2: If $d_N \neq 0$, then there exists t with $1 \leq t < N$ such that $k_t < k_{t+1} = \cdots = k_N$. Set $f_{N+1}(T) = f_N(T) - d_N d_t^{-1} T^{N-t} f_t(T)$ and $k_{N+1} = \max\{k_N, N + 1 - k_N\}$.

Remark 9.3.3   (i) The polynomial $f_n(T)$ generating the sequence $\mathsf{s} = s_0 s_1 \ldots s_{n-1}$ is not unique. However, the degree is unique as it is the least one.

(ii) Let $\ell_i(\mathsf{s})$ denote the linear complexity of the first i terms. Since $f_{m+1}(T)$ can generate the first m terms as well, we must have $\ell_m(\mathsf{s}) \leq \ell_{m+1}(\mathsf{s})$.

(iii) From the above algorithm, we know that if $\ell_m(\mathsf{s}) < \ell_{m+1}(\mathsf{s})$, then we must have $\ell_m(\mathsf{s}) < (m+1)/2$ since $\ell_m(\mathsf{s}) + \ell_{m+1}(\mathsf{s}) = m + 1$.

(iv) According to the Berlekamp-Massey algorithm, if $\ell_m(\mathsf{s}) > (m+1)/2$ for some $m > 0$, then $\ell_{m+1}(\mathsf{s}) \leq \max\{\ell_m(\mathsf{s}), m + 1 - \ell_m(\mathsf{s})\}$. This implies that $\ell_m(\mathsf{s}) = \ell_{m+1}(\mathsf{s})$ since $\ell_m(\mathsf{s}) \leq \ell_{m+1}(\mathsf{s})$.

Example 9.3.4 Consider the binary sequence $\mathsf{s} = 0010001$ of length 7. We have initial values $d_1 = d_0 = s_1 = s_0 = 0$, $d_2 = s_2 = 1$; $f_1(T) = f_2(T) = 1$ and $k_1 = k_2 = 0$.

Thus, according to the above algorithm, we have $f_3(T) = 1 + T^3$ and $k_3 = 3$. Then we have $d_3 = 1 \cdot s_3 + 1 \cdot s_0 = 0$, hence $f_4(T) = f_3(T) = 1 + T^3$ and $k_4 = k_3 = 3$.

Continuing, we obtain $d_4 = 1 \cdot s_4 + 1 \cdot s_1 = 0$, hence $f_5(T) = f_4(T) = 1 + T^3$ and $k_5 = k_4 = 3$.

We then have $d_5 = 1 \cdot s_5 + 1 \cdot s_2 = 1$, hence $f_6(T) = f_5(T) + T^{5-2} f_2(T) = 1$ and $k_6 = \max\{k_5, 5 + 1 - k_5\} = 3$.

Finally, we get $d_6 = 1$ and hence $f_7(T) = f_6(T) + T^{6-2} f_2(T) = 1 + T^4$, $k_7 = \max\{k_6, 6 + 1 - k_6\} = 4$.

Thus, a shortest LFSR generating **s** is $< 1 + T^4, 4 >$.

Remark 9.3.5 (i) Apart from the Berlekamp-Massey algorithm, the complexity of a sequence can be determined through its generating function and continued fractions. The reader may refer to [95] for details.

(ii) The result in Remark 9.3.3(iii) can be obtained using continued fractions as well.

Example 9.3.6 The linear complexity of the sequence $00 \ldots 001$ of length n is n by the Berlekamp-Massey algorithm. One of the shortest LFSRs generating this sequence is $1 + T^n$.

9.3.2 Constructions of d-Perfect Sequences from Algebraic Curves

In this subsection, we introduce the notions of perfect and d-perfect sequences, and discuss a construction of d-perfect sequences from algebraic curves.

For an infinite sequence $\mathbf{s} = s_0 s_1 s_2 \ldots$, we denote by $\ell_n(\mathbf{s})$ the linear complexity of the first n terms. Then we obtain a sequence $\{\ell_n(\mathbf{s})\}_{n=1}^{\infty}$ of nonnegative integers. This sequence is called the **linear complexity profile** of **s**.

To make it hard to recover the first terms of **s**, we want $\ell_n(\mathbf{s})$ to be large. However, according to Remark 9.3.3(iv), if $\ell_m(\mathbf{s}) > (m+1)/2$ for some m, then $\ell_n(\mathbf{s})$ remains unchanged from $n = m$ until an integer $k > m$ where $\ell_k(\mathbf{s}) = \ell_m(\mathbf{s}) < (k+1)/2$. This observation tells us that it is impossible to construct a sequence **s** such that $\ell_n(\mathbf{s})$ is bigger than $(n+1)/2$ for every n. Thus, it is natural to find sequences with linear complexity profile equal to $\{\lceil n/2 \rceil\}_{n=1}^{\infty}$. This gives the following definition introduced by Rueppel [128].

Definition 9.3.7 An infinite sequence **s** is called **perfect** if $\ell_n(\mathbf{s}) = \lceil n/2 \rceil$ for all $n \geq 1$.

Perfect sequences were generalized by Niederreiter [113] to d-perfect sequences, which are defined as follows.

Definition 9.3.8 For an integer $d \geq 1$, an infinite sequence **s** is called d-**perfect** if $\ell_n(\mathbf{s}) \geq (n + 1 - d)/2$ for all $n \geq 1$. A d-perfect sequence is also called an **almost perfect sequence**.

Lemma 9.3.9 *Let* **s** *be an infinite sequence. Then* **s** *is d-perfect if and only if* $(n + 1 - d)/2 \le \ell_n(\mathbf{s}) \le (n + 1 + d)/2$ *for all* $n \ge 1$.

Proof. It is sufficient to show that, for a d-perfect sequence **s**, we have $\ell_n(\mathbf{s}) \le (n + 1 + d)/2$ for all $n \ge 1$. Suppose that this is false. Let m be the smallest positive integer such that $\ell_m(\mathbf{s}) > (m + 1 + d)/2$. Then we must have $\ell_{m-1}(\mathbf{s}) < \ell_m(\mathbf{s})$. Thus, we have $\ell_{m-1}(\mathbf{s}) = m - \ell_m(\mathbf{s}) < (m - 1 - d)/2$. This is a contradiction. $\qquad\square$

Remark 9.3.10 From the above lemma, we can easily see that a perfect sequence is none other than a 1-perfect sequence.

Before introducing a construction of d-perfect sequences, we need some preparation on algebraic curves.

Let $\mathcal{X}$ be an algebraic curve over $\mathbb{F}_q$ and, for an $\mathbb{F}_q$-rational point P of $\mathcal{X}$, we choose a local parameter t of P in $\mathbb{F}_q(\mathcal{X})$.

Now we choose a sequence $\{t_r\}_{r=-\infty}^{\infty}$ of elements in $\mathbb{F}_q(\mathcal{X})$ such that

$$\nu_P(t_r) = r$$

for all integers r (for instance, we can let $t_r = t^r$). For a given function $f \in \mathbb{F}_q(\mathcal{X}) \setminus \{0\}$, we can find an integer v such that $\nu_P(f) \ge v$. Hence

$$\nu_P\left(\frac{f}{t_v}\right) \ge 0.$$

Put

$$a_v = \left(\frac{f}{t_v}\right)(P),$$

i.e., a_v is the value of the function f/t_v at P. Then a_v is an element of $\mathbb{F}_q$. Note that the function $f/t_v - a_v$ satisfies

$$\nu_P\left(\frac{f}{t_v} - a_v\right) \ge 1,$$

hence we know that

$$\nu_P\left(\frac{f - a_v t_v}{t_{v+1}}\right) \ge 0.$$

Put

$$a_{v+1} = \left(\frac{f - a_v t_v}{t_{v+1}}\right)(P).$$

Then a_{v+1} belongs to $\mathbb{F}_q$ and $\nu_P(f - a_v t_v - a_{v+1} t_{v+1}) \ge v + 2$.

Assume that we have obtained a sequence $\{a_r\}_{r=v}^{m}$ $(m > v)$ of elements of $\mathbb{F}_q$ such that

$$\nu_P\left(f - \sum_{r=v}^{k} a_r t_r\right) \ge k + 1 \tag{9.2}$$

for all $v \le k \le m$. Put

$$a_{m+1} = \left(\frac{f - \sum_{r=v}^{m} a_r t_r}{t_{m+1}} \right)(P).$$

Then $a_{m+1} \in \mathbb{F}_q$ and $\nu_P(f - \sum_{r=v}^{m+1} a_r t_r) \ge m+2$. Continuing our construction of a_r in this fashion, we obtain an infinite sequence $\{a_r\}_{r=v}^{\infty}$ of elements of $\mathbb{F}_q$ such that

$$\nu_P \left(f - \sum_{r=v}^{m} a_r t_r \right) \ge m + 1 \tag{9.3}$$

for all $m \ge v$. We summarize the above well-known construction in the formal expansion

$$f = \sum_{r=v}^{\infty} a_r t_r. \tag{9.4}$$

This is called the **local expansion** of f at P. The above local expansion (9.4) will be the core of our construction. We will use only the special case where $t_r = t^r$ for some local parameter t at P. For further background on algebraic curves and their function fields, we refer the reader to [116].

We now describe a construction of d-perfect sequences based on algebraic curves over finite fields. We fix some notations for this subsection:

- $\mathcal{X}/\mathbb{F}_q$ – an algebraic curve over $\mathbb{F}_q$;

- P – an $\mathbb{F}_q$-rational point on $\mathcal{X}$;

- t – a local parameter at P with $\deg(\mathrm{div}_\infty(t)) = 2$;

- f – a function in $\mathbb{F}_q(\mathcal{X}) \setminus \mathbb{F}_q(t)$ with $\nu_P(f) \ge 0$.

Consider the local expansion of f at P:

$$f = \sum_{n=0}^{\infty} a_n t^n,$$

where a_n $(n \ge 0)$ are elements of $\mathbb{F}_q$. We may define a sequence $\mathbf{a}(f)$, consisting of some coefficients of the above expansion, as follows:

$$\mathbf{a}(f) = (a_1, a_2, a_3, \ldots).$$

Proposition 9.3.11 *If $d \ge \deg(\mathrm{div}_\infty(f))$ and $\nu_P(f) \ge 0$, then the sequence $\mathbf{a}(f)$ constructed above is d-perfect.*

Proof. It is sufficient to prove that

$$\ell_n(\mathbf{a}(f)) \ge \frac{n - d + 1}{2}$$

for all $n \geq d$.

Suppose that there exist $r + 1 \leq n$ elements $\lambda_0, \ldots, \lambda_r$ of $\mathbb{F}_q$ with $\lambda_0 = 1$ such that

$$\lambda_0 a_{i+r-1} + \lambda_1 a_{i+r-2} + \cdots + \lambda_{r-1} a_i + \lambda_r a_{i-1} = 0 \qquad (9.5)$$

for $i = 1, 2, \ldots, n - r$. Consider the function

$$L = (\lambda_r t^r + \lambda_{r-1} t^{r-1} + \cdots + \lambda_0) f$$
$$- \left(\lambda_0 a_0 + (\lambda_0 a_1 + \lambda_1 a_0)t + \cdots + (\lambda_0 a_{r-1} + \cdots + \lambda_{r-1} a_0) t^{r-1}\right).$$

Since $\lambda_0 = 1$ and $f \notin \mathbb{F}_q(t)$, we know that L is a nonzero element of $\mathbb{F}_q(\mathcal{X})$. By applying the recursion (9.5) and considering the local expansion of L at P, we obtain

$$\nu_P(L) \geq n + 1.$$

On the other hand, the pole divisor of L satisfies

$$\deg(\mathrm{div}_\infty(L)) \leq \deg(\mathrm{div}_\infty(f)) + \deg(\mathrm{div}_\infty(t^r)) \leq d + 2r.$$

Therefore

$$n + 1 \leq \nu_P(L) \leq \deg(\mathrm{div}_0(L)) = \deg(\mathrm{div}_\infty(L)) \leq d + 2r,$$

i.e.,

$$r \geq \frac{n + 1 - d}{2}.$$

This implies that $\mathbf{a}(f)$ is d-perfect. $\qquad\qquad\qquad\qquad\qquad \square$

9.3.3   Examples of d-Perfect Sequences

In this subsection, we systematically discuss examples of the construction in Subsection 9.3.2 from the projective line and elliptic curves. In particular, we determine all possible binary and ternary perfect sequences from the projective line.

A function t on a curve $\mathcal{X}$ is called a **degree-d function** or a **function of degree** d if $\deg(\mathrm{div}_\infty(t)) = d$.

For convenience, we rewrite Proposition 9.3.11 in the following theorem.

Theorem 9.3.12 *Let $\mathcal{X}/\mathbb{F}_q$ be a curve for which there exist an $\mathbb{F}_q$-rational point P on $\mathcal{X}$ and a degree-two function t in $\mathbb{F}_q(\mathcal{X})$ such that $\nu_P(t) = 1$. Let x be a function of degree d on $\mathcal{X}$ such that $\mathbb{F}_q(\mathcal{X}) = \mathbb{F}_q(x, t)$ with $(x/t)(P) = 1$. Then there exists a unique power series expansion of the form*

$$x(t) = t + a_2 t^2 + a_3 t^3 + a_4 t^4 + \cdots$$

for x which defines an embedding of the function field $\mathbb{F}_q(\mathcal{X})$ in $\mathbb{F}_q((t))$. The corresponding sequence $(1, a_2, a_3, a_4, \ldots)$ is d-perfect.

The series $x(t)$ can be rapidly computed via the effective form of Hensel's Lemma (see [93, Chapter II, Proposition 2]).

Theorem 9.3.13 (Hensel's Lemma) *Let x and t be as in the previous theorem, and let x have minimal polynomial $F(X)$ over $\mathbb{F}_q(t)$. Then the power series expansion for x in $\mathbb{F}_q((t))$ can be determined by setting $x_1(t) = t$ and, for all i,*

$$x_{i+1} = x_i - \frac{F(x_i)}{F'(x_i)},$$

where $F'(X)$ is the derivative of $F(X)$ with respect to X. The sequence (x_i) satisfies $\nu_P(F(x_i)) \geq 2^i$.

Case 1: Projective lines

To obtain perfect sequences using the above construction, we require a curve $\mathcal{X}/\mathbb{F}_q$, which has a degree-one function x.

Lemma 9.3.14 *If there is a degree-one function x on a curve $\mathcal{X}/\mathbb{F}_q$, then $\mathcal{X}$ has genus 0, i.e., $\mathcal{X}$ is the projective line. Moreover, the function field of $\mathcal{X}$ is the rational function field $\mathbb{F}_q(x)$.*

Proof. Let $\operatorname{div}_\infty(x) = P$ for an $\mathbb{F}_q$-rational point P on $\mathcal{X}$. Thus, $\dim \mathcal{L}(mP) \geq m + 1$ since $1, x, x^2, \ldots, x^m$ are elements in $\mathcal{L}(mP)$ and they are $\mathbb{F}_q$-linearly independent. By the Riemann-Roch Theorem, we have $\dim \mathcal{L}(2gP) = 2g + 1 - g = g + 1$, where g is the genus of $\mathcal{X}$. Hence, we obtain $g + 1 \geq 2g + 1$, i.e., $g \leq 0$. This implies that $g = 0$ as the genus of a curve is nonnegative.

Consider the extension $\mathbb{F}_q(\mathcal{X})/\mathbb{F}_q(x)$. By Lemma 1.4.5, we have $1 = \deg(\operatorname{div}_\infty(x)) = [\mathbb{F}_q(\mathcal{X}) : \mathbb{F}_q(x)]$. This means that $\mathbb{F}_q(\mathcal{X}) = \mathbb{F}_q(x)$. $\square$

We begin with the projective line $\mathcal{X}/\mathbb{F}_q$ and a function x generating $\mathbb{F}_q(\mathcal{X})$. Let P be the zero of x, and let t be a degree-two function on $\mathcal{X}$ such that $\nu_P(t) = 1$ and $(x/t)(P) = 1$. The form of t is described by means of the following proposition.

Proposition 9.3.15 *Let $\mathcal{X}/\mathbb{F}_q$ be a genus zero curve with $\mathbb{F}_q(\mathcal{X}) = \mathbb{F}_q(x)$, and let P be the zero of x. A degree-two function $t \in \mathbb{F}_q(x)$ satisfying $(x/t)(P) = 1$ has the form*

$$t = \frac{x + ax^2}{1 + bx + cx^2}$$

for some a, b, and c in $\mathbb{F}_q$, where a or c is nonzero, and $\gcd(1 + ax, 1 + bx + cx^2) = 1$.

The proof of the above proposition is straightforward.

If x_1 is a generator for the rational function field $\mathbb{F}_q(x)$, i.e., $\mathbb{F}_q(x_1) = \mathbb{F}_q(x)$, then x_1 is a degree-one function. Hence, it is of the form

$$x_1 = \frac{ax + b}{cx + d},$$

where a, b, c, and d are elements of the base field $\mathbb{F}_q$ with $ad - bc \neq 0$. Given any x_1 in $\mathbb{F}_q(x)$ and a point P on $\mathcal{X}$, we may replace x_1 with $x_1 - x_1(P)$. Thus, by scaling, we may assume that

$$\nu_P(x_1) = 1 \quad \text{and} \quad (x_1/x)(P) = 1.$$

Then $x_1(P) = b/d = 0$ and $(x_1/x)(P) = a/d = 1$. Thus, any degree-one function x_1 can be normalized such that it is of the form

$$x_1 = \frac{x}{1 + cx}.$$

By Theorem 9.3.12, the sequence corresponding to the power series expansion for x_1 in t is also 1-perfect. Hence, we have shown that the following proposition is generally true.

Proposition 9.3.16 *Suppose x_0 and x_1 are degree-one functions on a curve $\mathcal{X}/\mathbb{F}_q$ and t is a degree-two local parameter at a point P. Suppose, moreover, that $\nu_P(x_0) = \nu_P(x_1) = 1$ and $(x_1/x_0)(P) = 1$. Then x_1 is of the form*

$$x_1 = \frac{x_0}{1 + cx_0},$$

for some $c \in \mathbb{F}_q$, and the sequences associated to x_0 and x_1 are both 1-perfect.

From Proposition 9.3.15, we see that there are four possible choices for the degree-two local parameter t over the binary field. These are:

$$(0.0) : t = x + x^2, \quad (1.0) : t = \frac{x + x^2}{1 + x + x^2},$$

$$(0.1) : t = \frac{x}{1 + x^2}, \quad (1.1) : t = \frac{x}{1 + x + x^2}.$$

In this case, the construction for a power series solution $x = x(t)$ to each of the above equations using Hensel's Lemma is a special case of an inversion formula for power series, i.e., if we take Equation $(i.j)$ to define the power series $t(x)$ in $\mathbb{F}_2[x]$, then $t(x(t)) = t$ and $x(t(x)) = x$.

We note that the linear fractional transformation $A(u) = u/(1 + u)$ determines an automorphism of order two of the projective line $\mathbf{P}^1(\mathbb{F}_2)$. If we denote by $x_{ij}(t)$ the power series which is a root of Equation $(i.j)$, and set $x(t) = x_{00}(t)$, then

$$x_{ij}(t) = A^j(x(A^i(t))).$$

We find the following sequences associated to the t-expansions of x in the four cases:

$(0.0) \ (1,1,0,1,0,0,1,0,0,0,0,0,0,0,1,0,0,0,0,0,0,0,0,0,0,0,0,0,\ldots),$
$(0.1) \ (1,0,1,0,0,0,1,0,0,0,0,0,0,0,1,0,0,0,0,0,0,0,0,0,0,0,0,0,\ldots),$
$(1.0) \ (1,0,1,1,1,0,1,0,1,0,1,1,1,0,1,1,1,0,1,1,1,0,1,0,1,0,1,1,1,\ldots),$
$(1.1) \ (1,1,0,0,1,1,1,1,1,1,0,0,1,1,0,0,1,1,0,0,1,1,1,1,1,1,0,0,1,\ldots).$

These initial segments can be verified to follow the linear complexity of 1-perfect sequences.

The sequence (0.0) is the well-known 1-perfect sequence of the power series (see [113, 128, 129])

$$\sum_{i=0}^{\infty} t^{2^i} = t + t^2 + t^4 + t^8 + t^{16} + \cdots,$$

and the sequence (0.1) is that of the power series

$$\sum_{i=1}^{\infty} t^{2^i-1} = t + t^3 + t^7 + t^{15} + t^{31} + \cdots.$$

By making the substitution $t \mapsto t/(1+t)$, the corresponding sequences (1.0) and (1.1) can be seen to have the forms

$$\sum_{m=1}^{\infty} \sum_{i=0}^{\infty} t^{m2^i} \quad \text{and} \quad \sum_{m=1}^{\infty} \sum_{i=0}^{\infty} t^{m(2^i-1)},$$

respectively.

From Proposition 9.3.15, we see that there are 18 possible choices of degree-two local parameters t over $\mathbb{F}_3$, which we subdivide into a block of nine functions:

$$(0.0)\ t = \frac{x}{1+x+x^2}, \qquad (1.0)\ t = \frac{x}{1-x+x^2}, \qquad (2.0)\ t = \frac{x}{1+x^2},$$

$$(0.1)\ t = x+x^2, \qquad (1.1)\ t = \frac{x+x^2}{1+x+x^2}, \qquad (2.1)\ t = \frac{x+x^2}{1-x-x^2},$$

$$(0.2)\ t = \frac{x-x^2}{1-x+x^2}, \qquad (1.2)\ t = x-x^2, \qquad (2.2)\ t = \frac{x-x^2}{1+x-x^2},$$

and a second block of nine functions:

$$(0.0)'\ t = \frac{x}{1-x^2}, \qquad (1.0)'\ t = \frac{x}{1+x-x^2}, \qquad (2.0)'\ t = \frac{x}{1-x-x^2},$$

$$(0.1)'\ t = \frac{x+x^2}{1-x}, \qquad (1.1)'\ t = \frac{x+x^2}{1+x^2}, \qquad (2.1)'\ t = \frac{x+x^2}{1+x-x^2},$$

$$(0.2)'\ t = \frac{x-x^2}{1+x}, \qquad (1.2)'\ t = \frac{x-x^2}{1-x-x^2}, \qquad (2.2)'\ t = \frac{x-x^2}{1+x^2}.$$

The linear fractional transformation $A(u) = u/(1+u)$ determines an automorphism of $\mathbf{P}^1(\mathbb{F}_3)$ of order three. As in the binary case above, it is easy to verify that the roots of any two equations in the same block can be exchanged by some action of the group $< A >$. Since A generates the group of automorphisms of $\mathbf{P}^1(\mathbb{F}_3)$ which fix the point $P = [0,1]$ and the residue of functions at P, we may think of each of the two blocks of equations as comprising an equivalence class over $\mathbb{F}_3$.

The corresponding sequences for the first block are:

(0.0) $(1,1,2,1,0,0,0,1,2,1,1,2,1,0,0,0,0,0,0,0,0,0,0,0,1,2,1,1,2,\ldots)$,
(0.1) $(1,2,2,1,2,0,0,0,2,1,2,2,1,2,0,0,0,0,0,0,0,0,0,0,0,2,1,2,2,\ldots)$,
(0.2) $(1,0,1,1,0,0,0,0,1,1,0,1,1,0,0,0,0,0,0,0,0,0,0,0,0,1,1,0,1,\ldots)$,

(1.0) $(1,2,2,2,0,0,0,2,2,2,1,1,1,0,0,0,0,0,0,0,0,0,0,0,2,2,2,1,1,\ldots)$,
(1.1) $(1,0,1,2,0,0,0,0,1,2,0,2,1,0,0,0,0,0,0,0,0,0,0,0,0,1,2,0,2,\ldots)$,
(1.2) $(1,1,2,2,2,0,0,0,2,2,2,1,1,1,0,0,0,0,0,0,0,0,0,0,0,2,2,2,1,\ldots)$,

(2.0) $(1,0,1,0,2,0,2,0,2,0,0,0,0,0,0,0,2,0,2,0,2,0,1,0,1,0,1,0,0,0,\ldots)$,
(2.1) $(1,1,2,0,0,1,2,2,1,0,0,0,0,0,0,0,0,1,2,2,1,0,0,2,1,1,2,0,0,0,\ldots)$,
(2.2) $(1,2,2,0,0,2,2,1,1,0,0,0,0,0,0,0,0,2,2,1,1,0,0,1,1,2,2,0,0,0,\ldots)$,

and for the second block:

(0.0)′ $(1,0,2,0,2,0,1,0,2,0,0,0,0,0,0,0,2,0,1,0,2,0,2,0,1,0,2,0,0,0,\ldots)$,
(0.1)′ $(1,1,0,2,0,2,0,1,0,2,0,0,0,0,0,0,0,2,0,1,0,2,0,2,0,1,0,2,0,0,\ldots)$,
(0.2)′ $(1,2,0,1,0,1,0,2,0,1,0,0,0,0,0,0,0,1,0,2,0,1,0,1,0,2,0,1,0,0,\ldots)$,

(1.0)′ $(1,1,0,1,0,1,2,0,2,1,1,0,1,0,1,2,2,0,2,2,0,2,0,2,1,1,0,1,1,0,\ldots)$,
(1.1)′ $(1,2,0,0,1,1,2,1,1,0,1,2,2,0,0,1,0,2,2,1,0,0,2,2,1,2,0,0,1,2,\ldots)$,
(1.2)′ $(1,0,2,2,1,1,0,1,1,2,1,0,0,2,2,1,2,0,0,1,2,2,0,0,1,0,2,2,1,0,\ldots)$,

(2.0)′ $(1,2,0,2,0,2,2,0,2,2,1,0,1,0,1,1,2,0,2,1,0,1,0,1,1,2,0,2,1,0,\ldots)$,
(2.1)′ $(1,0,2,1,1,2,0,2,1,1,1,0,0,1,2,2,2,0,0,2,2,1,0,0,1,0,2,1,1,0,\ldots)$,
(2.2)′ $(1,1,0,0,1,2,2,2,1,0,1,1,2,0,0,2,0,1,2,2,0,0,2,1,1,1,0,0,1,1,\ldots)$.

All the above 18 sequences are perfect.

We have now found all the binary and ternary perfect sequences from the projective line.

Case 2: Elliptic curves

We now consider sequences derived from the series expansion of functions on the elliptic curve

$$E \ : \ y^2 + xy = x^3 + x$$

over the field $\mathbb{F}_2$ of two elements. Since every point on an elliptic curve can be translated to any other, we consider only expansions around the fixed point O at infinity.

On an elliptic curve, there exist no degree-one functions, so Theorem 9.3.12 provides no means of constructing sequences from functions on the curve which are provably 1-perfect. However, at the end of this subsection, we present a conjecture that certain 2-perfect sequences obtained from functions on this curve are in fact 1-perfect. In the interest of minimizing d, we consider only the series expansions of linearly independent functions of degree-two and local parameters for O. We classify these functions as follows.

The set of $\mathbb{F}_2$-rational points on E over $\mathbb{F}_2$ consists of the four points O, $(0,0)$, $(1,0)$, and $(1,1)$. Since the 2-torsion group contains only two elements, $E(\mathbb{F}_2)$ must be isomorphic to the group $\mathbb{Z}/4\mathbb{Z}$.

The automorphism $[-1]$ on E is the map $(x,y) \mapsto (x, x+y)$, which stabilizes $(0,0)$, so we identify $(0,0)$ as the 2-torsion point. Any degree-two function on E which has a zero of order one at O has exactly one other zero on E, which must be one of the $\mathbb{F}_2$-rational points $(0,0)$, $(1,0)$, or $(1,1)$.

The zeros and poles of the functions on E satisfy an additional relation. Let $\mathrm{div}(f)$ be the divisor of the function f, and let $\mathrm{Div}^0(E, \overline{\mathbb{F}}_2)$ be the set of degree zero divisors on E defined over an algebraic closure of $\mathbb{F}_2$. Then there exists an exact sequence

$$1 \longrightarrow \overline{\mathbb{F}}_2(E)^* \longrightarrow \mathrm{Div}^0(E, \overline{\mathbb{F}}_2) \longrightarrow E(\overline{\mathbb{F}}_2) \longrightarrow 0, \qquad (9.6)$$

where the map $\overline{\mathbb{F}}_2(E)^* \to \mathrm{Div}^0(E, \overline{\mathbb{F}}_2)$ takes a function to its principal divisor, and the map $\mathrm{Div}^0(E, \overline{\mathbb{F}}_2) \to E(\overline{\mathbb{F}}_2)$ is the group homomorphism which takes the degree zero divisor $P - O$ to the point P, for all $P \in E(\overline{\mathbb{F}}_2)$. From the exact sequence (9.6), we may classify the functions in $\mathbb{F}_2(E)$ by their divisors. Precisely, they correspond to divisors in the kernel of the map $\mathrm{Div}^0(E, \overline{\mathbb{F}}_2) \to E(\overline{\mathbb{F}}_2)$ which are invariant under the Galois group $\mathrm{Gal}(\overline{\mathbb{F}}_2/\mathbb{F}_2)$.

In Table 9.1, we list all the possible divisors of degree-two and degree-three functions with poles only at O, and the functions to which they correspond. In addition to the divisors of points in $E(\mathbb{F}_2)$, the two degree-two divisors $\mathcal{P}$ and $\mathcal{Q}$, corresponding to the Galois invariant pairs of points in $E(\mathbb{F}_4)$ disappearing on the ideals $(x^2 + x + 1, y + 1)$ and $(x^2 + x + 1, y + x + 1)$, may appear.

First, we determine the divisors of certain "building block" functions on E. We write $\mathrm{div}(f) = \mathrm{div}_0(f) - \mathrm{div}_\infty(f)$.

TABLE 9.1: Divisors of degree-two and degree-three functions with poles only at O.

f	$\mathrm{div}_0(f)$	$\mathrm{div}_\infty(f)$
x	$2(0,0)$	$2O$
$x+1$	$(1,0) + (1,1)$	$2O$
y	$(0,0) + 2(1,0)$	$3O$
$y+x$	$(0,0) + 2(1,1)$	$3O$
$y+1$	$(1,1) + \mathcal{P}$	$3O$
$y+x+1$	$(1,0) + \mathcal{Q}$	$3O$

By listing all possible degree-two divisors which can occur as $\mathrm{div}_\infty(f)$, we obtain in Table 9.2 the classification of all degree-two functions on E over $\mathbb{F}_2$, expressed as quotients of the functions in Table 9.1.

In Table 9.3, we give the minimal polynomial for the function f over the

TABLE 9.2: All degree-two functions on E over $\mathbb{F}_2$.

	f	$\mathrm{div}_0(f)$	$\mathrm{div}_\infty(f)$
(1)	x/y	$O + (0,0)$	$2(0,1)$
(2)	$x/(y+x)$	$O + (0,0)$	$2(1,1)$
(3)	$y/(x^2+x)$	$O + (1,0)$	$(0,0) + (1,1)$
(4)	$(x+1)/(y+1)$	$O + (1,0)$	$\mathcal{P}$
(5)	$(y+x)/(x^2+x)$	$O + (1,1)$	$(0,0) + (1,0)$
(6)	$(x+1)/(y+x+1)$	$O + (1,1)$	$\mathcal{Q}$

field $\mathbb{F}_2(t)$, where f and t are one of the six degree-two functions on $E/\mathbb{F}_2$ in Table 9.2. The entry $(i.j)$ corresponds to the pair (f,t), where f is entry (i) in the Table 9.2, and t is entry (j).

The sequence of coefficients for the series expansion of f with respect to t at O is proved to be 2-perfect in Theorem 9.3.12. For a given pair, the functions may in fact be 1-perfect. We give the experimentally determined value d_0 for which the series expansion is believed to be d_0-perfect. However, this value is provably only a lower bound. For a pair (f,t) generating a periodic sequence, we set d_0 equal to 0 in Table 9.3.

We note that, due to the automorphism $[-1]$ of the curve, the nontrivial minimal polynomials appear in pairs, corresponding to function pairs which are interchanged under the automorphism induced by $[-1]$.

From Table 9.3, we note that the sequence associated with the root $f(t)$ of valuation 1 in $\mathbb{F}_2((t))$ of any of the four polynomials

$$\begin{aligned}
&(1) \quad F(X) = (t+t^2)X^2 + (1+t)X + t \\
&(2) \quad F(X) = (1+t+t^2)X^2 + (1+t)X + t \\
&(3) \quad F(X) = tX^2 + (1+t)X + t + t^2 \\
&(4) \quad F(X) = (1+t+t^2)X^2 + (1+t)X + t + t^2
\end{aligned}$$

over $\mathbb{F}_2(t)$, is conjecturally 1-perfect. A proof using the geometry of the associated function field would be interesting and might point to additional avenues for constructions of 1-perfect sequences.

The sequences associated with the roots of the four polynomials above are:

(1) $(1,1,0,1,0,1,1,1,0,1,1,1,0,1,0,1,0,1,1,1,0,1,0,1,0,1,1,1,0,1,\ldots)$,
(2) $(1,0,1,0,0,1,0,1,1,1,1,0,1,0,0,1,0,0,1,1,0,0,1,0,0,1,0,1,1,1,\ldots)$,
(3) $(1,0,1,1,1,1,0,0,1,1,0,0,1,1,1,1,1,1,0,0,1,1,1,1,1,1,0,0,1,1,\ldots)$,
(4) $(1,1,0,0,1,0,0,0,0,1,0,1,1,1,1,1,1,1,1,1,0,1,0,0,1,0,0,1,1,0,0,\ldots)$.

In particular, we note that the four sequences above are different from the 1-perfect binary sequences constructed from the projective line. This points to the possibility of constructing further 1-perfect sequences from curves of higher genus, if the above sequences are indeed proved to be 1-perfect.

TABLE 9.3: Minimal polynomials of f over $\mathbb{F}_2(t)$, with f, t from Table 9.2.

	$F(X)$	d_0
(1.1)	$X + t$	0
(1.2)	$(1+t)X + t$	0
(1.3)	$(t+t^2)X^2 + (1+t)X + t$	1
(1.4)	$(1+t+t^2)X^2 + (1+t)X + t$	1
(1.5)	$tX^2 + (1+t)X + t + t^2$	1
(1.6)	$(1+t+t^2)X^2 + (1+t)X + t + t^2$	1

	$F(X)$	d_0
(2.1)	$(1+t)X + t$	0
(2.2)	$X + t$	0
(2.3)	$tX^2 + (1+t)X + t + t^2$	1
(2.4)	$(1+t+t^2)X^2 + (1+t)X + t + t^2$	1
(2.5)	$(t+t^2)X^2 + (1+t)X + t$	1
(2.6)	$(1+t+t^2)X^2 + (1+t)X + t$	1

	$F(X)$	d_0
(3.1)	$t^2X^2 + (1+t+t^2)X + t$	2
(3.2)	$X^2 + (1+t+t^2)X + t$	2
(3.3)	$X + t$	0
(3.4)	$(1+t)X + t$	0
(3.5)	$tX^2 + (1+t+t^2)X + t$	2
(3.6)	$(1+t+t^2)X^2 + (1+t+t^2)X + t$	2

	$F(X)$	d_0
(4.1)	$t^2X^2 + (1+t+t^2)X + t + t^2$	2
(4.2)	$(1+t^2)X^2 + (1+t+t^2)X + t + t^2$	2
(4.3)	$(1+t)X + t$	0
(4.4)	$X + t$	0
(4.5)	$(t+t^2)X^2 + (1+t+t^2)X + t + t^2$	2
(4.6)	$(1+t+t^2)X^2 + (1+t+t^2)X + t + t^2$	2

	$F(X)$	d_0
(5.1)	$X^2 + (1+t+t^2)X + t$	2
(5.2)	$t^2X^2 + (1+t+t^2)X + t$	2
(5.3)	$tX^2 + (1+t+t^2)X + t$	2
(5.4)	$(1+t+t^2)X^2 + (1+t+t^2)X + t$	2
(5.5)	$X + t$	0
(5.6)	$(1+t)X + t$	0

	$F(X)$	d_0
(6.1)	$(1+t^2)X^2 + (1+t+t^2)X + t + t^2$	2
(6.2)	$t^2X^2 + (1+t+t^2)X + t + t^2$	2
(6.3)	$(t+t^2)X^2 + (1+t+t^2)X + t + t^2$	2
(6.4)	$(1+t+t^2)X^2 + (1+t+t^2)X + t + t^2$	2
(6.5)	$(1+t)X + t$	0
(6.6)	$X + t$	0

9.4   Constructions of Multisequences

For practical applications, we have to consider the scenario where a joint sequence family is used (see [49]). By a joint sequence family, we mean a set of sequences, or a multisequence, over $\mathbb{F}_q$ and we need to find an LFSR that generates all the sequences in this family.

For a multisequence $\mathcal{S} = \{\mathbf{c}_1, \mathbf{c}_2, \ldots, \mathbf{c}_m\}$ of dimension m (where dimension means the number of sequences in this family) and of length n, i.e., every sequence has length n, we define its linear complexity as follows.

Definition 9.4.1 The **linear complexity** of a **multisequence** set $\mathcal{S} = \{\mathbf{c}_i = (c_{i1}, c_{i2}, \ldots, c_{in})\}_{i=1}^{m}$ of length n is defined to be the smallest order k of an LFSR $< f(T), k >$ generating all of the sequences $\mathbf{c}_i$, $i = 1, 2, \ldots, m$.

For an LFSR $< f(T), k >$ and a multisequence $\mathcal{S} = \{\mathbf{c}_i\}_{i=1}^{m}$ of length n, we simply say that $< f(T), k >$ generates $\mathcal{S}$ if it generates each of the sequences $\mathbf{c}_1, \mathbf{c}_2, \ldots, \mathbf{c}_m$.

All the sequences we have discussed above in this section are of finite length. We next turn to sequences of infinite length.

Consider m sequences

$$
\begin{aligned}
\mathbf{a}_1 &= (a_{11}, a_{12}, a_{13}, \ldots) \in \mathbb{F}_q^{\infty} \\
\mathbf{a}_2 &= (a_{21}, a_{22}, a_{23}, \ldots) \in \mathbb{F}_q^{\infty} \\
&\vdots \\
\mathbf{a}_m &= (a_{m1}, a_{m2}, a_{m3}, \ldots) \in \mathbb{F}_q^{\infty},
\end{aligned}
$$

and let $\mathcal{A}$ be the multisequence $\{\mathbf{a}_i\}_{i=1}^{m}$. We also denote by $\mathcal{A}_n$ the multisequence

$$
\{(a_{i1}, a_{i2}, \ldots, a_{in})\}_{i=1}^{m}
$$

of length n.

Definition 9.4.2 The **linear complexity profile** of a multisequence $\mathcal{A} = \{\mathbf{a}_1, \mathbf{a}_2, \ldots, \mathbf{a}_m\}$ is defined by the sequence of integers

$$
\{\ell_n(\mathcal{A})\}_{n=1}^{\infty},
$$

where $\ell_n(\mathcal{A})$ denotes the linear complexity of $\mathcal{A}_n = \{(a_{i1}, a_{i2}, \ldots, a_{in})\}_{i=1}^{m}$, for all $n \geq 1$.

Analogous to the case of single sequences, we may also define almost perfect multisequences.

Definition 9.4.3 A multisequence $\mathcal{A} = \{\mathbf{a}_1, \mathbf{a}_2, \ldots, \mathbf{a}_m\}$ is called **almost perfect** if

$$
\ell_n(\mathcal{A}) \geq \frac{m(n+1)}{m+1} + O(1)
$$

for all $n \geq 1$, where $O(1)$ is a function independent of n.

Furthermore, we can define d-perfect multisequences in a similar manner.

Definition 9.4.4 A multisequence $\mathcal{A} = \{\mathbf{a}_1, \mathbf{a}_2, \ldots, \mathbf{a}_m\}$ is called d-**perfect** for a positive integer d if

$$\ell_n(\mathcal{A}) \geq \left\lceil \frac{m(n+1) - d}{m+1} \right\rceil$$

for all $n \geq 1$. In particular, $\mathcal{A}$ is called **perfect** if $\mathcal{A}$ is an m-perfect multisequence, i.e.,

$$\ell_n(\mathcal{A}) \geq \left\lceil \frac{mn}{m+1} \right\rceil$$

for all $n \geq 1$.

Remark 9.4.5   (i) It can be seen in Section 9.3 that our definitions of perfect multisequences are quite natural and consistent with Niederreiter's and Rueppel's definitions in the case where $m = 1$.

(ii) We will see in Remark 9.4.9 that, if $\mathcal{A}$ is a d-perfect multisequence of dimension m, then d is at least m.

For a multisequence $\mathcal{A} = \{\mathbf{a}_i = (a_{i1}, a_{i2}, a_{i3}, \ldots)\}_{i=1}^m$, put

$$\mathbf{s}_j = (a_{1j}, a_{2j}, \ldots, a_{mj})^T \in \mathbb{F}_q^m \tag{9.7}$$

for all $j \geq 1$, where ^{T} stands for the transpose of a vector. Then we immediately have the following lemma from the definitions.

Lemma 9.4.6 *The LFSR* $< f(T) = \sum_{i=0}^{k} \lambda_i T^i, k >$ *with* $\lambda_0 = 1$ *generates* $\mathcal{A}_n = \{(a_{i1}, a_{i2}, \ldots, a_{in})\}_{i=1}^m$ *if and only if*

$$\sum_{i=0}^{k} \lambda_{k-i} \mathbf{s}_{i+u} = \mathbf{0} \in \mathbb{F}_q^m$$

for all $1 \leq u \leq n - k$, *i.e.,*

$$\begin{pmatrix} \mathbf{s}_1 & \mathbf{s}_2 & \cdots & \mathbf{s}_{k+1} \\ \mathbf{s}_2 & \mathbf{s}_3 & \cdots & \mathbf{s}_{k+2} \\ \vdots & \vdots & \cdots & \vdots \\ \mathbf{s}_{n-k} & \mathbf{s}_{n-k+1} & \cdots & \mathbf{s}_n \end{pmatrix} \begin{pmatrix} \lambda_k \\ \lambda_{k-1} \\ \vdots \\ \lambda_0 \end{pmatrix} = \mathbf{0} \in \mathbb{F}_q^{(n-k)m}.$$

In order to give an equivalent condition for perfect multisequences, we need another lemma.

Lemma 9.4.7 *Let* $\mathcal{A} = \{\mathbf{a}_1, \mathbf{a}_2, \ldots, \mathbf{a}_m\}$ *be a multisequence. If*

$$\ell_n(\mathcal{A}) \geq \left\lceil \frac{mn}{m+1} \right\rceil$$

for all $n \geq 1$, then the rank of the $um \times um$ matrix over $\mathbb{F}_q$

$$
S_{um} \overset{\text{def}}{=} \begin{pmatrix} s_1 & s_2 & \cdots & s_{um} \\ s_2 & s_3 & \cdots & s_{um+1} \\ \vdots & \vdots & \cdots & \vdots \\ s_u & s_{u+1} & \cdots & s_{um+u-1} \end{pmatrix}
$$

is equal to um for all $u \geq 1$, where s_j is defined as in (9.7).

Proof. Suppose that the rank of S_{um} is less than um, i.e., there exists a nonzero solution $(\alpha_0, \alpha_1, \ldots, \alpha_{um-1}) \in \mathbb{F}_q^{um}$ such that

$$
\begin{pmatrix} s_1 & s_2 & \cdots & s_{um} \\ s_2 & s_3 & \cdots & s_{um+1} \\ \vdots & \vdots & \cdots & \vdots \\ s_u & s_{u+1} & \cdots & s_{um+u-1} \end{pmatrix} \begin{pmatrix} \alpha_0 \\ \alpha_1 \\ \vdots \\ \alpha_{um-1} \end{pmatrix} = \mathbf{0} \in \mathbb{F}_q^{um}.
$$

Let $0 \leq r \leq um - 1$ be the integer satisfying $\alpha_r \neq 0, \alpha_{r+1} = \alpha_{r+2} = \cdots = \alpha_{um-1} = 0$. Then

$$
\begin{pmatrix} s_1 & s_2 & \cdots & s_{r+1} \\ s_2 & s_3 & \cdots & s_{r+2} \\ \vdots & \vdots & \cdots & \vdots \\ s_u & s_{u+1} & \cdots & s_{r+u} \end{pmatrix} \begin{pmatrix} \alpha_0 \\ \alpha_1 \\ \vdots \\ \alpha_r \end{pmatrix} = \mathbf{0} \in \mathbb{F}_q^{um}.
$$

Hence, by Lemma 9.4.6, we obtain

$$
\ell_{r+u}(\mathcal{A}) \leq r. \tag{9.8}
$$

Write

$$
r + u = w(m+1) + k
$$

for some $w \geq 0$ and $0 \leq k \leq m$. Observe that $w \leq u - 1$ since $r + u \leq um - 1 + u = u(m+1) - 1$. Therefore

$$
\ell_{r+u}(\mathcal{A}) \geq \left\lceil \frac{(r+u)m}{m+1} \right\rceil = \left\lceil \frac{(m+1)mw + km}{m+1} \right\rceil
$$

$$
= mw + k + \left\lceil \frac{-k}{m+1} \right\rceil = mw + k = r + u - w \geq r + 1.
$$

The above inequality contradicts (9.8). This completes the proof. $\square$

We have the following result on perfect multisequences.

Theorem 9.4.8 *A multisequence* $\mathcal{A} = \{\mathbf{a}_1, \mathbf{a}_2, \ldots, \mathbf{a}_m\}$ *of dimension* m *is perfect if and only if*

$$
\ell_n(\mathcal{A}) = \left\lceil \frac{mn}{m+1} \right\rceil
$$

for all $n \geq 1$.

Proof. One direction is clear. Now we assume that $\mathcal{A}$ is perfect.

It is obvious that $\ell_n(\mathcal{A}) \leq n$ for all $n \geq 1$, since the LFSR $< T^n, n >$ generates every sequence of length n. On the other hand, we have

$$\ell_n(\mathcal{A}) \geq \left\lceil \frac{mn}{m+1} \right\rceil = n$$

for all $1 \leq n \leq m$. Therefore, we obtain $\ell_n(\mathcal{A}) = n = \lceil (mn)/(m+1) \rceil$ for all $1 \leq n \leq m$.

We now assume that $n \geq m+1$. Write

$$n = u(m+1) + v, \quad u \geq 1, \ 0 \leq v \leq m.$$

Then we get

$$\ell_n(\mathcal{A}) \geq \left\lceil \frac{nm}{m+1} \right\rceil = \left\lceil \frac{u(m+1)m + vm}{m+1} \right\rceil = mu + v. \tag{9.9}$$

By Lemma 9.4.7, the vector

$$(\mathbf{s}_{mu+v+1}^T, \mathbf{s}_{mu+v+2}^T, \ldots, \mathbf{s}_{mu+v+u}^T)^T \in \mathbb{F}_q^{um}$$

can be represented as an $\mathbb{F}_q$-linear combination of

$$\begin{pmatrix} \mathbf{s}_1 \\ \mathbf{s}_2 \\ \vdots \\ \mathbf{s}_u \end{pmatrix}, \begin{pmatrix} \mathbf{s}_2 \\ \mathbf{s}_3 \\ \vdots \\ \mathbf{s}_{u+1} \end{pmatrix}, \ldots, \begin{pmatrix} \mathbf{s}_{um} \\ \mathbf{s}_{um+1} \\ \vdots \\ \mathbf{s}_{um+u-1} \end{pmatrix},$$

i.e., there exist $\lambda_1, \lambda_2, \ldots, \lambda_{um} \in \mathbb{F}_q$ such that

$$\begin{pmatrix} \mathbf{s}_{um+v+1} \\ \mathbf{s}_{um+v+2} \\ \vdots \\ \mathbf{s}_{um+v+u} \end{pmatrix} = \sum_{i=1}^{um} \lambda_{um-i+1} \begin{pmatrix} \mathbf{s}_i \\ \mathbf{s}_{i+1} \\ \vdots \\ \mathbf{s}_{i+u-1} \end{pmatrix}.$$

Hence, by Lemma 9.4.6, the LFSR

$$< 1 - \sum_{i=1}^{um} \lambda_i T^i, mu + v >$$

generates the sequences $(a_{11}, a_{12}, \ldots, a_{1n}), (a_{21}, a_{22}, \ldots, a_{2n}), \ldots, (a_{m1}, a_{m2}, \ldots, a_{mn})$. This implies that

$$\ell_n(\mathcal{A}) = \ell_{u(m+1)+v}(\mathcal{A}) \leq mu + v. \tag{9.10}$$

Combining (9.9) with (9.10) gives the desired result. $\qquad \square$

Remark 9.4.9 We claim that d is at least m if there exists a d-perfect multisequence $\mathcal{A}$ of dimension m. Suppose $d < m$, then

$$\ell_n(\mathcal{A}) \geq \frac{nm + m - d}{m + 1} > \frac{mn}{m + 1}$$

for all $n \geq 1$. Therefore, $\mathcal{A}$ is a perfect multisequence. Hence, according to Theorem 9.4.8,

$$\ell_n(\mathcal{A}) = \left\lceil \frac{mn}{m + 1} \right\rceil \tag{9.11}$$

for all $n \geq 1$. On the other hand, we have

$$\ell_{m+1}(\mathcal{A}) \geq \left\lceil \frac{(m + 1)m + m - d}{m + 1} \right\rceil = m + 1, \tag{9.12}$$

which is in contradiction to (9.11).

Next, we construct d-perfect multisequences from algebraic curves.

Let P be a closed point of degree m. We consider the set of functions of $\mathbb{F}_q(\mathcal{X})$ which are regular at every point in P, i.e.,

$$\mathcal{O}_{\mathsf{P}} \stackrel{\text{def}}{=} \{f \in \mathbb{F}_q(\mathcal{X}) : \nu_P(f) \geq 0 \text{ for all } P \in \mathsf{P}\},$$

and the set of functions of $\mathbb{F}_q(\mathcal{X})$ which vanish at every point of P, i.e.,

$$\mathcal{P}_{\mathsf{P}} \stackrel{\text{def}}{=} \{h \in \mathbb{F}_q(\mathcal{X}) : h(P) = 0 \text{ for all } P \in \mathsf{P}\}.$$

Then $\mathcal{O}_{\mathsf{P}}$ is a local ring and $\mathcal{P}_{\mathsf{P}}$ is its unique maximal ideal. The quotient field is called the **residue class field** at P, denoted by F_{P}. This is a finite field isomorphic to $\mathbb{F}_{q^m}$. For a function $x \in \mathcal{O}_{\mathsf{P}}$, we denote by $x(\mathsf{P})$ the residue class of x in F_{P}. For details about the residue class fields at closed points, the reader may refer to [151, 117].

Throughout this section, we use the following notations and assumptions:

- $\mathcal{X}/\mathbb{F}_q$ – an algebraic curve over $\mathbb{F}_q$;

- P – a closed point of degree m of $\mathcal{X}$;

- $x_1, x_2, \ldots, x_m$ – m elements of $\mathcal{O}_{\mathsf{P}}$ such that $x_1(\mathsf{P}), x_2(\mathsf{P}), \ldots, x_m(\mathsf{P})$ form an $\mathbb{F}_q$-basis of F_{P};

- t – a local parameter of P with $\deg(\text{div}_\infty(t)) = m + 1$;

- y – an element of $\mathcal{O}_{\mathsf{P}}$ satisfying $y \notin \bigoplus_{i=1}^m \mathbb{F}_q(t)x_i$.

Remark 9.4.10 The condition $\deg(\text{div}_\infty(t)) = m + 1$ implies that $\mathbb{F}_q(\mathcal{X})$ is an $\mathbb{F}_q(t)$-linear space of dimension $m + 1$ (see [151, Theorem I.4.11]). Hence, the set $\{x_1, x_2, \ldots, x_m\}$ generates a proper $\mathbb{F}_q(t)$-linear subspace of $\mathbb{F}_q(\mathcal{X})$, i.e., $\mathbb{F}_q(\mathcal{X}) \setminus \bigoplus_{i=1}^m \mathbb{F}_q(t)x_i$ is not empty.

As in Section 9.3, for a function $y \in \mathcal{O}_\mathsf{P}$, we can also consider the local expansion of y at P. The residue class $y(\mathsf{P})$ of y in $\mathcal{O}_\mathsf{P}$ can be represented as an $\mathbb{F}_q$-linear combination of $x_1(\mathsf{P}), x_2(\mathsf{P}), \ldots, x_m(\mathsf{P})$. Let $a_{10}, a_{20}, \ldots, a_{m0} \in \mathbb{F}_q$ satisfy

$$y(\mathsf{P}) = \sum_{i=1}^{m} a_{i0} x_i(\mathsf{P}).$$

The above equality is equivalent to

$$\nu_\mathsf{P}\left(y - \sum_{i=1}^{m} a_{i0} x_i \right) \geq 1.$$

Hence $(y - \sum_{i=1}^{m} a_{i0} x_i)/t \in \mathcal{O}_\mathsf{P}$. Let $a_{11}, a_{21}, \ldots, a_{m1} \in \mathbb{F}_q$ satisfy

$$\left(\frac{y - \sum_{i=1}^{m} a_{i0} x_i}{t} \right)(\mathsf{P}) = \sum_{i=1}^{m} a_{i1} x_i(\mathsf{P}),$$

i.e.,

$$\nu_\mathsf{P}\left(\frac{y - \sum_{i=1}^{m} a_{i0} x_i}{t} - \sum_{i=1}^{m} a_{i1} x_i \right) \geq 1.$$

This is equivalent to

$$\nu_\mathsf{P}\left(y - \sum_{i=1}^{m} a_{i0} x_i - \left(\sum_{i=1}^{m} a_{i1} x_i \right) t \right) \geq 2.$$

Hence $(y - \sum_{i=1}^{m} a_{i0} x_i - (\sum_{i=1}^{m} a_{i1} x_i) t)/t^2 \in \mathcal{O}_\mathsf{P}$.

By induction, we obtain a sequence of vectors $\{(a_{1j}, a_{2j}, \ldots, a_{mj})\}_{j=0}^{\infty}$ such that

$$\nu_\mathsf{P}\left(y - \sum_{j=0}^{n} \left(\sum_{i=1}^{m} a_{ij} x_i \right) t^j \right) \geq n + 1$$

for all $n \geq 0$. We express this fact by the formal series

$$y = \sum_{j=0}^{\infty} \left(\sum_{i=1}^{m} a_{ij} x_i \right) t^j. \tag{9.13}$$

The above series is called a **local expansion** of y at P. The coefficients of the local expansion (9.13) will be used to construct perfect multisequences.

Put

$$\mathbf{a}_i(y) = (a_{i1}, a_{i2}, a_{i3}, \ldots) \in \mathbb{F}_q^{\infty}$$

for any $1 \leq i \leq m$ and define the multisequence

$$\mathcal{A}(y) = \{\mathbf{a}_i(y)\}_{i=1}^{m}. \tag{9.14}$$

For two divisors $D = \sum_\mathsf{P} m_\mathsf{P} \mathsf{P}$ and $G = \sum_\mathsf{P} n_\mathsf{P} \mathsf{P}$ on $\mathcal{X}$, we define a divisor $D \vee G \stackrel{\text{def}}{=} \sum_\mathsf{P} \max\{m_\mathsf{P}, n_\mathsf{P}\} \mathsf{P}$.

Theorem 9.4.11 *Let $A(y) = \{\mathbf{a}_i(y)\}_{i=1}^m$ be constructed as in (9.14). Then $A(y)$ is d-perfect, where $d = \deg(\operatorname{div}_\infty(y) \vee \operatorname{div}_\infty(x_1) \vee \operatorname{div}_\infty(x_2) \vee \cdots \vee \operatorname{div}_\infty(x_m))$. In particular, $\{\mathbf{a}_i(y)\}_{i=1}^m$ is a perfect multisequence if $d = m$.*

Proof. Denote by x the vector $(x_1, x_2, \ldots, x_m)^T \in \mathbb{F}_q(\mathcal{X})^m$. For another vector $\mathsf{z} = (z_1, z_2, \ldots, z_m)^T \in \mathbb{F}_q(\mathcal{X})^m$, we denote by $\mathsf{x} \cdot \mathsf{z}$ the inner product $\sum_{i=1}^m x_i z_i$. Then the local expansion of y can be written in the form

$$y = \sum_{j=0}^\infty (\mathsf{x} \cdot \mathsf{s}_j) t^j,$$

where $\mathsf{s}_j = (a_{1j}, a_{2j}, \ldots, a_{mj})^T$.

Suppose that an LFSR $< \sum_{i=0}^k \lambda_i T^i, k >$ with $\lambda_0 = 1$ generates the multisequence of length n

$$A_n(y) = \{(a_{i1}, a_{i2}, \ldots, a_{in})\}_{i=1}^m,$$

i.e.,

$$\sum_{i=0}^k \lambda_{k-i} \mathsf{s}_{i+u} = \mathbf{0} \in \mathbb{F}_q^m$$

for all $1 \le u \le n - k$ by Lemma 9.4.6.

Consider the function

$$
\begin{aligned}
L &= (\lambda_k t^k + \lambda_{k-1} t^{k-1} + \cdots + \lambda_0) y - [\lambda_0(\mathsf{x} \cdot \mathsf{s}_0) + \\
&\quad (\lambda_0(\mathsf{x} \cdot \mathsf{s}_1) + \lambda_1(\mathsf{x} \cdot \mathsf{s}_0)) t + \cdots + (\lambda_0(\mathsf{x} \cdot \mathsf{s}_k) + \cdots + \lambda_k(\mathsf{x} \cdot \mathsf{s}_0)) t^k] \\
&= \sum_{j=k+1}^\infty \lambda_k(\mathsf{x} \cdot \mathsf{s}_{j-k}) t^j + \sum_{j=k+1}^\infty \lambda_{k-1}(\mathsf{x} \cdot \mathsf{s}_{j-k+1}) t^j + \cdots + \sum_{j=k+1}^\infty \lambda_0(\mathsf{x} \cdot \mathsf{s}_j) t^j \\
&= \sum_{j=k+1}^\infty \sum_{l=0}^k \lambda_{k-l}(\mathsf{x} \cdot \mathsf{s}_{j-k+l}) t^j = \sum_{j=k+1}^\infty \mathsf{x} \cdot \left(\sum_{l=0}^k \lambda_{k-l} \mathsf{s}_{j-k+l} \right) t^j \\
&= \sum_{j=n+1}^\infty \mathsf{x} \cdot \left(\sum_{l=0}^k \lambda_{k-l} \mathsf{s}_{j-k+l} \right) t^j = \sum_{j=n+1}^\infty \sum_{l=0}^k \lambda_{k-l}(\mathsf{x} \cdot \mathsf{s}_{j-k+l}) t^j.
\end{aligned}
$$

First of all, we can see $L \ne 0$ since $\lambda_0 = 1$ and $y \notin \bigoplus_{i=1}^m \mathbb{F}_q(t) x_i$. Considering the zero divisor of L gives

$$\deg(\operatorname{div}_0(L)) \ge \nu_\mathsf{P}(L) \cdot \deg(\mathsf{P}) \ge (n+1)m, \tag{9.15}$$

whereas considering the pole divisor of L gives

$$
\begin{aligned}
\deg(\operatorname{div}_\infty(L)) &\le \deg\left(\operatorname{div}_\infty(t^k) + \operatorname{div}_\infty(y) \vee \operatorname{div}_\infty(x_1) \vee \cdots \vee \operatorname{div}_\infty(x_m) \right) \\
&\le (m+1)k + d. \tag{9.16}
\end{aligned}
$$

Combining (9.15) with (9.16) yields

$$(n+1)m \leq \deg(\operatorname{div}_0(L)) = \deg(\operatorname{div}_\infty(L)) \leq (m+1)k + d.$$

Hence,

$$k \geq \frac{m(n+1) - d}{m+1}.$$

This implies

$$\ell_n(\mathcal{A}(y)) \geq \frac{m(n+1) - d}{m+1}$$

for all $n \geq 1$.

Since $\ell_n(\mathcal{A}(y))$ are integers, our result follows.

If $d = m$, then $\ell_n(\mathcal{A}(y)) \geq (mn)/(m+1)$ for all $n \geq 1$. It follows from Theorem 9.4.8 and Remark 9.4.9 that $\{\mathbf{a}_i(y)\}_{i=1}^m$ is a perfect multisequence. This completes the proof. □

Remark 9.4.12 Theorem 9.4.11 does not mean that the multisequence $\mathcal{A}(y)$ is not perfect for $d > m$ (see examples in Section 9.3 for $m = 1$).

Example 9.4.13 (Binary multisequence of dimension $m = 2$) Let $\mathcal{X}$ be the projective line over $\mathbb{F}_2$ and let P be the unique zero of $x^2 + x + 1$. Then $t = x(x^2 + x + 1)$ is a local parameter of P.

Put $y = x^2$ and $x_1 = 1, x_2 = x$. Then $x_1(\mathsf{P}), x_2(\mathsf{P})$ form an $\mathbb{F}_2$-basis of F_P and $d = \deg(\operatorname{div}_\infty(y) \vee \operatorname{div}_\infty(x_1) \vee \operatorname{div}_\infty(x_2)) = 2 = \deg(\mathsf{P})$. The local expansion of y at P is

$$y = (x_1 + x_2) + (x_1 + x_2)t + x_2 t^2 + (x_1 + x_2)t^3 + x_1 t^4 + x_1 t^6 + (x_1 + x_2)t^8 + \cdots.$$

Set

$$\mathbf{a}_1(y) = (1, 0, 1, 1, 0, 1, 0, 1, \ldots)$$
$$\mathbf{a}_2(y) = (1, 1, 1, 0, 0, 0, 0, 1, \ldots).$$

By Theorem 9.4.11, $\{\mathbf{a}_1(y), \mathbf{a}_2(y)\}$ is a perfect multisequence of dimension 2.

Example 9.4.14 (Binary multisequence of dimension $m = 3$) Let $\mathcal{X}$ be the projective line over $\mathbb{F}_2$ and let P be the unique zero of $x^3 + x + 1$. Then $t = x(x^3 + x + 1)$ is a local parameter of P.

Put $y = x^3$ and $x_1 = 1, x_2 = x, x_3 = x^2$. Then $x_1(\mathsf{P}), x_2(\mathsf{P}), x_3(\mathsf{P})$ form an $\mathbb{F}_2$-basis of F_P and $d = \deg(\operatorname{div}_\infty(y) \vee \operatorname{div}_\infty(x_1) \vee \operatorname{div}_\infty(x_2) \vee \operatorname{div}_\infty(x_3)) = 3 = \deg(\mathsf{P})$. The local expansion of y at P is

$$y = (x_1 + x_2) + (x_1 + x_3)t + (x_1 + x_2 + x_3)t^2 + x_1 t^3 + (x_1 + x_2)t^4 + x_1 t^5$$
$$+ x_1 t^6 + (x_1 + x_3)t^8 + \cdots.$$

Set

$$\mathbf{a}_1(y) = (1, 1, 1, 1, 1, 1, 0, 1, \ldots)$$
$$\mathbf{a}_2(y) = (0, 1, 0, 1, 0, 0, 0, 0, \ldots)$$
$$\mathbf{a}_3(y) = (1, 1, 0, 0, 0, 0, 0, 1, \ldots).$$

By Theorem 9.4.11, $\{\mathbf{a}_1(y), \mathbf{a}_2(y), \mathbf{a}_3(y)\}$ is a perfect multisequence of dimension 3.

Example 9.4.15 (Ternary multisequence of dimension $m = 2$) Let $\mathcal{X}$ be the projective line over $\mathbb{F}_3$ and let P be the unique zero of x^2+1. Then $t = x(x^2+1)$ is a local parameter of P.

Put $y = x^2$ and $x_1 = 1, x_2 = x$. Then $x_1(\mathsf{P}), x_2(\mathsf{P})$ form an $\mathbb{F}_3$-basis of F_P and $d = \deg(\mathrm{div}_\infty(y) \vee \mathrm{div}_\infty(x_1) \vee \mathrm{div}_\infty(x_2)) = 2 = \deg(\mathsf{P})$. The local expansion of y at P is

$$y = 2x_1 + 2x_2t + 2x_1t^2 + x_2t^3 + 2x_1t^4 + 2x_1t^6 + 2x_2t^9 + \cdots.$$

Set

$$\mathbf{a}_1(y) = (0, 2, 0, 2, 0, 2, 0, 0, \ldots)$$
$$\mathbf{a}_2(y) = (2, 0, 1, 0, 0, 0, 0, 0, \ldots).$$

By Theorem 9.4.11, $\{\mathbf{a}_1(y), \mathbf{a}_2(y)\}$ is a perfect multisequence of dimension 2.

Example 9.4.16 (Binary 4-perfect multisequence of dimension $m = 2$) Let $\mathcal{X}$ be the elliptic curve over $\mathbb{F}_2$ defined by $z^2 + z = x^3 + 1$ and let P be the unique common zero of z and $x^2 + x + 1$. Then $t = z$ is a local parameter of P.

Put $y = x^2$ and $x_1 = 1, x_2 = x$. Then $x_1(\mathsf{P}), x_2(\mathsf{P})$ form an $\mathbb{F}_2$-basis of F_P and $d = \deg(\mathrm{div}_\infty(y) \vee \mathrm{div}_\infty(x_1) \vee \mathrm{div}_\infty(x_2)) = 4$. The local expansion of y at P is

$$y = x_1 + x_2 + x_2t + x_1t^2 + x_2t^3 + 0 \cdot t^4 + \cdots.$$

Set

$$\mathbf{a}_1(y) = (0, 1, 0, 0, \ldots)$$
$$\mathbf{a}_2(y) = (1, 0, 1, 0, \ldots).$$

By Theorem 9.4.11, $\{\mathbf{a}_1(y), \mathbf{a}_2(y)\}$ is a 4-perfect multisequence of dimension 2.

9.5 Sequences with Low Correlation and Large Linear Complexity

Sequences with both low correlation and large linear complexity are required for stream ciphers [128, 129]. The randomness of a sequence is measured by its correlation, while the replication complexity of a key cipher sequence depends on the linear complexity of the sequence. In this section, we make use of algebraic curves to construct sequences with both low correlation and large linear complexity. Sequences produced by other methods often satisfy only one of these two requirements.

We focus only on binary sequences in this section, though correlation can be defined for sequences over any finite field. The reader may refer to [49] for the correlation of nonbinary sequences.

Definition 9.5.1 Let $\mathbf{s} = s_0 s_1 \ldots$ be a periodic binary sequence with least period p. For $\ell \geq 0$, the **autocorrelation** of $\mathbf{s}$ is defined by

$$C_{\mathbf{s}}(\ell) \overset{\text{def}}{=} \sum_{i=0}^{p-1} (-1)^{s_i + s_{i+\ell}}.$$

For a sequence $\mathbf{s}$ of period p, it is clear that $C_{\mathbf{s}}(p + \ell) = C_{\mathbf{s}}(\ell)$ for all $\ell \geq 0$ and $C_{\mathbf{s}}(0) = p$. If $|C_{\mathbf{s}}(\ell)|$ is small, it implies that $\mathbf{s}$ is far from both its ℓ-shift $\mathbf{s}_\ell = s_\ell s_{\ell+1} \ldots$ and the complement of its ℓ-shift $\bar{\mathbf{s}}_\ell = 1 + s_\ell, 1 + s_{\ell+1}, \ldots$. The converse is also true. This means that the autocorrelation is a measurement for the randomness of a periodic binary sequence. For some other applications such as optical communication, CDMA systems, etc., the interest is in families of binary sequences with low correlation.

Definition 9.5.2 Let $\mathcal{S}$ be a finite set of periodic binary sequences. Assume that all the sequences have least period equal to p. For $\ell \geq 0$, then the **correlation** between the ith sequence $\mathbf{s}_i = s_0^{(i)} s_1^{(i)} \ldots$ and the jth sequence $\mathbf{s}_j = s_0^{(j)} s_1^{(j)} \ldots$ at shift ℓ is given by

$$C_{\mathbf{s}_i, \mathbf{s}_j}(\ell) \overset{\text{def}}{=} \sum_{t=0}^{p-1} (-1)^{s_t^{(i)} + s_{t+\ell}^{(j)}}.$$

The **correlation of the sequence family** is defined to be

$$C_{\mathcal{S}} \overset{\text{def}}{=} \max_{i \neq j \text{ or } \ell \not\equiv 0 \pmod{p}} \{C_{\mathbf{s}_i, \mathbf{s}_j}(\ell)\}.$$

A commonly used method of generating low-correlation sequences is described below.

For $e \geq 1$ and $d \geq 1$, let $\mathcal{P}_d$ denote all polynomials over $\mathbb{F}_{2^e}$ of degree at most d. Consider the set of sequences

$$\mathcal{S} \overset{\text{def}}{=} \{\text{Tr}((f(\alpha^i)))_{i=0}^{\infty} : f \in \mathcal{P}_d\}, \tag{9.17}$$

where Tr stands for the trace map from $\mathbb{F}_{2^e}$ to $\mathbb{F}_2$ and α is a primitive element of $\mathbb{F}_{2^e}$. It is clear that $2^e - 1$ is a period for every sequence in this family. Take a subset $\mathcal{P}$ of $\mathcal{S}$ such that all sequences in $\mathcal{P}$ have least period $2^e - 1$ and no two are cyclically equivalent (note that two sequences are called cyclically equivalent if one is a cyclic shift of the other). The reader may refer to [122] for details about such a family of sequences. Many well-known families of sequences can be produced in this manner. For instance, if we take $d = 3$ and let e be odd, then this family yields the optimal Gold sequences (see [122] for the definition of Gold sequences). Furthermore, a generalization of this method to Galois rings produces some families of sequences with low correlation as well (see [122]).

On the other hand, there are two disadvantages for the family of sequences constructed in (9.17):

(i) the linear complexities of the sequences in this family are not large enough, hence they are not suitable for stream ciphers;

(i) we cannot produce sequences of arbitrary period as the period $2^e - 1$ in this family is always fixed.

In this section, we employ algebraic curves to construct families of sequences that overcome these two disadvantages. We first need some further technical notions and results related to algebraic curves before we introduce the construction of sequences.

Let $q = 2^e$ and let $\mathcal{X}$ be an algebraic curve over $\mathbb{F}_q$ with function field $\mathbb{F}_q(\mathcal{X})$. An element z of $\mathbb{F}_q(\mathcal{X})$ is called **degenerate** if it can be written in the form $\alpha + h^2 - h$ for some $\alpha \in \mathbb{F}_q$ and $h \in \mathbb{F}_q(\mathcal{X})$. Otherwise, it is called **nondegenerate**. Apparently, if there exists a point P of $\mathcal{X}$ such that $\nu_P(z)$ is odd, then z is nondegenerate since $\nu_P(\alpha + h^2 - h)$ is always even for all $\alpha \in \mathbb{F}_q$ and $h \in \mathbb{F}_q(\mathcal{X})$.

For a given nondegenerate element z of $\mathbb{F}_q(\mathcal{X})$, we add one more equation $y^2 + y = z$ to the system of equations defining $\mathcal{X}$. The new curve is denoted by $\mathcal{X}_z$. In fact, the function field $\mathbb{F}_q(\mathcal{X}_z)$ of $\mathcal{X}_z$ is the field obtained by adjoining a root of $y^2 + y - z$ to $\mathbb{F}_q(\mathcal{X})$. We have the following easy result.

Lemma 9.5.3 *Let $\mathcal{X}$ be an algebraic curve over $\mathbb{F}_q$, where $q = 2^e$, and let z be a nondegenerate element of $\mathbb{F}_q(\mathcal{X})$. Then the field $\mathbb{F}_q(\mathcal{X}_z)$ is a Galois extension over $\mathbb{F}_q(\mathcal{X})$ of degree 2.*

The genus of the curve $\mathcal{X}_z$ can be estimated as follows.

Lemma 9.5.4 *Let $\mathcal{X}$ be an algebraic curve over $\mathbb{F}_q$, where $q = 2^e$, and let z be a nondegenerate element of $\mathbb{F}_q(\mathcal{X})$. Then the genus of $\mathcal{X}_z$ is at most $2g(\mathcal{X}) + d - 1$, where $g(\mathcal{X})$ is the genus of $\mathcal{X}$ and d is the degree of the divisor $\mathrm{div}_\infty(z)$.*

The reader may refer to [178] for the proofs of Lemmas 9.5.3 and 9.5.4.

Next, we study the Hamming weight of the trace vectors associated with nondegenerate elements of $\mathbb{F}_q(\mathcal{X})$ and some $\mathbb{F}_q$-rational points of $\mathcal{X}$.

The following result can be found in [151, Proposition VIII.2.8] when $\mathcal{X}$ is the projective line.

Lemma 9.5.5 *Let $P_1, \ldots, P_n$ be distinct $\mathbb{F}_q$-rational points of $\mathcal{X}$. Let z be a nondegenerate element of $\mathbb{F}_q(\mathcal{X})$ such that $\nu_{P_i}(z) \geq 0$ for all $i = 1, \ldots, n$. Let $\mathfrak{P}_z$ be the set of $\mathbb{F}_q$-rational points of $\mathcal{X}_z$ lying above those $\mathbb{F}_q$-rational points of $\mathcal{X}$ that are outside $\{P_1, \ldots, P_n\}$ (by a point Q of $\mathcal{X}_z$ lying above a point P*

of $\mathcal{X}$, we mean that, for any $f \in \mathbb{F}_q(\mathcal{X})$ with $\nu_P(f) \geq 0$, $f(Q) = 0$ whenever $f(P) = 0$). Then the Hamming weight of the vector

$$(\mathrm{Tr}(z(P_1)), \ldots, \mathrm{Tr}(z(P_n)))$$

is

$$n - \frac{N(\mathcal{X}_z) - |\mathfrak{P}_z|}{2},$$

where Tr is the trace function from $\mathbb{F}_q$ to $\mathbb{F}_2$ and $N(\mathcal{X}_z)$ denotes the number of $\mathbb{F}_q$-rational points of $\mathcal{X}_z$.

Proof. First of all, note the fact that, for an element $\alpha \in \mathbb{F}_q$, $\mathrm{Tr}(\alpha) = 0$ if and only if there exists $\beta \in \mathbb{F}_q$ such that $\alpha = \beta^2 - \beta$ (see [95]). Thus, for a fixed index i, $\mathrm{Tr}(z(P_i)) = 0$ if and only if $z(P_i) = a^2 - a$ for some $a \in \mathbb{F}_q$. This is equivalent to saying that there are two points of $\mathcal{X}_z$ lying over P_i.

Let r be the number of points of $\{P_1, \ldots, P_n\}$ for which there are two points lying over each. Then the Hamming weight of $(\mathrm{Tr}(z(P_1)), \ldots, \mathrm{Tr}(z(P_n)))$ is $n - r$ and it is clear that $2r + |\mathfrak{P}_z| = N(\mathcal{X}_z)$. The desired result follows. $\square$

The following lemma provides a sufficient condition under which a set of elements of $\mathbb{F}_q(\mathcal{X})$ are $\mathbb{F}_q$-linearly independent.

Lemma 9.5.6 *Let* $z_1, \ldots, z_n$ *be* n *elements of* $\mathbb{F}_q(\mathcal{X})$. *Suppose that there are* n *distinct points* $P_1, \ldots, P_n$ *such that* $\nu_{P_i}(z_j) < 0$ *if and only if* $i = j$, *for all* $1 \leq i, j \leq n$. *Then* $z_1, \ldots, z_n$ *are* $\mathbb{F}_q$-*linearly independent.*

Proof. Suppose that there exist n elements $\lambda_i \in \mathbb{F}_q$, $i = 1, \ldots, n$, with $\lambda_k \neq 0$ (for some k) such that $\sum_{i=1}^n \lambda_i z_i = 0$. Then $\lambda_k z_k = -\sum_{i \neq k} \lambda_i z_i$. However, $\nu_{P_k}(\lambda_k z_k) < 0$, while $\nu_{P_k}(-\sum_{i \neq k} \lambda_i z_i) \geq 0$. The result follows from this contradiction. $\square$

For an algebraic curve $\mathcal{X}$ in the projective space $\mathbf{P}^N$ over $\mathbb{F}_q$, an **automorphism** of $\mathcal{X}$ is a bijective map from $\mathcal{X}$ to itself defined by $P \mapsto [f_0(P), f_1(P), \ldots, f_N(P)]$, where f_i are homogeneous polynomials in $\mathbb{F}_q[x_0, \ldots, x_N]$ of the same degree. For an automorphism σ of $\mathcal{X}$, it induces an automorphism of the function field $\mathbb{F}_q(\mathcal{X})$ by sending $f(x_0, \ldots, x_N)$ to $f(\sigma(x_0), \ldots, \sigma(x_N))$. All the automorphisms of $\mathcal{X}$ form a group which is denoted by $\mathrm{Aut}(\mathcal{X}/\mathbb{F}_q)$.

Recall that a closed point P on $\mathcal{X}$ is a set of conjugate points. If we consider divisors, it is more convenient to denote the closed point P by $\mathsf{P} = \sum_P P$. In this case, we have that $\sigma(\mathsf{P}) = \sum_P \sigma(P)$ is again a closed point of $\mathcal{X}$ for any automorphism $\sigma \in \mathrm{Aut}(\mathcal{X}/\mathbb{F}_q)$, and P and $\sigma(\mathsf{P})$ have the same degree. For a closed point P and a function $f \in \mathbb{F}_q(\mathcal{X})$, it is easy to see that $\nu_Q(f)$ is the same for all the points Q in P. Therefore, we may define $\nu_{\mathsf{P}}(f) = \nu_P(f)$, where P is a point in the set of conjugate points defining P.

The reader may refer to [144, 117] for more details on rational maps, automorphisms of curves, and closed points.

We have the following useful results on automorphisms (see [52, 151]).

Lemma 9.5.7 *Let σ be an automorphism of $\mathcal{X}$. Let P be a point of $\mathcal{X}$ (not necessarily $\mathbb{F}_q$-rational) and let f be a function in $\mathbb{F}_q(\mathcal{X})$. Then*

(i) $\nu_{\sigma(P)}(\sigma(f)) = \nu_P(f)$;

(ii) $\sigma(f)(\sigma(P)) = f(P)$ *if* $\nu_P(f) \geq 0$.

We are now ready to present the construction of sequences using algebraic curves.

Construction of Sequences

Let $\mathcal{X}$ be an algebraic curve over $\mathbb{F}_q$. Choose an $\mathbb{F}_q$-rational point P of $\mathcal{X}$ and an automorphism $\sigma \in \mathrm{Aut}(\mathcal{X}/\mathbb{F}_q)$. Then $\sigma^i(P)$ are also $\mathbb{F}_q$-rational points for all $i = 0, 1, \ldots$. Since there are only finitely many $\mathbb{F}_q$-rational points by the Hasse-Weil bound (see Corollary 1.5.4(iii)), we have some least $n \geq 0$ such that $\sigma^{n+i}(P) = \sigma^i(P)$ for all $i \in \mathbb{Z}$. Put $P_i = \sigma^i(P)$ for all $i \geq 0$ and choose an element $z \in \mathbb{F}_q(\mathcal{X})$ such that $\nu_{P_i}(z) \geq 0$ for all $i \geq 0$. Define a sequence

$$\mathbf{s}_z \stackrel{\text{def}}{=} \{\mathrm{Tr}(z(P_i))\}_{i=0}^{\infty}.$$

It is clear that n is a period of $\mathbf{s}_z$. The following result provides a sufficient condition under which n is the least period of $\mathbf{s}_z$.

Lemma 9.5.8 *Let P_i and z satisfy the conditions in the above construction. Assume that $\mathrm{div}_\infty(z) = m\mathsf{P}$ for a closed point P and some odd m. If $\mathsf{P}, \sigma(\mathsf{P}), \ldots, \sigma^{n-1}(\mathsf{P})$ are all distinct and $d = \deg(\mathrm{div}_\infty(z)) = m\deg(\mathsf{P})$ satisfies*

$$q + 1 + 2(2g(\mathcal{X}) + 2d - 1)\sqrt{q} < 2n,$$

then the least period of $\mathbf{s}_z$ is exactly n.

Proof. Suppose that there exists an integer $0 \leq k < n$ such that k is a period of $\mathbf{s}_z$. Consider the function

$$f = z - \sigma^k(z).$$

Then f is a nondegenerate element since

$$\mathrm{div}_\infty(f) = \mathrm{div}_\infty(z - \sigma^k(z)) = m\mathsf{P} + m\sigma^k(\mathsf{P}).$$

Now consider the evaluation of f at P_{i+k}:

$$\begin{aligned}
f(P_{i+k}) &= z(P_{i+k}) - \sigma^k(z)(P_{i+k}) \\
&= z(P_{i+k}) - z(\sigma^{-k}(P_{i+k})) \\
&= z(P_{i+k}) - z(P_i).
\end{aligned}$$

Thus, $\mathrm{Tr}(f(P_{i+k})) = \mathrm{Tr}(z(P_{i+k}) - z(P_i)) = 0$ as k is a period of $\mathbf{s}_z$. Hence, the vector $(\mathrm{Tr}(f(P_1)), \ldots, \mathrm{Tr}(f(P_n)))$ is the zero vector. By Lemma 9.5.5, we get

$$n - \frac{N(\mathcal{X}_f) - |\mathfrak{P}_f|}{2} = 0.$$

This gives

$$N(\mathcal{X}_f) = 2n + |\mathfrak{P}_f| \geq 2n. \tag{9.18}$$

By the Hasse-Weil bound (Corollary 1.5.4(iii)), we have

$$N(\mathcal{X}_f) \leq q + 1 + 2g(\mathcal{X}_f)\sqrt{q}.$$

On the other hand, by Lemma 9.5.4, we have

$$g(\mathcal{X}_f) \leq 2g(\mathcal{X}) + \deg(\mathrm{div}_\infty(f)) - 1 = 2g(\mathcal{X}) + 2d - 1.$$

Together, the above two inequalities give

$$N(\mathcal{X}_f) \leq q + 1 + 2(2g(\mathcal{X}) + 2d - 1)\sqrt{q}. \tag{9.19}$$

Combining (9.18) with (9.19) yields

$$2n \leq q + 1 + 2(2g(\mathcal{X}) + 2d - 1)\sqrt{q},$$

contradicting the assumption. Hence, the least period of $\mathbf{s}_z$ is equal to n. $\quad\square$

In view of Lemma 9.5.8, it may be possible to obtain in this manner sequences with period different from $2^e - 1$. Indeed, some such examples are given in Subsection 9.5.2.

Now we turn to study the linear complexity and correlation for the sequences constructed above. We first consider the linear complexity.

Theorem 9.5.9 *Let P_i and z satisfy the conditions in the above construction. Assume that $\mathrm{div}_\infty(z) = m\mathsf{P}$ for a closed point P and some odd m. If $\mathsf{P}, \sigma(\mathsf{P}), \ldots, \sigma^{n-1}(\mathsf{P})$ are all distinct, then the linear complexity of $\mathbf{s}_z$ satisfies*

$$\ell(\mathbf{s}_z) \geq \min\left\{ n, \frac{2n - q - 1 - 2(2g(\mathcal{X}) + d - 1)\sqrt{q}}{2d\sqrt{q}} \right\},$$

where $d = \deg(\mathrm{div}_\infty(z)) = m\deg(\mathsf{P})$.

Proof. Denote $\ell(\mathbf{s}_z)$ by s. If $s = n$, we have nothing to prove. Hence, we may assume that $s < n$. Then, there exist $s + 1$ binary numbers $\lambda_0, \lambda_1, \ldots, \lambda_s$ such that $\lambda_0 = \lambda_s = 1$ and

$$\sum_{i=0}^{s} \lambda_i \mathrm{Tr}(z(P_{i+v})) = 0$$

for all $v \geq 0$. Setting

$$u \stackrel{\mathrm{def}}{=} \sum_{i=0}^{s} \lambda_i \sigma^{-i}(z),$$

then u is nondegenerate since $\operatorname{div}_\infty(\sigma^{-i}(z)) = \sigma^{-i}(m\mathsf{P})$ for any $0 \le i \le n-1$. Moreover, $\deg(\operatorname{div}_\infty(u)) = (s+1)\deg(\operatorname{div}_\infty(z)) = (s+1)d$.

Since

$$\sum_{i=0}^{s} \lambda_i \operatorname{Tr}(z(P_{i+v})) = \operatorname{Tr}\left(\sum_{i=0}^{s} \lambda_i z(P_{i+v})\right)$$

$$= \operatorname{Tr}\left(\sum_{i=0}^{s} \lambda_i z(\sigma^i(P_v))\right)$$

$$= \operatorname{Tr}\left(\sum_{i=0}^{s} \lambda_i \sigma^{-i}(z)(P_v)\right)$$

$$= \operatorname{Tr}(u(P_v)),$$

it follows that

$$(\operatorname{Tr}(u(P_1)), \operatorname{Tr}(u(P_2)), \ldots, \operatorname{Tr}(u(P_n))) = \mathbf{0}.$$

By Lemma 9.5.5, we have

$$0 = n - \frac{N(\mathcal{X}_u) - |\mathfrak{P}_u|}{2} \ge n - \frac{1}{2}N(\mathcal{X}_u).$$

Hence, by the Hasse-Weil bound (Corollary 1.5.4(iii)) and Lemma 9.5.4,

$$2n \le N(\mathcal{X}_u) \le q + 1 + 2g(\mathcal{X}_u)\sqrt{q} \le q + 1 + 2(2g(\mathcal{X}) + d(s+1) - 1)\sqrt{q}.$$

The desired result now follows. $\qquad\square$

The above theorem indicates that the linear complexity of $\mathbf{s}_z$ is good if the period n is relatively large compared with q and $2g(\mathcal{X})\sqrt{q}$. Next, we look at the correlation of such sequences.

Theorem 9.5.10 *Let* z_1, z_2 *be two elements of* $\mathbb{F}_q(\mathcal{X})$ *with* $d_i = \deg(\operatorname{div}_\infty(z_i))$ *and* $\nu_{P_j}(z_i) \ge 0$ *for all* $1 \le i \le 2$ *and* $0 \le j \le n-1$ *(it is allowed that* $z_1 = z_2$*). Suppose that* $z_1 + \sigma^{-w}(z_2)$ *is nondegenerate for some* $w \in \mathbb{Z}$*. Then the correlation* $C_{\mathbf{s}_{z_1}, \mathbf{s}_{z_2}}(w)$ *satisfies*

$$|C_{\mathbf{s}_{z_1}, \mathbf{s}_{z_2}}(w)| \le 2(2g(\mathcal{X}) + d - 1)\sqrt{q} + |q + 1 - n| + 2(N(\mathcal{X}) - n),$$

where d *is the degree of the divisor* $\operatorname{div}_\infty(z_1 + \sigma^{-w}(z_2))$.

Proof. Put $u = z_1 + \sigma^{-w}(z_2)$. By the definition of the correlation, we have

$$C_{\mathbf{s}_{z_1}, \mathbf{s}_{z_2}}(w) = \sum_{i=1}^{n} (-1)^{\mathrm{Tr}(z_1(P_i)) + \mathrm{Tr}(z_2(P_{i+w}))}$$

$$= \sum_{i=1}^{n} (-1)^{\mathrm{Tr}(z_1(P_i) + z_2(P_{i+w}))}$$

$$= \sum_{i=1}^{n} (-1)^{\mathrm{Tr}(z_1(P_i) + \sigma^{-w}(z_2)(P_i))}$$

$$= \sum_{i=1}^{n} (-1)^{\mathrm{Tr}(u(P_i))}$$

$$= n - 2\mathrm{wt}(\mathrm{Tr}(u(P_1)), \mathrm{Tr}(u(P_2)), \ldots, \mathrm{Tr}(u(P_n)))$$

$$= n - 2 \left(n - \frac{N(\mathcal{X}_u) - |\mathfrak{P}_u|}{2} \right)$$

$$= N(\mathcal{X}_u) - n - |\mathfrak{P}_u|.$$

By the Hasse-Weil bound (Corollary 1.5.4(iii)) and Lemma 9.5.4, we have

$$|N(\mathcal{X}_u) - (q+1)| \leq 2g(\mathcal{X}_u)\sqrt{q} \leq 2(2g(\mathcal{X}) + d - 1)\sqrt{q}.$$

It is also clear that the size of $\mathfrak{P}_u$ is at most $2(N(\mathcal{X}) - n)$.
    Hence,

$$\begin{aligned}
|C_{\mathbf{s}_{z_1}, \mathbf{s}_{z_2}}(w)| &= |N(\mathcal{X}_u) - n - |\mathfrak{P}_u|| \\
&\leq |N(\mathcal{X}_u) - (q+1)| + |q+1-n| + |\mathfrak{P}_u| \\
&\leq 2(2g(\mathcal{X}) + d - 1)\sqrt{q} + |q+1-n| + 2(N(\mathcal{X}) - n).
\end{aligned}$$

The proof is complete. $\qquad\square$

We proceed now to discuss two examples from the above construction of sequences.

9.5.1 Construction Using a Projective Line

We continue to assume that q is even in this subsection. We fix some notations for this subsection:

ϵ – a fixed primitive element of $\mathbb{F}_q$;
$F = \mathbb{F}_q(x)$ – the rational function field of a projective line $\mathcal{X}$;
ϕ – the automorphism of $\mathcal{X}/\mathbb{F}_q$: $x \mapsto \epsilon x$;
P – the unique zero of $x - 1$.

Let P_i be the unique zero of $\phi^i(x-1) = \epsilon^i x - 1 = \epsilon^i(x - \epsilon^{-i})$ for all $i \in \mathbb{Z}$, and put $n = q - 1$. Then $P_i, P_{i+1}, \ldots, P_{i+n-1}$ are pairwise distinct for any fixed $i \in \mathbb{Z}$. Moreover, $P_j = P_{j+n}$ for all $j \in \mathbb{Z}$ since ϕ^n is the identity.

Let $P_d(F)$ be the set of all closed points of degree $d \geq 2$. It is not difficult to verify that there is a one-to-one correspondence between $P_d(F)$ and the set of all monic irreducible polynomials of degree d of $\mathbb{F}_q[x]$. Therefore, the size $I_q(d)$ of $P_d(F)$ is equal to the number of monic irreducible polynomials of degree d of $\mathbb{F}_q[x]$ and is given by (cf. [95, Theorem 3.25, page 93])

$$I_q(d) = \frac{1}{d} \sum_{b|d} \mu\left(\frac{d}{b}\right) q^b,$$

where $\mu(\cdot)$ is the Möbius function.

Since ϵ is a primitive element of $\mathbb{F}_q$, for $P \in P_d(F)$, it is clear that $\phi^i(P) = \phi^{i+n}(P)$ for all $i \in \mathbb{Z}$.

Lemma 9.5.11 *Let* $P \in P_d(F)$. *If* $d \geq 2$ *and* $\gcd(d, q - 1) = 1$, *then* $\phi^i(P), \phi^{i+1}(P), \ldots, \phi^{i+n-1}(P)$ *are pairwise distinct for any fixed* $i \in \mathbb{Z}$.

Proof. It is sufficient to show that $\phi^k(P) \neq P$ for any $1 \leq k \leq n - 1$. Let $f(x)$ be the monic irreducible polynomial with $\text{div}_0(f) = P$. Then $\phi^k(P)$ is the unique zero of $\phi^k(f(x)) = f(\phi^k(x)) = f(\epsilon^k x)$, i.e., $\text{div}_0(f(\epsilon^k x)) = \phi^k(P)$. In order to prove that $\phi^k(P) \neq P$, we need to show that the roots of $f(x)$ are not roots of $\phi^k(f(x)) = f(\epsilon^k x)$.

Let $\alpha \in \mathbb{F}_{q^d}$ be a root of $f(x)$. We want to show that α is not a root of $f(\epsilon^k x)$. This is equivalent to showing that $\epsilon^k \alpha$ is not a root of $f(x)$. Suppose that $\epsilon^k \alpha$ is a root of $f(x)$. Since all the roots of $f(x)$ are $\alpha, \alpha^q, \ldots, \alpha^{q^{d-1}}$, there exists an integer t with $1 \leq t \leq d - 1$ such that $\epsilon^k \alpha = \alpha^{q^t}$, i.e., $\epsilon^k = \alpha^{q^t - 1}$. This yields

$$1 = (\epsilon^k)^{q-1} = \alpha^{(q^t - 1)(q-1)}. \tag{9.20}$$

Since $\gcd(d, q - 1) = 1$, we have $\gcd((q^d - 1)/(q - 1), q - 1) = 1$. Knowing $\alpha^{q^d - 1} = 1$, we obtain from (9.20) that $\alpha^{q^t - 1} = 1$, i.e., $\alpha^{q^t} = \alpha$. This contradicts the fact that $f(x)$ is an irreducible polynomial of degree d since $1 \leq t \leq d - 1$. $\square$

By the above lemma, we find that for $d \geq 2$ with $\gcd(d, q - 1) = 1$, the action of the cyclic group $<\phi>$ of order $n = q - 1$ on $P_d(F)$ divides $P_d(F)$ into $r \overset{\text{def}}{=} I_q(d)/n$ equivalence classes. Each class contains n closed points of degree d. We choose only one point from each class. Thus, we obtain r closed points of degree d

$$P_1, P_2, \ldots, P_r.$$

It is clear that for $1 \leq i \neq j \leq r$,

$$P_j \notin \{\phi^s(P_i) : s \in \mathbb{Z}\} = \{P_i, \phi(P_i), \ldots, \phi^{n-1}(P_i)\}.$$

For each $1 \leq i \leq r$, let $f_i(x)$ be the monic irreducible polynomial of degree d of $\mathbb{F}_q[x]$ with unique zero P_i, i.e., $\text{div}_0(f_i) = P_i$. Put

$$z_i \overset{\text{def}}{=} \frac{1}{f_i(x)}.$$

Then P_i is the unique pole of z_i (i.e., $\mathrm{div}_\infty(z_i) = P_i$) and $\nu_{P_i}(z_i) = -1$. Consider the family of binary sequences

$$\mathcal{S}_d = \{\mathbf{s}_{z_i} : i = 1, 2, \dots, r = I_q(d)/(q-1)\}$$

with

$$\mathbf{s}_{z_i} = \{\mathrm{Tr}(z_i(P_j))\}_{j=0}^{\infty} = \{\mathrm{Tr}(z_i(\epsilon^j))\}_{j=0}^{\infty},$$

where P_j is the unique zero of $x - \epsilon^j$.

Theorem 9.5.12 *Let* $2 \le d \le \frac{1}{2}((q-7)/(2\sqrt{q}) + 1)$ *and* $\gcd(d, q-1) = 1$. *Let* $\mathcal{S}_d$ *be the family of binary sequences as constructed above. Then* $\mathcal{S}_d$ *is of size* $I_q(d)/(q-1)$ *and each sequence in* $\mathcal{S}_d$ *is of period* $n = q-1$. *Moreover,*

$$\ell_{\min}(\mathcal{S}_d) \overset{\text{def}}{=} \min\{\ell(\mathbf{s}) : \mathbf{s} \in \mathcal{S}_d\} \ge \frac{q - 3 - 2(d-1)\sqrt{q}}{2d\sqrt{q}}$$

and

$$C_{\mathcal{S}_d} \le 2(2d-1)\sqrt{q} + 6.$$

Proof. By the condition $d \le \frac{1}{2}((q-7)/(2\sqrt{q}) + 1)$, we get $q + 1 + 2(2g(\mathcal{X}) + 2d - 1)\sqrt{q} < 2n$. It follows from Proposition 9.5.8 that the least period of each sequence in $\mathcal{S}_d$ is $n = q - 1$. Let $\mathbf{s}_{z_i} \in \mathcal{S}_d$ for some $1 \le i \le r = I_q(d)/(q-1)$. Then P_i is the unique pole of z_i and $P_i, \phi(P_i), \dots, \phi^{n-1}(P_i)$ are pairwise distinct. By Theorem 9.5.9, we have

$$\ell(\mathbf{s}_{z_i}) \ge \frac{2n - q - 1 - 2(2g(\mathcal{X}) + d - 1)\sqrt{q}}{2d\sqrt{q}} = \frac{q - 3 - 2(d-1)\sqrt{q}}{2d\sqrt{q}}$$

for all $1 \le i \le r$. This means that

$$\ell_{\min}(\mathcal{S}_d) \ge \frac{q - 3 - 2(d-1)\sqrt{q}}{2d\sqrt{q}}.$$

Now let $\mathbf{s}_{z_i}$ and $\mathbf{s}_{z_j}$ be two sequences of $\mathcal{S}_d$ (it is allowed that $j = i$). For $w \in \mathbb{Z}$, consider the function

$$u \overset{\text{def}}{=} z_i + \phi^{-w}(z_j).$$

The closed point P_i is the unique pole of z_i and $\phi^{-w}(P_j)$ is the unique pole of $\phi^{-w}(z_j)$. We consider two cases.
Case 1: $i \ne j$. We must have $P_i \ne \phi^{-w}(P_j)$ since

$$P_j \notin \{\phi^s(P_i) : s \in \mathbb{Z}\}.$$

Thus, P_i is not a pole of $\phi^{-w}(z_j)$ and $\nu_{P_i}(u) = \min\{\nu_{P_i}(z_i), \nu_{P_i}(\phi^{-w}(z_j))\} = -1$. Therefore, u is nondegenerate.
Case 2: $i = j$ and $0 < w < n$. Again, we have $P_i \ne \phi^{-w}(P_i)$, thus the same argument as in Case 1 shows that u is nondegenerate.

For both cases, $z_i + \phi^{-w}(z_j)$ is nondegenerate. By Theorem 9.5.10, we obtain

$$\begin{aligned}
|C_{\mathbf{s}_{z_i}, \mathbf{s}_{z_j}}(w)| &\leq 2(2g(\mathcal{X}) + 2d - 1)\sqrt{q} + |q + 1 - n| + 2(N(\mathcal{X}) - n) \\
&= 2(2d - 1)\sqrt{q} + |q + 1 - (q - 1)| + 2(q + 1 - (q - 1)) \\
&= 2(2d - 1)\sqrt{q} + 6.
\end{aligned}$$

Since $C_{\mathcal{S}_d} < n$, it is clear that the sequences $\mathbf{s}_{z_1}, \mathbf{s}_{z_2}, \ldots, \mathbf{s}_{z_r}$ are pairwise distinct. Hence, the size of $\mathcal{S}_d$ is equal to $r = I_q(d)/(q-1)$. □

We rewrite Theorem 9.5.12 into the following form by specifying q to be 2^e.

Theorem 9.5.13 *Let* $e \geq 3$, *let* $2 \leq d \leq (2^e - 7 + 2^{e/2+1})/2^{e/2+2}$, *and* $\gcd(d, 2^e - 1) = 1$. *Then there exists a family* $\mathcal{S}_d$ *of binary sequences such that*

(a) $|\mathcal{S}_d| = I_{2^e}(d)/(2^e - 1)$;

(b) *each sequence in* $\mathcal{S}_d$ *is of period* $2^e - 1$;

(c) $\ell_{\min}(\mathcal{S}_d) \geq (2^e - 3 - (d-1)2^{e/2+1})/(d2^{e/2+1})$;

(d) $C_{\mathcal{S}_d} \leq (2d - 1)2^{e/2+1} + 6$.

If we take $d = 2$ in Theorem 9.5.13, we obtain the following:

Corollary 9.5.14 (i) *Let* $e \geq 6$, *then there exists a family* $\mathcal{S}_2$ *of binary sequences such that*

(a) $|\mathcal{S}_2| = 2^{e-1}$;

(b) *each sequence in* $\mathcal{S}_2$ *is of period* $2^e - 1$;

(c) $\ell_{\min}(\mathcal{S}_2) \geq 2^{e/2-2} - \frac{1}{2} - \frac{3}{2^{e/2+2}}$;

(d) $C_{\mathcal{S}_2} \leq 6(2^{e/2} + 1)$.

(ii) *Let* e *be a positive integer and let* d *be a prime satisfying* $\gcd(d, 2^e - 1) = 1$ *and* $d \leq ((2^e - 7)/2^{e/2+1} + 1)/2$. *Then there exists a family* $\mathcal{S}_d$ *of binary sequences such that*

(a) $|\mathcal{S}_d| = (2^{ed} - 2^e)/(d(2^e - 1))$;

(b) *each sequence in* $\mathcal{S}_d$ *is of period* $2^e - 1$;

(c) $\ell_{\min}(\mathcal{S}_d) \geq (2^e - 3 - (d-1)2^{e/2+1})/(d \times 2^{e/2+1})$;

(d) $C_{\mathcal{S}_d} \leq (2d - 1)2^{e/2+1} + 6$.

Proof. Note that

$$\frac{I_{2^e}(2)}{2^e - 1} = 2^{e-1}.$$

The desired result in (i) follows from Theorem 9.5.10 when we take $d = 2$. Furthermore, for a prime d, we have

$$\frac{I_{2^e}(d)}{2^e - 1} = \frac{2^{ed} - 2^e}{d(2^e - 1)}.$$

The desired result in (ii) follows from Theorem 9.5.10. □

9.5.2 Construction Using Elliptic Curves

We continue to assume that q is even in this subsection. First, we review some results on elliptic curves [52, 144].

Let $\mathcal{E}/\mathbb{F}_q$ be an elliptic curve defined over $\mathbb{F}_q$ with at least one $\mathbb{F}_q$-rational point O. Let $\mathcal{E}(\mathbb{F}_q)$ be the set of all $\mathbb{F}_q$-rational points on $\mathcal{E}$. We can take O as the zero element of $\mathcal{E}(\mathbb{F}_q)$. The number of $\mathbb{F}_q$-rational points of $\mathcal{E}$ is always between $q + 1 - 2\sqrt{q}$ and $q + 1 + 2\sqrt{q}$ by the Hasse-Weil bound (Corollary 1.5.4(iii)). Furthermore, for any $d \geq 1$, the number of closed points of degree d is determined by the number of $\mathbb{F}_q$-rational points. More precisely, suppose that $\mathcal{E}$ has $q + 1 + t$ rational places, then the number $B_q(d, t)$ of closed points of degree d of $\mathcal{E}$ is determined by

$$B_q(d, t) \overset{\text{def}}{=} \frac{1}{d} \sum_{b \mid d} \mu(\frac{d}{b})(q^b + 1 - \omega_1^b - \omega_2^b),$$

where ω_1, ω_2 are the two roots of the quadratic equation $X^2 + tX + q = 0$. In particular,

$$B_q(2, t) = \frac{q^2 + q - t - t^2}{2} \quad \text{and} \quad B_q(3, t) = \frac{q^3 - q - 3qt - t + t^3}{3}.$$

Lemma 9.5.15 (see [52, pages 194–195]) *Let $\mathcal{E}/\mathbb{F}_q$ be an elliptic curve with at least one $\mathbb{F}_q$-rational point O. Then, for any rational point P of $\mathcal{E}$, there exists a unique automorphism σ_P of $\mathrm{Aut}(\mathcal{E}/\mathbb{F}_q)$ such that, for any closed point P of degree d, $\sigma_P(\mathsf{P}) + O - \mathsf{P} - [d]P$ is a principal divisor. In particular, σ_O is the identity, and $\sigma_P^i = \sigma_{[i]P}$ for all $i \in \mathbb{Z}$.*

Remark 9.5.16 All the $\mathbb{F}_q$-rational points of $\mathcal{E}$ form a finite abelian group (with group operation $\oplus$) that is isomorphic to the divisor class group of degree zero of $\mathcal{E}$. If we take O as the zero element of the group and Q is an $\mathbb{F}_q$-rational point, then

$$\sigma_P(Q) = P \oplus Q.$$

An elliptic curve is called **cyclic** if the $\mathbb{F}_q$-rational points of this curve form a cyclic group. The next result on the existence of cyclic elliptic curves can be deduced from Theorem 3.3.10 with $p = 2$.

Lemma 9.5.17 *Let $q = 2^e$. Let t be an integer satisfying one of the following three conditions:*

(i) *t is an odd integer between $-2\sqrt{q} = -2^{e/2+1}$ and $2\sqrt{q} = 2^{e/2+1}$;*

(ii) *$t = 0$;*

(iii) *$t = \sqrt{q} = 2^{e/2}$ if e is even, and $t = \sqrt{2q} = 2^{(e+1)/2}$ if e is odd.*

Then there exists a cyclic elliptic curve over $\mathbb{F}_q$ with $1 + q + t$ $\mathbb{F}_q$-rational points.

Lemma 9.5.18 *Let $\mathcal{E}$ be a cyclic elliptic curve over $\mathbb{F}_q$. Let R be a generator of $\mathcal{E}(\mathbb{F}_q)$ and let P be a closed point of degree d of $\mathcal{E}$. Suppose the order n of $\mathcal{E}(\mathbb{F}_q)$ is relatively prime to d. Then $\sigma_R^j(\mathsf{P}) = \sigma_R^{j+n}(\mathsf{P})$, for all $j \in \mathbb{Z}$, and $\sigma_R^i(\mathsf{P}), \sigma_R^{i+1}(\mathsf{P}), \ldots, \sigma_R^{i+n-1}(\mathsf{P})$ are pairwise distinct for any fixed $i \in \mathbb{Z}$.*

Proof. For any $j \in \mathbb{Z}$,

$$\sigma_R^{j+n}(\mathsf{P}) = \sigma_R^j(\sigma_R^n(\mathsf{P})) = \sigma_R^j(\sigma_{[n]R}(\mathsf{P})) = \sigma_R^j(\sigma_O(\mathsf{P})) = \sigma_R^j(\mathsf{P}).$$

In order to prove that $\sigma_R^i(\mathsf{P}), \sigma_R^{i+1}(\mathsf{P}), \ldots, \sigma_R^{i+n-1}(\mathsf{P})$ are pairwise distinct for any fixed $i \in \mathbb{Z}$, we only need to show that $\sigma_R^\ell(\mathsf{P}) = \mathsf{P}$ only if $\ell \equiv 0 \pmod{n}$. Suppose $\sigma_R^\ell(\mathsf{P}) = \mathsf{P}$, i.e., $\sigma_{[\ell]R}(\mathsf{P}) = \mathsf{P}$. Then

$$\sigma_{[\ell]R}(\mathsf{P}) + O - \mathsf{P} - [d][\ell]R = O - [d \cdot \ell]R$$

is a principal divisor. Therefore, $d \cdot \ell \equiv 0 \pmod{n}$ since R is a generator of $\mathcal{E}(\mathbb{F}_q)$. Since n is relatively prime to d, this implies $\ell \equiv 0 \pmod{n}$. □

Let $\mathcal{E}$ be a cyclic elliptic curve of order $n \stackrel{\mathrm{def}}{=} q+1+t$ and let R be a generator of $\mathcal{E}(\mathbb{F}_q)$. Put $P_i = [i]R$ for all $i \in \mathbb{Z}$. For $d \geq 2$, let $\mathsf{P}_d(\mathcal{E})$ be the set of all places of $\mathcal{E}$ of degree d. Assume $\gcd(d, n) = 1$. According to Lemma 9.5.18, the action of $< \sigma_R >$ on $\mathsf{P}_d(\mathcal{E})$ divides $\mathsf{P}_d(\mathcal{E})$ into $r \stackrel{\mathrm{def}}{=} B_q(d,t)/n = B_q(d,t)/(q+1+t)$ equivalence classes. Each class contains n closed points of degree d. We choose only one closed point from each class. Thus, we obtain r closed points of degree d

$$\mathsf{P}_1, \mathsf{P}_2, \ldots, \mathsf{P}_r.$$

It is clear that for $1 \leq i \neq j \leq r$,

$$\mathsf{P}_j \notin \{\sigma^s(\mathsf{P}_i) \; : \; s \in \mathbb{Z}\} = \{\mathsf{P}_i, \sigma(\mathsf{P}_i), \ldots, \sigma^{n-1}(\mathsf{P}_i)\}.$$

For each $1 \leq i \leq r$, as $\dim_{\mathbb{F}_q} \mathcal{L}(\mathsf{P}_i) = \deg(\mathsf{P}_i) + 1 - g(\mathcal{E}) = d > 1$, we can find an element

$$z_i \in \mathcal{L}(\mathsf{P}_i) \setminus \mathbb{F}_q.$$

It is obvious that P_i is the unique pole of z_i and $\nu_{P_i}(z_i) = -1$. Consider the family of binary sequences

$$\mathcal{T}_d = \{\mathbf{s}_{z_i} \; : \; i = 1, 2, \ldots, r = B_q(d, t)/(q+1+t)\},$$

where

$$\mathbf{s}_{z_i} = \{\mathrm{Tr}(z_i(P_j))\}_{j=0}^{\infty} = \{\mathrm{Tr}(z_i([j]R))\}_{j=0}^{\infty}.$$

Theorem 9.5.19 *Let t be an integer satisfying one of the three conditions in Lemma 9.5.17. Let $2 \le d \le (q+1+2t-2\sqrt{q})/(4\sqrt{q})$ and $\gcd(d, q+1+t) = 1$. Let $\mathcal{T}_d$ be the family of binary sequences as constructed above. Then $\mathcal{T}_d$ is of size $B_q(d, t)/(q+1+t)$ and each sequence in $\mathcal{T}_d$ is of period $n \overset{\text{def}}{=} q+1+t$. Moreover,*

$$\ell_{\min}(\mathcal{T}_d) \ge \frac{q+1+2t-2(d+1)\sqrt{q}}{2d\sqrt{q}}$$

and

$$C_{\mathcal{T}_d} \le 2(2d+1)\sqrt{q} + |t|.$$

Proof. By Lemma 9.5.17, there exists a cyclic elliptic curve $\mathcal{E}/\mathbb{F}_q$ with $q+1+t$ $\mathbb{F}_q$-rational places. Employing arguments similar to those in the proof of Theorem 9.5.12 and using the results of Lemma 9.5.18, and Theorems 9.5.9 and 9.5.10, we can obtain the desired results. $\square$

We rewrite Theorem 9.5.19 in the following form by specifying q to be 2^e.

Theorem 9.5.20 *Let $e \ge 3$ be an integer and let t be an integer satisfying one of the three conditions in Lemma 9.5.17. Let $2 \le d \le (2^e + 1 + 2t - 2^{e/2+1})/(2^{e/2+2})$ and $\gcd(d, 2^e + 1 + t) = 1$. Then there exists a family $\mathcal{T}_d$ of binary sequences such that*

(a) $|\mathcal{T}_d| = B_{2^e}(d, t)/(2^e + 1 + t)$;

(b) *each sequence in $\mathcal{T}_d$ is of period $2^e + 1 + t$;*

(c) $\ell_{\min}(\mathcal{T}_d) \ge (2^e + 1 + 2t - (d+1)2^{e/2+1})/(d2^{e/2+1})$;

(d) $C_{\mathcal{T}_d} \le (2d+1)2^{e/2+1} + |t|$.

Corollary 9.5.21 (i) *Let $e \ge 8$ and let t satisfy condition (ii) or (iii) in Lemma 9.5.17. Then there exists a family $\mathcal{T}_2$ of binary sequences such that*

(a) $|\mathcal{T}_2| = (2^e - t)/2$;

(b) *each sequence in $\mathcal{T}_2$ is of period $2^e + 1 + t$;*

(c) $\ell_{\min}(\mathcal{T}_2) \ge 2^{e/2-2} - \frac{3}{2} - \frac{1+2t}{2^{e/2+2}}$;

(d) $C_{\mathcal{T}_2} \le 10 \cdot 2^{e/2} + |t|$.

(ii) *Let $e \geq 9$ and let t satisfy one of the three conditions in Lemma 9.5.17. In addition, suppose $\gcd(3, t+1+(-1)^e) = 1$. Then there exists a family $\mathcal{T}_3$ of binary sequences such that*

(a) $|\mathcal{T}_3| = (2^{2e} - 2^e - t2^e + t^2 - t)/3;$

(b) *each sequence in $\mathcal{T}_3$ is of period $2^e + 1 + t;$*

(c) $\ell_{\min}(\mathcal{T}_3) \geq 2^{e/2-2} - 2 - \frac{1+2t}{2^{e/2+2}};$

(d) $C_{\mathcal{T}_3} \leq 14 \cdot 2^{e/2} + |t|.$

Proof. Note that

$$\frac{B_q(2,t)}{q+1+t} = \frac{q-t}{2} = \frac{2^e - t}{2}$$

and that $q + t + 1$ is an odd number, when t satisfies condition (ii) or (iii) of Lemma 9.5.17. Taking $d = 2$ in Theorem 9.5.20 gives the results of part (i).

Note that

$$\frac{B_q(3,t)}{q+1+t} = \frac{q^2 - q - qt + t^2 - t}{3} = \frac{(2^{2e} - 2^e - t2^e + t^2 - t)}{3}$$

and that $\gcd(3, q+1+t) = \gcd(3, t+1+(-1)^e)$. Taking $d = 3$ in Theorem 9.5.20 gives the results of part (ii). $\qquad\square$

Bibliography

[1] M. Abdalla, Y. Shavitt, and A. Wool. Towards making broadcast encryption practical. In *Financial Cryptography, Lecture Notes in Computer Science 1648*, pages 140–152. Springer, 1999.

[2] N. Alon. Explicit construction of exponential sized families of k-independent sets. *Discrete Mathematics*, 58:191–193, 1986.

[3] J. Anzai, N. Matsuzaki, and T. Matsumoto. A quick group key distribution scheme with entity revocation. In *Advances in Cryptology – ASIACRYPT '99, Lecture Notes in Computer Science 1716*, pages 333–347. Springer, 1999.

[4] C. Asmuth and J. Bloom. A modular approach to key safeguarding. *IEEE Transactions on Information Theory*, 29:208–210, 1983.

[5] M. Atici, S. S. Magliveras, D. R. Stinson, and W. D. Wei. Some recursive constructions for perfect hash families. *Journal of Combinatorial Designs*, 4:353–363, 1996.

[6] O. Barkol, Y. Ishail, and E. Weinreb. On d-multiplicative secret sharing. *Journal of Cryptology*, 23:580–593, 2010.

[7] A. Bassa, A. Garcia, and H. Stichtenoth. A new tower over cubic finite fields. *Moscow Mathematical Journal*, 8:401–418, 2008.

[8] A. Beimel. *Secure Schemes for Secret Sharing and Key Distribution*. PhD thesis, Technion, Israel Institute of Technology, 1996.

[9] A. Beimel. Secret sharing scheme: A survey. In *Coding and Cryptology, Lecture Notes in Computer Science 6639*, pages 11–46. Springer, 2011.

[10] A. Beimel and B. Chor. Communication in key distribution schemes. *IEEE Transactions on Information Theory*, 40:19–28, 1996.

[11] J. Benaloh and J. Leichter. Generalized secret sharing and monotone functions. In *Advances in Cryptology – CRYPTO '88, Lecture Notes in Computer Science 403*, pages 27–35. Springer, 1988.

[12] D. J. Bernstein and T. Lange. Faster addition and doubling on elliptic curves. In *Advances in Cryptology – ASIACRYPT '07, Lecture Notes in Computer Science 4833*, pages 29–50. Springer, 2007.

[13] Th. Beth, D. Jungnickel, and H. Lenz. *Design Theory*. Bibliographisches Institut, Zurich, 1985.

[14] J. Bierbrauer, T. Johansson, G. Kabatianskii, and B. Smeets. On families of hash functions via geometric codes and concatenation. In *Advances in Cryptology – CRYPTO '93, Lecture Notes in Computer Science 773*, pages 331–342. Springer, 1994.

[15] S. R. Blackburn. Combinatorics and threshold cryptology. In *Combinatorial Designs and Their Applications*, pages 49–70. Chapman & Hall / CRC, 1999.

[16] S. R. Blackburn. Frameproof codes. *SIAM Journal on Discrete Mathematics*, 16:499–510, 2003.

[17] S. R. Blackburn, M. Burmester, Y. Desmedt, and P. R. Wild. Efficient multiplicative sharing schemes. In *Advances in Cryptology – EURO-CRYPT '96, Lecture Notes in Computer Science 1070*, pages 107–118. Springer, 1996.

[18] S. R. Blackburn and P. R. Wild. Optimal linear perfect hash families. *Journal of Combinatorial Theory, Series A*, 83:233–250, 1998.

[19] I. Blake, G. Seroussi, and N. Smart. *Elliptic Curves in Cryptography*. London Mathematical Society Lecture Note Series. Cambridge University Press, 1999.

[20] I. Blake, G. Seroussi, and N. Smart. *Advances in Elliptic Curve Cryptography*. London Mathematical Society Lecture Note Series. Cambridge University Press, 2004.

[21] G. R. Blakley. Safeguarding cryptographic keys. In *National Computer Conference: AFIPS 1979*, pages 313–317, 1979.

[22] G. R. Blakley and C. Meadows. Security of ramp schemes. In *Advances in Cryptology – CRYPTO '84, Lecture Notes in Computer Science 196*, pages 242–268. Springer, 1985.

[23] R. Blom. An optimal class of symmetric key generation systems. In *Advances in Cryptology – EUROCRYPT '84, Lecture Notes in Computer Science 209*, pages 335–338. Springer, 1985.

[24] C. Blundo and A. Cresti. Space requirements for broadcast encryption. In *Advances in Cryptology – EUROCRYPT '94, Lecture Notes in Computer Science 950*, pages 287–298. Springer, 1995.

[25] C. Blundo, L. A. Frota Mattos, and D. R. Stinson. Trade-offs between communication and storage in unconditionally secure schemes for broadcast encryption and interactive key distribution. In *Advances in Cryptology – CRYPTO '96, Lecture Notes in Computer Science 1109*, pages 387–400. Springer, 1996.

[26] C. Blundo, A. De Santis, A. Herzberg, S. Kutten, U. Vaccaro, and M. Yung. Perfectly-secure key distribution for dynamic conferences. In *Advances in Cryptology – CRYPTO '92, Lecture Notes in Computer Science 740*, pages 471–486. Springer, 1993.

[27] D. Boneh and J. Shaw. Collusion-secure fingerprinting for digital data. *IEEE Transactions on Information Theory*, 44:1897–1905, 1998.

[28] D. Boneh and A. Silverberg. Applications of multilinear forms to cryptography. *Contemporary Mathematics*, 324:71–90, 2003.

[29] I. Bouchemakh and K. Engel. The order-interval hypergraph of a finite poset and the König property. *Discrete Mathematics*, 170:51–61, 1997.

[30] E. F. Brickell. A few results in message authentication. *Congressus Numerantium*, 43:141–154, 1984.

[31] E. F. Brickell. A problem in broadcast encryption. In *5th Vermont Summer Workshop on Combinatorics and Graph Theory*, 1991.

[32] E. F. Brickell, G. Di Crescenzo, and Y. Frankel. Sharing block ciphers. In *Information Security and Privacy, Lecture Notes in Computer Science 1841*, pages 457–470. Springer, 2000.

[33] R. Canetti, J. Garay, G. Itkis, D. Micciancio, M. Naor, and B. Pinkas. Multicast security: A taxonomy and some efficient constructions. In *INFOCOM '99*, pages 708–716, 1999.

[34] R. Canetti, T. Malkin, and K. Nissim. Communication-storage tradeoffs for multicast encryption. In *Advances in Cryptology – EUROCRYPT '99, Lecture Notes in Computer Science 1592*, pages 459–474. Springer, 1999.

[35] J. L. Carter and M. N. Wegman. Universal classes of hash functions. *Journal of Computer and System Science*, 18:143–154, 1979.

[36] I. Chang, R. Engel, D. Kandlur, D. Pendarakis, and D. Saha. Key management for secure internet multicast using Boolean function minimization techniques. In *INFOCOM '99*, pages 689–698, 1999.

[37] L. S. Charlap and D. P. Robbins. An elementary introduction to elliptic curves. Technical report, Center for Communications Research, Princeton, 1988.

[38] H. Chen and R. Cramer. Algebraic geometric secret sharing schemes and secure multi-party computations over small fields. In *Advances in Cryptology – CRYPTO '06, Lecture Notes in Computer Science 4117*, pages 521–536. Springer, 2006.

[39] H. Chen, R. Cramer, S. Goldwasser, R. de Haan, and V. Vaikuntanathan. Secure computation from random error correcting codes. In *Advances in Cryptology – EUROCRYPT '07, Lecture Notes in Computer Science 4515*, pages 291–310. Springer, 2007.

[40] H. Chen, S. Ling, C. Padró, H. Wang, and C. Xing. Key predistribution schemes and one-time broadcast encryption schemes from algebraic geometry codes. In *IMA International Conference, Lecture Notes in Computer Science 5921*, pages 263–277. Springer, 2009.

[41] B. Chor, A. Fiat, and M. Naor. Tracing traitors. In *Advances in Cryptology – CRYPTO '94, Lecture Notes in Computer Science 839*, pages 480–491. Springer, 1994.

[42] H. Cohen and G. Frey. *Handbook of Elliptic and Hyperelliptic Curve Cryptography*. Discrete Mathematics and Its Applications 34. Chapman & Hall / CRC, 2005.

[43] C. J. Colbourn and J. H. Dinitz. *CRC Handbook of Combinatorial Designs*. CRC Press, 1996.

[44] T. Cover and J. Thomas. *Elements of Information Theory, 2nd Edition*. Wiley-Interscience, 2005.

[45] R. Cramer, I. Damgård, and U. Maurer. General secure multi-party computation from any linear secret-sharing scheme. In *Advances in Cryptology – EUROCRYPT '00, Lecture Notes in Computer Science 1807*, pages 316–334. Springer, 2000.

[46] Z. J. Czech, G. Havas, and B. S. Majewski. Perfect hashing. *Theoretical Computer Science*, 182:1–143, 1997.

[47] M. Deuring. Die typen der multiplikatorenringe elliptischer funktionenkörper. *Abhandlungen aus dem Mathematischen Seminar der Universität Hamburg*, 14:197–272, 1941.

[48] C. Diem. *On Arithmetic and the Discrete Logarithm Problem in Class Groups of Curves*. Habil. thesis, Leipzig, 2009.

[49] C. Ding, G. Xiao, and W. Shan. *The Stability Theory of Stream Ciphers*. Springer, 1991.

[50] A. G. D'yachkov and V. V. Rykov. Bounds on the length of disjunctive codes. *Problems of Information Transmission*, 18:7–13.

[51] M. Dyer, T. Fenner, A. Frieze, and A. Thomason. On key storage in secure networks. *Journal of Cryptology*, 8:189–200, 1995.

[52] M. Eichler. *Introduction to the Theory of Algebraic Numbers and Functions*. Academic, New York, 1966.

[53] A. Enge. *Elliptic Curves and Their Applications to Cryptography: An Introduction*. Springer, 1999.

[54] K. Engel. Interval packing and covering in the Boolean lattice. *Combinatorics, Probability and Computing*, 5:373–384, 1996.

[55] P. Erdős, P. Frankl, and Z. Füredi. Families of finite sets in which no set is covered by the union of r others. *Israel Journal of Mathematics*, 51:79–89, 1985.

[56] S. Fehr. Efficient construction of the dual span program. Master's thesis, Swiss Federal Institute of Technology (ETH), Zurich, 1999. http://homepages.cwi.nl/~fehr/publications.html.

[57] A. Fiat and M. Naor. Broadcast encryption. In *Advances in Cryptology – CRYPTO '93, Lecture Notes in Computer Science 773*, pages 480–491. Springer, 1994.

[58] A. Fiat and T. Tassa. Dynamic traitor tracing. In *Advances in Cryptology – CRYPTO '99, Lecture Notes in Computer Science 1666*, pages 354–371. Springer, 1999.

[59] M. L. Fredman and J. Komlòs. On the size of separating systems and families of perfect hash functions. *SIAM Journal on Algebraic and Discrete Methods*, 5:61–68, 1984.

[60] D. Freeman, M. Scott, and E. Teske. A taxonomy of pairing-friendly elliptic curves. *Journal of Cryptology*, 23:224–280, 2010.

[61] W. Fulton. *Algebraic Curves: An Introduction to Algebraic Geometry*. Benjamin, New York, 1969.

[62] Z. Füredi. On r-cover-free families. *Journal of Combinatorial Theory, Series A*, 73:172–173, 1996.

[63] E. M. Gabidulin. Theory of codes with maximum rank distance. *Problems of Information Transmission*, 21:1–12, 1985.

[64] A. Gál. *Combinatorial Methods in Boolean Function Complexity*. PhD thesis, University of Chicago, 1995.

[65] A. Garcia and H. Stichtenoth. A tower of Artin-Schreier extensions of function fields attaining the Drinfeld-Vlăduţ bound. *Inventiones Mathematicae*, 121:211–222, 1995.

[66] A. Garcia and H. Stichtenoth. On the asymptotic behaviour of some towers of function fields over finite fields. *Journal of Number Theory*, 61:248–273, 1996.

[67] S. Garg, C. Gentry, and S. Halevi. Attribute based encryption for general circuits. *Preprint*, 2012.

[68] S. Garg, C. Gentry, and S. Halevi. Candidate multilinear maps from ideal lattices and applications. *Cryptology ePrint Archive*, 610, 2012.

[69] E. N. Gilbert, F. J. MacWilliams, and N. J. A. Sloane. Codes which detect deception. *The Bell System Technical Journal*, 33:405–424, 1974.

[70] V. D. Goppa. Codes on algebraic curves. *Soviet Mathematics - Doklady*, 24:170–172, 1981.

[71] V. D. Goppa. Codes on algebraic curves (Russian). *Doklady Akademii Nauk SSSR*, 259:1289–1290, 1981.

[72] M. Grassl. Bounds on the minimum distance of linear codes. Online available at http://www.codetables.de.

[73] H.-D. O. F. Gronau and R. S. Mullin. On super-simple 2-$(v, 4\lambda)$ designs. *Journal of Combinatorial Mathematics and Combinatorial Computing*, 11:113–121, 1992.

[74] D. Hankerson, A. Menezes, and S. Vanstone. *Guide to Elliptic Curve Cryptography*. Springer, 2004.

[75] R. Hartshorne. *Algebraic Geometry*. Springer, New York, 1977.

[76] J. W. P. Hirschfeld, G. Korchmaros, and F. Torres. *Algebraic Curves over a Finite Field*. Princeton University Press, 2008.

[77] J. Hoffstein, J. Pipher, and J. H. Silverman. *Introduction to Mathematical Cryptography*. Springer, 2008.

[78] Y. Ihara. Some remarks on the number of rational points of algebraic curves over finite fields. *Journal of the Faculty of Science, University of Tokyo, Section IA, Mathematics*, 28:721–724, 1981.

[79] M. Ito, A. Saito, and T. Nishizeki. Secret sharing scheme realizing general access structure. *Journal of Cryptology*, 6:15–20, 1993.

[80] W.-A. Jackson and K. Martin. Cumulative arrays and geometric secret sharing schemes. In *Advances in Cryptology – AUSCRYPT '92, Lecture Notes in Computer Science 718*, pages 48–55. Springer, 1993.

[81] T. Johansson. *Contributions to Unconditionally Secure Authentication*. PhD thesis, Lund University, 1994.

[82] T. Johansson. Authentication codes for non-trusting parties obtained from rank metric codes. *Designs, Codes and Cryptography*, 6:205–218, 1995.

[83] T. Johansson, G. Kabatianskii, and B. Smeets. On the relation between A-codes and codes correcting independent errors. In *Advances in Cryptology – EUROCRYPT '93, Lecture Notes in Computer Science 765*, pages 1–11. Springer, 1993.

[84] H. W. Lenstra, Jr. Factoring integers with elliptic curves. *Annals of Mathematics*, 126:649–643, 1987.

[85] G. Kabatianskii, B. Smeets, and T. Johansson. On the cardinality of systematic authentication codes via error-correcting codes. *IEEE Transactions on Information Theory*, 42:566–578, 1996.

[86] M. Karchmer and A. Wigderson. On span programs. In *8th Annual Symposium on Structure in Complexity Theory*, pages 102–111, 1993.

[87] E. D. Karnin, J. W. Greene, and M. E. Hellman. On secret sharing systems. *IEEE Transactions on Information Theory*, 29:35–41, 1983.

[88] W. H. Kautz and R. C. Singleton. Nonrandom binary superimposed codes. *IEEE Transactions on Information Theory*, 10:363–377, 1964.

[89] H.-K. Kim and V. Lebedev. On optimal superimposed codes. *Journal of Combinatorial Designs*, 12:79–91, 2003.

[90] H.-K. Kim, V. Lebedev, and D. Y. Oh. Some new results on superimposed codes. *Journal of Combinatorial Designs*, 13:276–285, 2004.

[91] R. Kumar, S. Rajagopalan, and A. Sahai. Coding constructions for blacklisting problems without computational assumptions. In *Advances in Cryptology – CRYPTO '99, Lecture Notes in Computer Science 1666*, pages 609–623. Springer, 1999.

[92] K. Kurosawa and K. Okada. Combinatorial lower bounds for secret sharing schemes. *Information Processing Letters*, 60:301–304, 1996.

[93] S. Lang. *Algebraic Number Theory, 2nd Edition*. Springer, 1986.

[94] V. Lebedev. New asymptotic upper bound on the rate of (w, r) cover free codes. *Problems of Information Transmission*, 39:75–89, 2003.

[95] R. Lidl and H. Niederreiter. *Finite Fields*. Cambridge University Press, Cambridge, 1997.

[96] S. Ling and C. Xing. *Coding Theory: A First Course*. Cambridge University Press, 2004.

[97] M. Luby and J. Staddon. Combinatorial bounds for broadcast encryption. In *Advances in Cryptology – EUROCRYPT '93, Lecture Notes in Computer Science 765*, pages 512–526. Springer, 1998.

[98] X. Ma and R. Wei. On a bound of cover-free families. *Designs, Codes and Cryptography*, 32:303–321, 2004.

[99] K. Martin and S.-L. Ng. The combinatorics of generalised cumulative arrays. *Journal of Mathematical Cryptology*, 1:13–32, 2007.

[100] K. Martin, J. Pieprzyk, R. Safavi-Naini, H. Wang, and P. R. Wild. Threshold MACs. In *5th International Conference on Information Security and Cryptology, Lecture Notes in Computer Science 2587*, pages 237–252. Springer, 2003.

[101] K. Martin, R. Safavi-Naini, H. Wang, and P. R. Wild. Distributing the encryption and decryption of a block cipher. *Designs, Codes and Cryptography*, 36:263–287, 2005.

[102] J. L. Massey. Minimal codewords and secret sharing. In *6th Joint Swedish-Russian Workshop on Information Theory*, pages 269–279, 1993.

[103] T. Matsumoto and H. Imai. On the key predistribution system: A practical solution to the key distribution problem. In *Advances in Cryptology – CRYPTO '87, Lecture Notes in Computer Science 293*, pages 185–193. Springer, 1988.

[104] R. McEliece and D. Sarwate. On sharing secrets and Reed-Solomon codes. *Communications of the ACM*, 24(9):582–584, 1981.

[105] R. J. McEliece. A public-key cryptosystem based on algebraic coding theory. *Deep Space Network Progress Report, Jet Propulsion Laboratory*, pages 22–44, 1978.

[106] K. Mehlhorn. *Data Structures and Algorithms, Volume 1*. Springer, Berlin, 1984.

[107] A. Menezes, T. Okamoto, and S. Vanstone. Reducing elliptic curve logarithms to logarithms in a finite field. *IEEE Transactions on Information Theory*, 39:1639–1646, 1993.

[108] S. Micali and R. Sidney. A simple method for generating and sharing pseudo-random functions, with applications to clipper-like escrow systems. In *Advances in Cryptology – CRYPTO '95, Lecture Notes in Computer Science 963*, pages 185–195. Springer, 1995.

[109] C. J. Mitchell and F. C. Piper. Key storage in secure networks. *Discrete Applied Mathematics*, 21:215–228, 1988.

[110] D. Mumford. *Abelian Varieties*. Oxford University Press, 1970.

[111] M. Naor and B. Pinkas. Efficient trace and revoke schemes. In *Financial Cryptography, Lecture Notes in Computer Science 1962*. Springer, 2000.

[112] H. Niederreiter. Knapsack-type cryptosystems and algebraic coding theory. *Problems of Control and Information Theory*, 15:159–166, 1986.

[113] H. Niederreiter. Sequences with almost perfect linear complexity profile. In *Advances in Cryptology – EUROCRYPT '88, Lecture Notes in Computer Science 304*, pages 37–51. Springer, 1988.

[114] H. Niederreiter and M. Vielhaber. Linear complexity profiles: Hausdorff dimension for almost perfect profiles and measures for general profiles. *Journal of Complexity*, 13:353–383, 1997.

[115] H. Niederreiter and C. Xing. Low-discrepancy sequences and global function fields with many rational places. *Finite Fields and Their Applications*, 2:241–273, 1996.

[116] H. Niederreiter and C. Xing. *Rational Points on Curves over Finite Fields: Theory and Practice*. Cambridge University Press, 2001.

[117] H. Niederreiter and C. Xing. *Algebraic Geometry in Coding Theory and Cryptography*. Princeton University Press, 2009.

[118] W. Ogata and K. Kurosawa. Some basic properties of general nonperfect secret sharing schemes. *Journal of Universal Computer Science*, 4:690–704, 1998.

[119] C. Padró, I. Gracia, S. Martín, and P. Morillo. Linear key predistribution schemes. *Designs, Codes and Cryptography*, 25:281–298, 2002.

[120] C. Padró, I. Gracia, S. Martín, and P. Morillo. Linear broadcast encryption schemes. *Discrete Applied Mathematics*, 128:223–238, 2003.

[121] D. Pei. Information-theoretic bounds for authentication codes and block designs. *Journal of Cryptology*, 8:177–188, 1995.

[122] V. Pless and W. C. Huffman. *Handbook of Coding Theory*. Elsevier, 1998.

[123] H. Randriam. Hecke operators with odd determinant and binary frameproof codes beyond the probabilistic bound. In *ITW 2010 Dublin, IEEE Information Theory Workshop*, 2010.

[124] H. Randriam. Bilinear complexity of algebras and the Chudnovsky-Chudnovsky interpolation method. *Complexity*, 28:489–517, 2012.

[125] H. Randriam. (2, 1)-separating systems beyond the probabilistic bound. *Israel Journal of Mathematics*, 2012. DOI:10.1007/s11856-012-0126-9.

[126] U. Rosenbaum. A lower bound on authentication after having observed a sequence of messages. *Journal of Cryptology*, 6:135–156, 1993.

[127] H.-G. Rück. A note on elliptic curves over finite fields. *Mathematics of Computation*, 49:301–304, 1987.

[128] R. A. Rueppel. *Analysis and Design of Stream Ciphers*. Springer-Verlag, 1986.

[129] R. A. Rueppel. Stream ciphers. In *Contemporary Cryptology: The Science of Information Integrity*. IEEE Press, New York, 1992.

[130] M. Ruszinkó. On the upper bound of the size of the r-cover-free families. *Journal of Combinatorial Theory, Series A*, 66:302–310, 1994.

[131] R. Safavi-Naini and H. Wang. New constructions for multicast re-keying schemes using perfect hash families. In *7th ACM Conference on Computer and Communication Security*, pages 228–234. ACM Press, 2000.

[132] R. Safavi-Naini, H. Wang, and D. S. Wong. Resilient LKH: Secure multicast key distribution schemes. *International Journal of Foundations of Computer Science*, 17:1205–1221, 2006.

[133] A. Sahai and B. Waters. Attribute-based encryption for circuits from multilinear maps. *Cryptology ePrint Archive*, 592, 2012.

[134] D. V. Sarwate. A note on universal hash functions. *Information Processing Letters*, 10:41–45, 1980.

[135] T. Satoh and K. Araki. Fermat quotients and the polynomial time discrete log algorithm for anomalous elliptic curves. *Commentarii Mathematici Universitatis Sancti Pauli*, 47:81–92, 1998.

[136] J. Seberry, C. Charnes, J. Pieprzyk, and R. Safavi-Naini. Crypto topics and applications II, secret sharing, threshold cryptography, signature schemes, quantum key distributions. In *Algorithms and Theory of Computation Handbook*, pages 1–30. Chapman & Hall / CRC, 2010.

[137] I. A. Semaev. Evaluations of discrete logarithms in a group of p-torsion points of an elliptic curve with characteristic p. *Mathematics of Computation*, 67:353–356, 1998.

[138] J.-P. Serre. Nombres de points des courbes algébriques sur $\mathbb{F}_q$. *Séminaire de Théorie des Nombres 1982–1983*, Exp. 22, Université de Bordeaux I, Talence, 1983.

[139] J.-P. Serre. Sur le nombre des points rationnels d'une courbe algébrique sur un corps fini. *Comptes Rendus de l'Académie des Sciences, Paris, Series I, Mathematics*, 296:397–402, 1983.

[140] J.-P. Serre. Résumé des cours de 1983–1984. *Annuaire du Collège de France*, pages 79–83, 1984.

[141] J.-P. Serre. *Rational Points on Curves over Finite Fields*. Notes for lectures at Harvard University, 1985.

[142] A. Shamir. How to share a secret. *Communications of the ACM*, 22:612–613, 1979.

[143] K. W. Shum, I. Aleshnikov, P. V. Kumar, H. Stichtenoth, and V. Deolalikar. A low-complexity algorithm for the construction of algebraic-geometric codes better than the Gilbert-Varshamov bound. *IEEE Transactions on Information Theory*, 47:2225–2240, 2001.

[144] J. H. Silverman. *The Arithmetic of Elliptic Curves*. Springer, 2009.

[145] G. J. Simmons. A game theory model of digital message authentication. *Congressus Numerantium*, 34:413–423, 1982.

[146] G. J. Simmons. Authentication theory/coding theory. In *Advances in Cryptology – CRYPTO '84, Lecture Notes in Computer Science 196*, pages 411–431. Springer, 1984.

[147] G. J. Simmons. How to (really) share a secret. In *Advances in Cryptology – CRYPTO '88, Lecture Notes in Computer Science 403*, pages 390–448. Springer, 1990.

[148] G. J. Simmons. A survey of information authentication. In *Contemporary Cryptology: The Science of Information Integrity*, pages 379–419. IEEE Press, 1992.

[149] N. P. Smart. The discrete logarithm problem on elliptic curves of trace one. *Journal of Cryptology*, 12:193–196, 1999.

[150] J. Staddon, D. R. Stinson, and R. Wei. Combinatorial properties of frameproof and traceability codes. *IEEE Transactions on Information Theory*, 47:1042–1049, 2001.

[151] H. Stichtenoth. *Algebraic Function Fields and Codes*. Springer, Berlin, 1993.

[152] D. R. Stinson. Combinatorial characterization of authentication codes. *Designs, Codes and Cryptography*, 2:175–187, 1992.

[153] D. R. Stinson. Universal hashing and authentication codes. *Designs, Codes and Cryptography*, 4:377–346, 1994.

[154] D. R. Stinson. On the connection between universal hashing, combinatorial designs and error-correcting codes. *Congressus Numerantium*, 114:7–27, 1996.

[155] D. R. Stinson. On some methods for unconditionally secure key distribution and broadcast encryption. *Designs, Codes and Cryptography*, 12:215–243, 1997.

[156] D. R. Stinson. *Cryptography: Theory and Practice, 3rd Edition*. CRC Press, 2002.

[157] D. R. Stinson, T. van Trung, and R. Wei. Secure frameproof codes, key distribution patterns, group testing algorithms and related structures. *Journal of Statistical Planning and Inference*, 86:595–617, 2000.

[158] D. R. Stinson and R. Wei. Combinatorial properties and constructions of traceability schemes and frameproof codes. *SIAM Journal on Discrete Mathematics*, 11:41–53, 1998.

[159] D. R. Stinson and R. Wei. Generalized cover-free families. *Discrete Mathematics*, 279:463–477, 2004.

[160] D. R. Stinson, R. Wei, and L. Zhu. New constructions for perfect hash families and related structures using combinatorial designs and codes. *Journal of Combinatorial Designs*, 8:189–200, 2000.

[161] J. Tate. Endomorphisms of abelian varieties over finite fields. *Inventiones Mathematicae*, 2:134–144, 1966.

[162] L. Tombak and R. Safavi-Naini. Authentication codes in plaintext and content-chosen attacks. *Designs, Codes and Cryptography*, 6:83–99, 1995.

[163] M. A. Tsfasman and S. G. Vlăduţ. *Algebraic-Geometric Codes*. Kluwer, Dordrecht, 1991.

[164] G. van der Geer and M. van der Vlugt. Tables of curves with many points. *Mathematics of Computation*, 69:797–810, 2000.

[165] M. van Dijk, C. Gehrmann, and B. Smeets. Unconditionally secure group authentication. *Designs, Codes and Cryptography*, 14:281–296, 1998.

[166] J. H. van Lint and R. M. Wilson. *A Course in Combinatorics*. Cambridge University Press, 1992.

[167] S. G. Vlăduţ and V. G. Drinfeld. Number of points of an algebraic curve. *Functional Analysis and Its Applications*, 17:53–54, 1983.

[168] D. M. Wallner, E. J. Harder, and R. C. Agee. *Key Management for Multicast: Issues and Architectures*, 1999. Internet Draft (http://tools.ietf.org/html/draft-wallner-key-arch-01).

[169] H. Wang and J. Pieprzyk. Shared generation of pseudo-random function with cumulative maps. In *CT-RSA '03, Lecture Notes in Computer Science 2612*, pages 281–294. Springer, 2003.

[170] H. Wang and C. Xing. Explicit constructions of perfect hash families from algebraic curves over finite fields. *Journal of Combinatorial Theory, Series A*, 93:112–124, 2001.

[171] H. Wang, C. Xing, and R. Safavi-Naini. Linear authentication codes: Bounds and constructions. *IEEE Transactions on Information Theory*, 49:866–872, 2003.

[172] L. C. Washington. *Elliptic Curves: Number Theory and Cryptography*. Chapman & Hall / CRC, 2003.

[173] W. C. Waterhouse. Abelian varieties over finite fields. *Annales Scientifiques de l'Ecole Normale Supérieure*, 2:521–560, 1996.

[174] R. Wei. On cover-free families. *Preprint*, 2006.

[175] C. K. Wong, M. Gouda, and S. S. Lam. Secure group communication using key graphs. In *SIGCOMM '98*, pages 68–79, 1998.

[176] C. Xing. Algebraic geometry codes with asymptotic parameters better than Gilbert-Varshamov bound and Tsfasman-Vlăduț-Zink bound. *IEEE Transactions on Information Theory*, 47:347–352, 2001.

[177] C. Xing. Asymptotic bounds on frameproof codes. *IEEE Transactions on Information Theory*, 48:2991–2995, 2002.

[178] C. Xing, V. J. Kumar, and C. Ding. Low-correlation, large linear span sequences from function fields. *IEEE Transactions on Information Theory*, 49:1439–1446, 2003.

[179] C. Xing, H. Wang, and K. Y. Lam. Constructions of authentication codes from algebraic curves over finite fields. *IEEE Transactions on Information Theory*, 46:886–892, 2000.

[180] C. Xing and S. L. Yeo. Algebraic curves with many points over the binary field. *Journal of Algebra*, 311:775–780, 2007.

[181] Z. Zhang, M. Liu, Y. M. Chee, S. Ling, and H. Wang. Strongly multiplicative and 3-multiplicative linear secret sharing schemes. In *Advances in Cryptology – ASIACRYPT '09, Lecture Notes in Computer Science 5350*, pages 19–36. Springer, 2008.

[182] T. Zink. Degeneration of Shimura surfaces and a problem in coding theory. In *Fundamentals of Computation Theory, Lecture Notes in Computer Science 199*, pages 503–511, 1985.

Index

Printed and bound by CPI Group (UK) Ltd, Croydon, CR0 4YY

21/10/2024

01777085-0011